Physics without Metaphysics?

Raphael Neelamkavil

Physics without Metaphysics?

Categories of Second Generation Scientific Ontology

With an Appraisal by Prof. Saju Chackalackal

Bibliographic Information published by the Deutsche Nationalbibliothek
The Deutsche Nationalbibliothek lists this publication in the Deutsche Nationalbibliografie; detailed bibliographic data is available in the internet at http://dnb.d-nb.de.

Library of Congress Cataloging-in-Publication Data
Neelamkavil, Raphael, 1964-
 Physics without metaphysics? : categories of second generation scientific ontology / Raphael Neelamkavil ; with an appraisal by Prof. Saju Chackalackal.
    pages cm
 Includes bibliographical references and index.
 ISBN 978-3-631-66431-5 (print) – ISBN 978-3-653-05589-4 (e-book)
 1.Categories (Philosophy) 2. Causality (Physics) 3. Quantum theory. I. Title.
 BD331.N44 2015
 111–dc23
                            2015006238

1st Edition 2006, Dharmaram Publications, Bangalore, India
ISBN: 81-86861-89-0, © Raphael Neelamkavil

ISBN 978-3-631-66431-5 (Print)
E-ISBN 978-3-653-05589-4 (E-Book)
DOI 10.3726/978-3-653-05589-4

© Peter Lang GmbH
Internationaler Verlag der Wissenschaften
Frankfurt am Main 2015
All rights reserved.
PL Academic Research is an Imprint of Peter Lang GmbH.

Peter Lang – Frankfurt am Main · Bern · Bruxelles · New York ·
Oxford · Warszawa · Wien

All parts of this publication are protected by copyright. Any utilisation outside the strict limits of the copyright law, without the permission of the publisher, is forbidden and liable to prosecution. This applies in particular to reproductions, translations, microfilming, and storage and processing in electronic retrieval systems.

This publication has been peer reviewed.

www.peterlang.com

Preface

I develop in the present work *a new set of system-building categories and a new truth-probabilistic scientific ontology* from these categories. They have taken shape from the vantage of my critiques: (1) of the basic question of causality in Quantum Mechanics (QM), (2) of the question of the limiting velocity in Special Theory of Relativity (STR), (3) of some purely ontological and epistemological questions in the Philosophy of Physics and Causality, in Analytic and Possible-Worlds Metaphysics, Epistemology etc., and (4) of some seminal concepts in Aristotle, Kant, Einstein, Bohr, Whitehead, Gödel, Quine, Strawson, Husserl, Heidegger, Armstrong, Musgrave, van Fraassen, Field, Carroll, Stenlund, Aspect etc. The new systemic and scientific-ontological categories proposed here *invite the attention of philosophers of physics, analytic ontologists, phenomenologists and metaphysicians*. The QM- and STR aspects of causality are studied elaborately in my doctoral work, published under the title: *Causal Ubiquity in Quantum Physics: A Superluminal and Local-Causal Physical Ontology* (Frankfurt: Peter Lang, 2014, 361 pp.).

My inspiration for the present volume has primarily been decades-long efforts to critically understand anew the Philosophies of Being, Nature, Knowledge and the Divine and to base them on a cosmology-compatible ontological system. To begin with, such personal readings and enthusiastic discussions with Father Jacob Achandy many times every month, for three-and-a-half years from the age of 13, nurtured in me passion for Cosmology, Philosophy of Physics / Cosmos and Metaphysics. I admit also the vantage obtained thereafter from detailed and critical reading of (1) Russell's *Introduction to Mathematical Philosophy* (1984), Whitehead's *Process and Reality* (1984–'85) and related works, Husserl's *Formal and Transcendental Logic* (2004) and other works, Heidegger's *Contributions to Philosophy* and Wittgenstein's *Philosophical Investigations* (2005) and related works, and (2) the theories of Physics, Astrophysics and Cosmology (from 1981), beginning with articles from encyclopediae, and then moving to works by fine authors.

As I offer here a set of systemic ontological categories, I have in view the development of a system of Einaic Ontology (always insisting on the broadest sense of *Einai*, "the To Be of Reality-in-total" as the ontological categorial base), a second generation scientific ontology synthesizing as much as possible in Reality. I project for the future the work of Einaic Philosophy with sub-disciplines (local ontologies), already attempted in 2001 in a book-length material for private circulation without elaborating on the categories. As I go about the present work on systemic metaphysical categories, I clarify the concept of categories from various systematic angles and from various thinkers. The exercise in the various chapters does at times repeat what is already said in other chapters but each time it aims at developing different related concepts – this is unavoidable, since the theme is just three ontological categories.

The whole work dwells much on my "ways of being of processes" interpretation of ontological universals that are the basis for the connection between matter and form, and staticity and fluency in Reality. To Be, which belongs to Reality-in-total, is the fundamentally causal-processual and relational universality of ontological universals / essences active in particular processes in their totality that is Reality. The special variety of ontologically maximally broadened universals giving shape to laws of nature using their conceptual expression as connotative universals, and to the notions of Reality, its To Be and the conscious-connotative Reality-in-general, have been my own intuitions. These were conceived during my studies and years of teaching Philosophy, through the experience of critical encounter with Heraclitus, Parmenides, Aristotle, Hume, Kant, Peirce, Wittgenstein, Whitehead, Quine, Gödel, Husserl, Heidegger, Cosmology, Philosophy of Physics, Analytic and Possible-Worlds Metaphysics and Epistemology, and Eastern Metaphysics.

My chancing (just before the publication of the 1st edition) on a one-page passage similar to my interpretation of universals as ways of processes, in David M. Armstrong, "Universals as Attributes", 80 (see Bibliography), gave me the confidence that I have something more systemic to contribute to it, though I am unable to vouch for the particularism and logical purism Armstrong and almost all analytic ontologists have espoused by historical necessity. Armstrong refers in the above article to David A. J. Seargent, who held that "properties are ways things are" in Chapter 4 of Seargent's *Plurality and Continuity: An Essay in G. F. Stout's Theory of Universals* (The Hague: Martinus Nijhoff, 1985), which, Armstrong says, presented this concept for the first time.

New horizons productive of exploratory theories are always available. Such might fence me in, giving a sense of grandeur about the truth claims made accessible using the new species of theoretical horizons. But I hope the new set of categories here can give *a fundamental categorial challenge* (1) to scientific-ontological systems which abide by one or more of the families of theories of truth claims, (2) to anti-metaphysical tendencies like postmodernism, and (3) to analytic ontologists' particularist generalizations of definitions of universal concepts (e.g., 'necessity' defined as in: 'If all swans are white, swanhood and whiteness are the bases of necessity', unjustifiably becomes the definition of 'necessity' for more general purposes, justified by the merits of simplicity, clarity, directness and substitutability of terms). Einaic Ontology contends that ontological notions like necessity, possibility, causality, universals, concept-formation etc. presuppose more broadly Reality-imbued conditions.

I view Einaic Ontology with its collusive categories as a matrix facilitating ever more truth-probable systems after systems that are flexible for continuous broadening and deepening of the categories, and of the resulting system that does not remain just *one* system with fixed particularist definitions of categories. The Einaic system would remain a matrix to facilitate ever better systems. Thus, my attempt is *to attract philosophy* beyond absolutism, relativism, pragmatism, instrumentalism, particularism, logicism etc. *back into the world of systems*, with the firm difference of epistemological probabilism of ontological meanings of all that we project in theory, which includes also truth-probabilism regarding sentential truths within systems.

The revised version of the book supplants and ontologically characterizes the scientific-epistemological and geometrical sub-category of Space-Time with *the physical-ontological sub-category of Extension-Motion*: where *space is the measure of extension including the extension achieved in motion, and time is the measure of motion except the extension covered thereby*. I have added here and there relevant remarks to evaluate the concept of *possible worlds* from the perspective of Einaic Ontology. Each chapter is self-subsisting in that each treats in passing also the categories developed in the others. The *overcoming of Kant's phenomena-noumena distinction, of QM acausalism and of STR limit-velocity* in Chapter 1 will be interesting for motivated students of philosophy and physics alike. To facilitate a *comparative study of the various positions as against mine, additional citations* from relevant authors are given in the text and in footnotes, with or without clarifications. The volume and extent of discussions here on the new system-building set of categories and the development of some seminal ways for a new philosophical system make such citations necessary for comparative relevance. The "simplicity" of a Master's thesis written in a short time is still visible in some sources used. Nevertheless, I consider that the corrected and augmented 2nd edition deserves presentation to a wide readership.

The present edition preserves the Appendix revised. My critical attitude, from 1984, to Heidegger's exclusive orientation of "Being" onto the human *Dasein* justifies my attempt in the Appendix to *try and reverse Heidegger's anthropic orientation of Being, by substituting Being with a cosmology-compatible, nomic-nominal and verbal-processual concept of To Be* that is coextensive with

Reality. While reading the Appendix it must be kept in mind that I reflect there from the point of view of the ontological and scientific-realistic arguments in the whole book.

I am indebted much to Prof. Saju Chackalackal, Dharmārām Vidyā Kṣetram (DVK), Bangalore, for the empowering guidance of my Master's thesis and the near-absolute freedom I was granted as he perceived in me responsibility and dedication for philosophical work. If he had been inflexible about the theme and content, or restricted the length of the thesis (he permitted me over 250 pages), my thoughts on these new categories or on the new scientific ontology would have been forced into slumber, or at least not have been published so early. Thanks also for his academic foreword, "An Appraisal of *Physics without Metaphysics?*" that appears (revised) in the present edition too.

I welcome critiques and assessments of my work per e-mail.

Raphael Neelamkavil wisdomcontemplation@gmail.com
Bochum, Germany, January 2015

An Appraisal of *Physics without Metaphysics?*

Prof. Saju Chackalackal

I

In the recent history of philosophy an obvious shift has taken place. While almost all philosophers attempted to address the perennial questions that intrigued humanity from its inception by way of formulating variously rationally consistent systems of thought – either by focusing on the object or the subject, as the case may be – a number of prominent thinkers have questioned the validity of such systematisations, as no system comes up neutrally. They are convinced, for example, that the so-called *a priori* is no more completely *a priori*. In fact, there are many factors that have colluded to give rise to anything that is said to be purely rational. Indeed, human reason itself is said to be the product of myriad influences, spread through the whole history of human existence. The trend, therefore, is to challenge and overthrow all systems that have been accepted as valid in providing answers to the perennial questions, and to move beyond the boundaries of systems with the hope of striking a better understanding or grasp of reality. Though this deal seems to be quite captivating, especially among the neophytes, the fundamental issue continues to haunt us: Can the human mind understand without a framework suitable to its nature? Or, in other words, is it possible for us to understand something in an absolute vacuum? Can reason try to understand the inner recesses of reality – which, in the traditional understanding, requires a move from physics to metaphysics – just by unsubscribing to all systematic schools of thought available to us down the centuries? Can rationalisation be nothing but anti-systematisation, leading to bits and pieces, whims and fancies, the whispers of the moment, and semantic-linguistic ratiocinations, intelligible or unintelligible?

As postmodernity and analysis became the catchwords in many philosophical and humanities circles, and many researches are being carried out to disprove the absolutist claims made by old and new systems of thought, I wonder whether we can totally be freed from the natural tendency for systematisation. Recent metaphysical trends in analytic philosophy show just this, although postmodernism, as a system of thought, is unable to support any metaphysics. In fact, a blanket rejection of systematisation will be suicidal to philosophical deliberations, as human mind cannot function outside a framework of its own, if at all it should make any sense. Recent trends in philosophical research clearly indicate the demise of vigorous postmodernism except in *some* French, Asian, African and South American Universities. It seems that postmodern and analytic thoughts have either missed or gone astray from their targets. It is true that the absolutist understanding of many a previous thinker or school of thought should be challenged; if not, it would be a blatant denial of the creativity of human reason. Moreover, Postmodernity brings to our attention that nobody can pronounce the final word on the understanding of truth; we do philosophise, but within frameworks which need to be constantly challenged and revised. Indeed, philosophising must remain an open project and an ongoing process, accessible to and extendable by the entire humanity.

The same dynamics could also be located in the constant struggle between physics and metaphysics, in their attempts to nullify each other, or to win over each other. The medieval glory of metaphysics was shelved during the Enlightenment, and physics had its field day: an unending saga of success enjoyed by the modern sciences and their applications in different areas of human need have made the contemporary humans shun any thrust on the metaphysical dimensions of reality. The questions such as 'what is reality', 'what are the dynamics of reality' etc. are usually answered within the parameters of physics; for many, only such answers are intelligible. Indeed, any answer that is beyond the terrains of verifiability, linguistic clarification and particularistically logical definitions is fashionably rejected as 'non-sense'. The trend, in general, seems to belittle and ridicule the value of that which cannot be given to us in observables and measurables: the directly practical principles of logic, mathematics and physics seem to reign supreme, and to a good number of physicists they are indisputable and absolute. Thus, physics, without recourse to any metaphysical understanding, apparently tries to master reality. To many, then, what is said by physicists, biologists, psychologists and other scientists is the final word on the nature of reality.

It is against these two trends – the freelancing philosophical arrogance of postmodernity, the rigorous particularism of analytic philosophy and the omniscience of physics in having copyrighted finality in understanding reality – that I find the relevance and daring nature of Raphael Neelamkavil's earnest research in 'Einaic Ontology', an attempt to re-capture the lost sense of the real, by taking recourse to philosophy and physics, and many other allied disciplines.

II

In *Physics without Metaphysics?* Neelamkavil successfully launches an articulation and defence of the ontology behind all sorts of philosophical endeavours, especially the philosophies of physics, astrophysics, and mathematics. It presupposes the history of ontological categories and scientific categories (space, time, cause, matter-energy/mass, etc.) from Plato and Aristotle through the modern theories in physical sciences to the twentieth century scientific ontology. His masterly focus, in particular on Aristotle, Kant, Bohr, Einstein, Armstrong, Strawson, Quine, and Heidegger, and, in general on thinkers in the philosophies of physics and mathematics, analytical epistemology and analytical ontology, fructifies in giving rise to the mutually collusive 'Einaic' categories of cosmology-epistemology-ontology.

The search in this undertaking begins by questioning the ability of purely classificational categories to do authentic scientific ontology, and with an admission that all that there are to Reality in ontological commitment are: (1) particular token entities (processes), (2) ontological particulars (species), (3) epistemologically connotative (of the conscious manner of noting together) universals (species names, qualities, truths, laws of nature, etc.), (4) ontological universals (*qualia* / ways of being) in processes, and (5) totalities of entities / processes. Of these, substances – processual tokens and totals – are transcendents; and universals – connotative species names, connotative qualities and ontological *qualia* – are transcendentals. He proposes to transcend the problem of particularist physics and ontology by using a new set of cosmological, epistemological and ontological categories that are maximally classificational, ideal, provincially singleton-case, and theoretically *a priori*. All of them, as Neelamkavil claims to show, are necessarily probabilistic and transcends particularism by the ever-widening nature of universals ingredient in tokens and totals.

Neelamkavil begins his articulation based on certain foundational assumptions in ontology: (1) the traditional categories equivalent to substance are to be maximized by the unique Transcendent domain of substances, i.e. Reality-in-total (termed often 'Reality'); (2) the connotative ("noting together") presentations of the various categorial attributes, universals, truths etc. in consciousness are to be maximized by the epistemologically connotative and Transcendently Transcendental universal of universals, i.e., the concept of Reality-in-general; and (3) ontological universals active at the processual-relational-essential aspect of beings are to be maximized by the processual-verbal (see the third paragraph in the General Introduction), nomic, nominal and Transcendental universal, the To Be of Reality. Maximization of transcendentals is by inductive generalization, and that of transcendents is by inductive totalization, justified by the ontology of infinities in mathematics. These dimensional (not fixed in definitions, ever open to bettering) categories make their mutual collusion and implication naturally generative of systems that are ever more truth-probable by reason of their idealistic coherence, theoretical pragmatism, probabilistic relativism and, finally, realistic correspondence of results with matters of fact.

The proposed categories have been argued out from the points of view of contemporary philosophy of physics, analytic philosophy of knowledge, and analytic plus some continental philosophy of being. Firstly, the work is informative of various contemporary analytic trends related to its field of study. Secondly, it delves into the radically cosmological, epistemological, and ontological questions typical of the interface of science and philosophy.

Against the Kantian epistemological approach, Neelamkavil attempts to overcome the phenomena-noumena divide by an ontological approach to the cosmological category of Reality-in-total, which is the uniquely continuous substance that includes all really possible (actual) worlds that are objectual-causally connected. To procure validity to this concept of substance, he moves to an analysis of the deep-seated scientific-instrumentalistic difficulties in the QM manner of cutting up its object into mere statistical phenomena, which one conceived to be a wave at one moment and a particle at another, without any ontological commitment admissible by instrumentalism. He shows how this dichotomy may be overcome scientifically and how Reality may be conceived as thoroughly continuous – a curious nicety, indeed. Soon he moves to the foundations of Quantum Physics and Special Theory of Relativity to fill up all possible values of energies and velocities in Reality as conceivable at the broadest possible realm of "all possible real worlds", by a short and succinct argument in favour of the actuality of all possible velocities (which he has studied in a detailed fashion in his doctoral dissertation, published *Causal Ubiquity in Quantum Physics: A Superluminal and Local-Causal Physical Ontology*. See Bibliography). By now he is on his way to derive the concept of Reality that is in all respects thoroughly continuous, which results in the Ontological Principle of Excluded Vacuous Middle. This process allows him to ontologically bridge the phenomena-noumena divide in the concept of the maximal substance, Reality-in-total. Thus, he proposes Reality-in-total as the ontological-cosmological ideal of all discourse on the Worlds, with all its objectual-causal roots that may possibly be in the Divine. Thus this work claims that physics cannot be done without scientific metaphysics.

The author proceeds, then, to show that the category of Reality-in-total does not stay alone. It needs the support of its theoretical, conceptual, ideal category. He comes out with a well-knit justification of the need of universals in epistemological, cosmological, and ontological discourses. He derives his justifications in a face-to-face encounter with twentieth century analytical thinkers, who do not, or, partially and particularistically do, favour the use of universals in any discourse.

He makes also a contribution to the philosophy of universals by bringing clarity to the concept of the universal, i.e., *by differentiating ontological universals (ways of being of processes) that are objectually present (by ontological commitment to processes and their ways of being "over there," not by reference) in processes in their relational realm from connotative-conscious (epistemological) universals that are probabilistically appropriated forms of the former.* Thus he conceives the connotative and inductively most generalized universal of all universals – Reality-in-general – as the epistemological ideal category of all ontological endeavours. He favours treating connotative universals and Reality-in-general as probabilistic because, ever after the Incompleteness Theorem of Gödel, we are not justified in fixing meanings of definitions and terms or validity of truth-statements as absolute. Instead, we must go on pushing axioms and the definitions of primitive and derivative notions, backwards into more general and succinct ones, in systems of any ontological order. So, connotative universals, which are presupposed (or ingredient) in concepts, are ever-widening and, hence, truth statements in Einaic systems are considered to be probabilistic in nature.

Finally, Neelamkavil moves to the purely ontological Transcendental that makes both Reality-in-total and Reality-in-general possible. He presents a concept of the Transcendental To Be, the supra-categorial category beyond the ways of being (*qualia*) and the to be of entities or processes. This seems to be an improvement beyond the traditions of the concept of Being. To Be is the maximization of all ontological universals by inductive generalization, which, in turn, is based on inductive totalization. But, this goes counter to ontological particularism, which, in none of its forms, goes beyond tokens and their immediately wider universals. Confident of the suitability of the new trans-classificational category of the To Be proper to Reality for ontological consumption, he proposes that we can no more do scientific ontology merely by classificational categories, for they are particularistic. Therefore, the author thinks it is tenable to claim that it is a handicap for science and philosophy at once, if they allow doing science and regional ontologies without the most generally probabilistic and self-transforming categories of scientific ontology. Particularism without Einaic Ontology has been the backbone of the hitherto practice of science and philosophy. The author highlights the problems of particularistic ontologies by making an in-depth study of the particularistic, linguistic, and ontological presuppositions in Quine, Strawson, etc. He contends that, without acceding to the most adequate dimensional categories of ontological thinking, the scientific categories of particular sciences and regional ontologies cannot be justified. Scientific ontology possibilizes science and, hence, without Einaic thinking, science is not what it can and ought to be. In short, only maximal categories can possibilize reality-in-particular and the discourse on actual entities or processes in the sciences. Moreover, by reason of the partiality of particularistic ontological universals with respect to the processes involved, each such universal refers to other and broader universals and, hence, ontological universals too are partial in their relation to token processes. The connotative reflections of ontological universals can thus be only epistemologically probabilistic and produce only probable truths. This fact couches all sciences and regional ontologies in Einaic Ontology with its truth-probabilism.

Such truth-probabilistic and universalistic inclination in theory allows this research to transform ontology, particularly scientific ontology, into Einaic Ontology, which is a pragmatic amalgamation of (1) Einaiology, which studies To Be in terms of Reality-in-total and Reality-in-general, and (2) General Ontology, which treats Reality and Reality-in-general in terms of To Be. Due to the collusive nature of the three categories, one can never do any one of these two sciences in isolation from the other. This makes Einaic Ontology not only viable, but also ideal and inevitable, universalistic in aim and probabilistically flexible in approach to truth – thus, frameworking the foundations of ontologies beyond metaphysical absolutism of ideal reifications and postmodern, analytic, sceptic, sophistic absolutism of haphazard relativisations.

In conversation with Neelamkavil I realize his claim to be that this sort of ontology is for him also a scientific ontology, since the category of Reality-in-total is the maximized cosmological category of the sciences, which is potent enough to make physical reality, processes and experiments possible. Although his work creates only the kernel for a new scientific ontology, it seems to hold the promise of further elaborations and the development of an entire system of philosophy in itself. Given the earnestness of the author visible in this venture, especially in this text, I am hopeful that many of us would live to see penetrating and extensive researches in the field of Einaic Philosophy, capable of shedding brighter light into the nature of reality, and to answer the perennial questions that keep us haunting in the realm of philosophical thinking.

III

I deem it important to draw the attention of the reader to the Appendix that juxtaposes (1) earlier Heidegger's verbal, aletheial Being (which "throws-open" Being in and through and for *Dasein*) based on *Dasein* and later Heidegger's enowning and projecting-open Being which bases *Dasein*, with (2) the slightly different, but systemic-probabilistic, concept of the nomic-nominal and processual-verbal To Be, which is simultaneously aletheial to man and enowning-projecting of and co-extensive with Reality in its maximality. Neelamkavil proposes this latter version of To Be for collusive, probabilistic, and systemic reasons. He seems to hold that both the earlier and later Heideggerian concepts of Being are still anthropic and epistemic, because they do not give *a priori* objectual validity (i.e., based purely on the mind-independent and trans-phenomenal *fact* of Reality and realities) beyond ordinary cognitive apriority, (1) to Being as the Transcendental beyond *Dasein*'s appropriation of it in himself and beings, (2) to Being's giving itself to *Dasein*, and (3) to Being's enowning of *Dasein* and Being's projecting open (*Entwurf des Seins*) of *Dasein* within and out of Being-thinking. Neelamkavil argues that such a Being is comparable to the connotative-epistemological category Reality-in-general, which is the giving itself of To Be in human consciousness' appropriation of it as the connotative universal of universals. Thus he shows that an adequate concept of Reality-in-total is lacking in Heidegger. I think, with this interesting suggestion, combined with his Einaic Ontological evaluation of earlier and later Heidegger and demonstration of the exact Einaic Ontological difference between earlier and later Heidegger, the Appendix would invite both appreciative and critical evaluations by those who are ready to leap beyond Heidegger.

IV

Although the subtlety of the analysis carried out in this research and the complicated and complicating terminological fiesta that abounds the work may create an impression of an 'arm-chaired' philosophical discourse, *Physics without Metaphysics?* is specifically oriented towards practical import as well. The changed scenario of philosophy – especially its need to speculate in collaboration with scholars who are involved in research in many other fields, e.g., in doing science – is well aware that many among the best of the minds are involved in and committed to scientific investigations. The new scientific theories that have come up in the last few decades have, in general, significantly altered human conception of Reality, though without an anchor to hold on to in the vicissitudes of the constant flux and rational unrest of unfathomableness. So, the practical intent of Neelamkavil's project is to draw insights into Reality from these theoreticians; and in attempting to go beyond them, he envisions the possibility of pioneering a novel way of doing ontology in the sure *dimension* of unfathomableness – not for its own sake, but for a genuine understanding of Reality, which is the rationalised goal of all human seeking. In this context he is categorical as far as his findings are concerned: Einaic Ontology is speculative as well as scientific in the sense that it makes possible and transcends the categories of the sciences. Moreover, he contends that the Einaic categories that he has justified in this work as the sure *dimensions* of all Reality and Thought, are applicable to their cosmological, epistemological, and ontological categorial concepts as well.

The novel perspective unveiled in this work is a promise. If the theories available in the known history of philosophy could not settle the issues that intrigued humanity with any definitive answer, and if the continued search for answers is the duty of every human being endowed with rationality, *Physics without Metaphysics?* and its path-breaking ideas in favour of Einaic Ontology are worth our serious consideration. They hold an impressive promise for doing philosophy that looks for ultimate answers, especially amidst the myriad theories that crop up every other day. Indeed, even for Neelamkavil – in accordance with Einaic probabilism concerning the new set of categories – this is not the final word in a philosophical settlement of the nature of reality. It is indeed a courageous and firm step that must be pursued further to unveil and traverse unforeseen horizons in encountering and understanding Reality.

Prof. Saju Chackalackal
Professor of Philosophy and President, Dharmārām Vidyā Kṣetram (DVK) and Professor of Philosophy, Christ University, Bangalore
Email: saju@chackalackal.com

Contents

General Introduction. Second Generation
Scientific-Ontological Categories 21
1. Mutually Collusive Non-classificational Categories 22
2. General Method and Rationale 28
3. Transcendentals and Transcendents 29
4. Methodology of Mutual Collusion 36
5. Aim 1: Scientific-Ontological Categories and System-Building 45
6. Aim 2: Synthesis of Nominal 'Being' and Verbal 'To Be' 47

Chapter 1. Ontological Categorial Transcendent of Cosmology 49

Introduction 49
1.1 Redeeming Cosmological Categories from Idealism 52
1.1.1 Entities and Predicates as Particularist Categories 52
1.1.2 Kant and Modern Physics on Phenomenal Categories 57
1.1.2.1 Understanding 'Phenomena' and Sensing 'Noumena' 57
1.1.2.2 Conflating Seeing-that, Seeing and Seeing-as 61
1.1.2.3 Phenomenalistic Idealization of Forms and Categories 64
1.1.3 A Category from Phenomena-Noumena Continuity 67
1.2 Microscopic Categorial Features of Quantum Mechanics 76
1.2.1 Apriority in Scientific Categories and Realism 76
1.2.2 Wave-Particle Duality and the Concept of Reality 87
1.2.3 Nonlocality and the Concept of Reality 113
1.2.4 Ontological Categorial Foundation of Quantum Mechanics 125
1.3 Categorial Features of Relativistic Theories 131
1.3.1 Questions in the Origin of the Special Theory of Relativity 131
1.3.1.1 Inertial Referential Velocity of Light 131
1.3.1.2 Principle of Relativity and Superluminal Velocities 136
1.3.2 Space, Time, Mass and Causality from Newton to Einstein 140
1.3.2.1 Classical and Reformed Views of Continuous Reality 140
1.3.2.2 Beyond Scientific Causality and Categories 142
1.4 Ontological Synthesis of Categories of the Cosmos 145
1.4.1 Reality-in-total: The Ultimate Transcendent Category 145
1.4.2 Mathematical Entities vs. Ultimacy in Denotability
 of Reality-in-total 150
Conclusion 162

Chapter 2. Ontological Categorial Transcendental of Epistemology 163

Introduction 163
2.1 Transcendental Categorial Orientation in Epistemic Actualities 165
2.1.1 Laws of Nature and the Epistemology of Essences 165
2.1.2 Transcendental Categorial Orientation in Particularism 184
2.2 Transcendental Categorial Dimension of Probabilistic Essences 196
2.2.1 Categorial Confluence of Causality, Laws and Essences 196
2.2.2 System-Building vs. Absolutism and Probabilism in Essences 219
2.2.2.1 Gödel's Theorem and System-Building 220
2.2.2.2 Gödel's Result vs. Probabilistic Universals and Categories 226
2.2.3 Transcendental Aspect of Particularist Probability-Makers 232
2.3 Reality-in-general: Synthesis of Categories of Knowing 235
Conclusion 237

Chapter 3. Ontolgoical Categorial Transcendental of Reality-in-total 241

Introduction 241
3.1 To Be: The Einaic Ontological Transcendental *Par Excellence* 242
3.2 Einaic Objectual Ontology beyond Quine 253
3.2.1 Relevance of Quine's Ontological Commitment and Holism 253
3.2.2 Objectual Ontology vs. Particularist Ontological Commitment 254
3.2.3 The 'Is' of Abstract Objects: Quine-Carnap Dialogue 265
3.2.4 From 'Is' to Language-Ladenness and To Be 274
3.3 Einaic and Semantic Ontology of Actualizing Possibility 289
3.3.1 The Ontological Transcendental vs. Semantic Holism 289
3.3.2 To Be as Possibilizing Actuality in Possibility 293
3.3.3 Dynamism of Einaic Semantic Connectivity 296
3.3.4 Einaic Logic and Ontology of Quantification 303
3.4 Trans-categorial Transcendental or Way of Being of Reality 310
Conclusion 313

General Conclusion. Prospects of the Transcendental-Transcendent Categorial Synthesis 315
1. Founding Second Generation Speculative Philosophy 315
2. Nature of Einaic, Second Generation Scientific, Categories 317
3. Einaic Ontology: Blend of Einaiology and General Ontology 324
4. Cosmological-Theological Prospects 330
5. New Einaic Sense and Universality of 'Category' 332

Appendix. Beyond Heidegger's Anthropologized Being: Nomic-Nominal, Verbal-Processual, Universal
"To Be" in Einaic Ontology 337

Bibliography 357

Index 369

General Introduction. Second Generation Scientific-Ontological Categories

The present monograph on the ontological foundations of physics and physical existence constructs the categories of Einaic Ontology and attempts to answer the question: Can physics be done without the dimension of foundational categories of metaphysics / ontology? The adjective 'Einaic' is from the Greek *Einai*, "To Be". Einaic Ontology studies the whole Reality and the highest conceptual ideals of thought in terms of the To Be of Reality-in-total (hereafter whenever possible, termed 'Reality'). The process of constructing a system that answers the above question gives rise to a set of mutually collusive and maximal case categories for second generation scientific ontologies. These categories are maximally collusive: each of them "plays together" with the others of them maximally.

To achieve this aim, the work (1) synthesizes Kant's phenomena and noumena, (2) discusses the cosmic ontology of Reality-in-total presupposed in the physical sciences through trans-phenomenalistic, trans-instrumentalistic interpretations of causality, continuity, values of velocity, objectuality, objectivity and other notions in QM and STR, (3) moves to a viable epistemology of connotative universals and their Gödelian connotative probabilism necessarily implied by the cosmic ontology of Reality, and then (4) develops the foundational categories of the ontology that pertains to Reality, and characterizes the categories by an episemological version of the implications of Gödelian probabilism and an ontological version of the implications of Quine's semantic holism. This provide the essential trans-scientific touch to ontology. It necessitates ontological and epistemological probabilism, an infinite, infinitesimal and objectual causalism, transcending of ontological and epistemological particularism and the ways-of-being interpretation of ontological universals. I name the new categories and the science that works on such a machinery as Einaic (Greek, *Einai*, "To Be") Ontology – which synthesizes the nomic-nominal ('nomic' from the Greek *nomos*, "rule") and objectual-causal concept of To Be with the processual-verbal concept. I show that its ingredients are Einaiology and General Ontology. **The processual-verbal** concept of To Be is what I mean always, when I speak of **the verbal meaning** of it. It reflects the very processual "thus-ness" or "that-it-is-so" of everything together. It is the most processual sense of all verbs and actions culminating in 'To Be'.

The synthesis of the nominal and verbal concepts of To Be and the transcendence of particularism impels me to analyse, in the Appendix, Heidegger's concept of Being and point out its major achievement (emphasis on man's conscious connection with the thinking of To Be allegedly *as* the "Be-ing" of human existence / "standing out") and its defect (forgetfulness of the nomic-nominal, processual-verbal aspects of the To Be of Reality).

1. Mutually Collusive Non-classificational Categories

By ontological commitment I mean the ontological *a priori* presupposition or seeing-that-there-is-*something* behind any (concrete or abstract) object/s in the process of conceptual grasping and/or propositional expression. A concrete object is a causal process; and an abstract object / ontological universal is an acausal and abstract *way of being of processes.* Instantiating the ways in consciousness, and given in an abstract conceptual formulation that always connotes ("notes together") the many similar causal processes, the same ontological commitment is *made* in consciousness, language and discourse. Quine distinguishes between the ontological commitment behind abstract and concrete terms:

> For I deplore that facile line of thought according to which we may freely use abstract terms, in all the ways terms are used, without thereby acknowledging the existence of any abstract objects. According to this counsel, abstract turns of phrases are mere linguistic usage innocent of metaphysical commitment to a peculiar realm of entities. For anyone with scruples about what objects he assumes, such counsel should be no less unsettling than reassuring; for it drops the distinction between irresponsible reification and its opposite.[1]

A responsible way of reification of abstract entities is to consider them as ways of being of beings / processes, that is, as general, irreducibly causal relationalities of processes, given connotatively in mental concepts based on the generalities of the processes. Thus, concrete entities (irreducibly and always token processes) are imbued with ontological abstract entities (based in concrete processes, and not in cognizing minds), i.e. ways of being of processes. If so, we cannot speak of an entity as what is spatiotemporally (i.e., according to measurements of extension-motion regions) the bare itself. It is bounded by ontological universals given in an array of similar ones, even in their particular expressions as instantiations of many universals. Every utterance has thus ontological commitment to there being universals, particulars / natural kinds,[2] tokens and the whole Reality, from the

---

1 Quine, *Word and Object*, 119–20.
2 By particulars I mean reality-in-particlar or natural kinds or species, and not particular instances or tokens of substance. "[…] [W]hat in modern science is taken to be a natural kind is a form of substance, not particular instances in which the form

causal potentialities of which universals issue. I study ontological commitment in Quine in a detailed manner in Chapter 3, and show its particularistic problems in comparison with ontological commitment in Einaic Ontology.

Regarding the problem of accounting for the universals in the mind I am opine that mind connotes universals in its epistemic processes. Thus, ontological universals, which are ways of being of processes as such, become the ameliorated connotative universals in the mind. By 'connotation' I mean epistemologically the connection between the ontological and conceptual universals by the conscious manner of "noting together" the relevant ontological universals. It refers epistemologically to the conscious activity of involving ontological universals in sensing, conceptualizing, referring and speech by ontological commitment. More clarity on the concept of connotation – within the limits of a few sentences – yielded while connecting it with the Whiteheadian concept of eternal objects, will be found under the title: "4. Methodology of Mutual Collusion". Eternal objects in purely conceptual prehension are connotatives. Connotation is the conceptual activity of 'noting together or referring together to' the many by means of involvement of respective universals – which never denote any thing but only are involved with other ontological universals in 'noting together'. This differs from denotation, which is direct reference to one or a few actual processes to which it pertains, of which it is. Just like objects / processes, universals too are denotables: we denote them with terms.

All that there are for ontological commitment and without which discourse is naught, are tokens (bare singular entities / processes, e.g. the book / a book), ontological particulars (actual *species* [Latin, "appearance" / "mirroring"], e.g. book, with its final instantiations in tokens), epistemologically connotative ("noting together") universals in consciousness (conscious notions, e.g. 'bookness', 'possibility', 'actuality', 'temporality', 'spatiality' etc.), ontological universals (connectives and "ways of being of beings" in processes, e.g. generalities / universals like possibility, actuality, temporality, spatiality etc.), particular species universals that are instantiative in species (e.g. bookness instantiating in book, which in turn instantiate in a specific token, which is a / the book) and totalities (ever-greater totalities of tokens on par with and at the level of the ontological universals that

---

is manifest. Thus gold is a natural kind while a particular lump of gold is not; and the species *tiger* is a natural kind, while an individual tiger or group of tigers is not. So, just as science with regard to change is concerned with laws and not their instantiations *per se*, science with regard to non-change is concerned with natural kinds and not their instantiations *pe se*." Craig Dilworth, *The Metaphysics of Science: An Account of Modern Science in Terms of Principles, Laws and Theories* (Dordrecht: Springer, 2006), 149.

encompass a certain level of actual entities, and specific actual entities that too are conglomerations of minute actualities).

If one accepts these ultimates as categories, one has an ontology proper to them, as in most of today's metaphysics. The question arises: Should we justify the level of primordiality of these notions as final, or, are there still deeper generalities capable of being taken as most general categories? Our conclusion – to anticipate the argument in this work – is that it is not enough to have recourse to this sort of classificational categories. We should look for their maximal cases, which make these and all their other specific cases possible. The present work concerns itself about the questions pertaining to this conclusion and attempts a systemic answer, so that the system will begin to explain issues that arise in this regard in an ever-more dynamic and systemic manner. The listed classificational ultimates of disourse are what the various ontological and scientific traditions have taken as categories. Einaic Ontology argues that the alleged primitiveness of these notions must be questioned. There are some foundational implications behind them, which are least classificational but most collusive of all notions that are categorial of systems and sub-systems. We seek here a system of systems based on most primitive, dynamically systemic categorial mould-notions that allow adequate explication of their own selfsameness. We should not need to dig deeper than these notions but only ever deeper in its definitional explications, because these themselves will then arise in ever more adequate forms – an expectation which, I believe, does not betray the process of philosophizing.

Any discourse has a train of theoretically and pragmatically presupposed and idealized primitive notions discoverable in the idealized theoretical recesses of the discourse or theory. All primitive notions are categorial (classificational) in nature. They are not instituted merely by the mind in the process of learning a language or describing a process. They have ontological universal foundations in causal processes in Reality, which are vaguely discovered and epistemologically idealized by the mind as being active "there" in the measuremenally spatio-temporal but ontologically extension-motion interfaces of particular processes and their totality.

Matter-energy, Space-time and Causality are the most commonly accepted scientific categories. Spacetime is measuremental and so is epistemological. Extension-Motion is their physical-ontological counterpart. As explained in the Preface, ***space is the measure of extension; and time is the measure of motion, excluding the extent covered by the measure of extension***. The present work analyses the particularism in the particularistic and classificational natures of prevalent cosmological, epistemological and ontological categories, and suggests three minimally classificational, mutually collusive and maximally representatve categories. In the process, it clarifies the nature of ontological universals as probabilistic in their conscious counterparts, and lays the foundation for a second generation scientific ontology that transcends the purely classificational and least collusive particularistic categories of the scientific[3] and traditional ontologies of the bygone, more than two millennia old classificational tendency of ontologies and sciences in the West and the East.

Synthesis warrants least classificational categories. Western philosophy has witnessed many sets of first-generation scientific and metaphysical classificational categories that purport to facilitate ontological thinking. Today scientific data and theories are at the base of all ontologies. The sciences severally admit presupposing some or other ontology. We examine whether they generally presuppose merely classificational primitive notions, or whether they presuppose the most collusive of primitive notions. The set of collusive second-generation ontological categories here envisage maximally systemic scientific ontology. Collusion is the characteristic of involvement and mutual implication of categories and other derivative concepts in systems. As the three categories here are, respectively, cosmological, epistemological and purely ontological, it is impossible to attempt a system of thought by use of these categories within a short space. Therefore, I limit myself here to justify the merits of the projected systemic categories over classificational ones. Each chapter develops one maximal category. By the end of the third it will be clear that these ontological-collusive categories will substitute classificational ones with greater ontological merits and make the scientific categories ontologically subordinate. These categories are collusive: each of them "plays with" all of them.

The real (*das Wirkliche*) fundament of all thought, as will be clear, is Reality-in-total, not particular realities or actual entities in isolation from Reality. 'Reality-in-total' denotes an actual, ontological, partially phenomenal thing-in-itself that is all that are: that is, actual entities in all their implications and whatever there are, including concepts as conscious events, and even the Divine if it exists. 'Reality' does not primarily have, as in some scientific realists, the epistemological sense as given to *Realität*, but the ontological sense attributed to *Wirklichkeit*.

---

3 By 'scientific' in 'scientific ontology' I mean 'pertaining to and inspired by positive science/s'. 'Scientific ontology' does not mean merely what a scientist admits or assumes as the physical-ontological backbone of his discipline, but what ever more wisely informed philosophical analysis and synthesis yield as presupposed and produced by the positive science in question.

Chapter 1 investigates categorial possibilities in the philosophy of physics and derives the cosmological category of Reality-in-total, as presupposed in the ontology of all discourse.[4] Chapter 2 explores some basic epistemological foundations of such ontology and derives the epistemological category of Reality-in-general, as presupposed in the epistemology of all discourse. 'Reality-in-general' facilitates to consciousness the generality of all generalities – universal of all universals – and presents the conceptual ideal of all thought, i.e. To Be. Chapter 3 transforms the concept of Being into the abstract, nominal, verbal and transcategorial category, To Be, and overcomes the particularist analytic ontology of the "is" or "there exists" as presupposed in any sort of "ontological commitment" (Quine) to physical and abstract objects in discourse. To Be does not mean the universal Being of Aristotle, Aquinas, Spinoza, Hegel etc., nor the Heideggerian continuous openness of the event-ing of Be-ing of all that is with respect to Being-thinking human. Instead, it is the absolutely verbal-processual aspect of Reality, in its universal and totally processual characteristics. It is also law-like (nomic) and nominal. Throughout this book 'to exist' is equivalent to the general attribute, 'to be', of any actual entity. The term 'to be' is not coextensive with the To Be of Reality.

Classificational categories follow the method of predication by predicables (qualities) and class terms (particulars) about tokens of any layer of totalities. Predicables and class terms refer to either transcendentals (in the sense of essential qualities or universals pertaining to things / processes and/or their conceptual reflections) or to transcendents (in the sense of tokens and particulars, which are sets of tokens with their universals). Any entity and any particular is a transcendent to any other such and to other transcendentals. Similarly, any attribute or quality is a transcendental to any other such and to transcendents: it indicates the involvement of transcendentals in both abstract and actual entities of all kinds. However, ontologically there are no pure transcendents without transcendentals for the following reasons: (1) Any transcendent exists by its 'to be' (in its particularity) and transcendental qualities; and the Transcendent Reality exists by its 'To Be'. A "to be", the qualitative universals and the "To Be" are transcendentals in their own unique manners. (2) Similarly, any transcendental is the case with respect to some transcendents. Transcendentals and transcendents are ontologically intertwined. Any predicable involves one or more

---

4   For the scientific-realist and epistemological sense of the term *Wirklichkeit*, see Wallner and Peschl, "Cognitive Science – An Experiment in Constructive Realism; Constructive Realism – An Experiment in Cognitive Science", 107–108.

such transcendentals. Hence, classification by means of predication is a transcendental and transcendent affair.

Classification always involves many entities and predicables, both mingled with particulars. That is, the domain of definition of classification by predicables is the many particulars proper to the predicable/s. There can be unique domains of definition of transcendentals and transcendents. As we shall see in the chapters, these are ideal cases of classification by predication. Thus, these can be the maximal cases of transcendents and transcendentals – (1) the ideally most inclusive Transcendent, which is the ontologically and cosmologically Total Entity that includes all that exist, (2) the broadest and deepest epistemological Transcendental, the ideal concept occurring in consciousness,[5] and (3) the broadest and deepest Transcendental on par with Reality, i.e. the ontological ideal for there being anything. These are unique realities.

The terms applied to them are classifications, i.e. maximal and unique ideals of (1) actuals (both tokens and totals), (2) transcendentals (both particular and the more universal of ontological universals) and (3) conceptual activity (both the particular and more universal of conceptual universals, which are all veiled appropriations of ontological universals). Such maximal cases are also categories by reason of their uniquely defined domains of definition of classification. They are not disqualified from being categories: somehow they too are classifications of uniquely defined singleton domains that include all modes of reality. Nor are they classifications *merely* in the sense of there being nothing else in their domains unrelated to classification by each such other term: they cannot be thought of without the others of the set. *The challenge that the present project is offered* is to transform such maximal and uniquely defined predications as coextensive to the very objects of such predication, and then to show that they qualify as the best scientific-ontological categories.

The term 'category' is from the Greek *katēgoría*, "accuse, affirm, predicate, speak, accuse or classify someone in the assembly / public place (as notified or noticed by the ruler)". In this original sense categories are basically classification terms. The Einaic categories are in that sense classifying terms. The maximality assignable to them reduces their classifying nature in comparison with the traditional non-maximal categories. Since the maximal nature of Einaic categories

---

5   By 'consciousness' I mean not merely the empirically experimentable mind of empirical psychology, where the distinction is between the conscious, sub-conscious and unconscious. I include in it also the pre-conscious and physical causal influences on and results of their causal past, and also the potential causal future.

reduces their classificational function to the minimum, we choose to term them non-classificational.

2. General Method and Rationale

This book does not proceed vertically into just one thinker or one theme in a thinker. It enters horizontally and cross-sectionally into a persistently single theme in and beyond some major philosophical disciplines and representative thinkers, most of them from the twentieth century. Its horizontal approach does not make it shallow.

Methodologically, I deem it appropriate to *analyse* here the theoretical recesses of the theme of categories in relation to its twentieth century scientific, epistemological and ontological revivals. I *presume* the ideas of some of the thinkers studied here; *evaluate* some of them; and *advance* from, without or beyond them – in order to enter upon the topic of this work. Justice cannot be done to depth of treatment in whole thoughts of individual authors. I avoid eclecticism (*ekklegein*, "selecting the best from a group of things") and make sure that the material used from different thinkers is used only to generate the the Einaic axial categories for scientific ontology by transcending their specific modes of procedure and conclusions. The three proposed Einaic categories are new in themselves, developed in juxtaposition and dialogue with philosophies who have produced other metaphysical categories. The method of procedure carries us beyond classificational categories.

The *rationale of this enterprise* is the intuition that although pragmatic and scientific purposes circumscribe meaning from within the immediately given, their justification requires an ontology that circumscribes (not just totalises) meanings from within their course to and from the whole, and is theoretically implied in every instance of meaning[6] that has its foundation on ontological processes and universals beyond them in the backdrop of the actuality of ever greater wholes (totalities). The concept of ever greater wholes is never completely whole and final. This opens the way for a set of categories capable of continuing to evolve more and more wholesomely and make each other evolve in a truth-probabilistic manner. Therefore, *I hold it possible to derive* a progressively more

---

6   For a similar contemporary semantic position not inclusive of the general ontological position held by this essay as the most fundamental for philosophies of science, epistemologies and ontologies, see Louise Antony, "Semantic Anorexia: On the Notion of 'Content' in Cognitive Science", 105–135, especially 114. The present work aims at more than such a semantic position.

science-enhancing, realism-enhancing, more exact, speculative, synthetic and yet ever more continuously truth-probabilistic *second generation scientific ontology* by creating a set of non-classificational, mutually colluding, particular-possibilizing, Transcendental and Transcendent categories – re-generalized beyond the categories, essences, Being, universals and realistic-empirical categories, respectively, in Kant, cosmology, epistemology and analytic ontology.

The possibilizer of this ontology is its highest nomic-nominal verbal-processual (both discussed later) trans-categorial condition, To Be, which is the purely ontological Transcendental (that connects everything with all possible others) of the two other proposed categories. Reality-in-total is not merely epistemologically taken as "objective", but as "objectual-causal", i.e., in its trans-mental entirety, with ontological commitment to *there being something* as the toal object, beyond but ontologically related to the cognitive activities of the mind. Reality-in-general is, conceptually / "connotatively" speaking, taken purely as the conceptually universal *conditio sine qua non* and maximal case of all thought – in which, connotatives are active in constituting concepts by mixing connotative universals with elements in conscious activities. 'Objectivity' has the overtone of adequacy of a concept or a point of view with respect to the actual. By 'objectual' I mean 'of actual objects (processes) in their mind-independent otherness'.

3. Transcendentals and Transcendents

Conceptual and extra-conceptual entities are all that there can be in ontology. *Conceptual entities are not non-entities*, but are transcendentals (generals) expressed based on conceptual events that have some basis on the ways of actual processes. These are based on conceptual events insofar as they are conceptual reflections of ways of being of processes in their inter-connectedness. *Ways of being of processes* are what we (agree to) call *ontological universals*. Hence, their conceptual reflections are connotative universals, by use of which we have concepts, words, sentences, their expression etc. Moreover, *ontological universals are not mere functional entities in discourse*: they are real ways of being of processes. *Universals may be denotative* (purely ontological, in processes) *and connotative* (active in concepts as derived variously from many processes). The connotative aspect is the reflection of ontological universals, and the denotative aspect is that which is qualitative in processes, without consideration of the blend of the ontological aspect with actual conscious elements and concept-formation, and it is functional in denoting by reference. The functional quality of denotative ontological concepts is supported by their ways-of-being quality and the connotative quality of their conscious reflections in concepts.

To clarify the concepts of 'object', 'event' and 'process': An *event* is so called based on the isolated activity or *process* of the *object* dealt with in naming it as entity. In the Kantian and Husserlian traditions one tends to understand 'object' as the mental counterpart, wherein there is something objective in the content of the act of perception or cognition. Even from this point of view, one tends to make 'object' universal-bounded, by arguing that this mental counterpart, present and active in the mind, is involved with other pre-reflective and reflective particulars, universals and concepts. For example, Brentano speaks of mental objects as objects: "Nothing distinguishes mental phenomena from physical phenomena more than the fact that something is immanent as an object in them." Having an object is "… having-of-something-objectively."[7] *This is not the concept of 'object' I use.* As Chapter 1 concentrates on redeeming cosmological categories from Kantian idealism and dualism of phenomena-noumena, and as we approach the concept of the infinitely and infinitesimally objectual concept of Reality by the end of that chapter, I hold that objects are mainly those processes that have direct or indirect causal effect on others or those ways of being of causal processes that can have ontologically universal-bounded objectification in the mind. Such objectification is some ameliorated reflection of processes (not objects) in conscious and conceptual processes, so that consciousness and speech give these processes referring capacity by fixed nominalization in consciously reflected, spatiotemporally sensed or imagined or rationalized, extension-motion processes.

An event also is universal-bounded. It is the object / actual entity / process – named *without direct recourse to the causal nature* of the processes within and without it – that is simultaneously in inner and outward causal activity and relation, in its own right. *Eventhood denotes* the activity and relation of the entity within and without, without explicit mention of its causal nature. Activity and relation mix the selfsameness of the active and related object with the essences of relationality with respect to the event. These essences are nothing but the universals involved in the particular, of which the event is in fact a causal instantiation: a process in extension-motion. In short, the concept of an event is always universal-bounded in an extra-mental manner. *These ontological universals are of causal processes, not of objects.* The universals involved therein are not lodged in the mind, but in the events themselves that they are related to with respect to those universals. 'Event' as a term does not carry with it its processual-causal

---

7   Brentano, *Psychology from an Empirical Standpoint*, 197. See also footnote 3 in Brentano, 197.

nature, but its relations with other processes or events. This is the sense in which I use the term 'event'. Though 'entity' and 'object' too are its representations without recourse to the causal nature of processes within it, the essential difference between 'event' on the one hand and 'entity' and 'object' on the other is that *an event is mentioned in its relationality without mention of its causal nature*, and 'entity' and 'object' are used to indicate its particularity, without reference to their process-nature and relational nature.

It is in place to distinguish between a universal and a transcendental. An universal is wider in range of denotation of what it means exactly in instantiation or possible instantiation in the many, and at the same time an ontological universal can be conceptually a connotative involving the many. A transcendental is an ontological pre-condition that involves a process and its relevant universals. All universals are transcendentals in the sense that they are conditions to ontological explication. The To Be of Reality is the only highest Transcendental as condition, and also a universal. A universal is (1) ontological – a way of being of processes and inherent in processes, not in one token but in the many – and (2) connotative – in that ontological universals are involved and invoked in conscious activities of all kinds. Wherever we use 'universal' and 'transcendental' interchangeably, the universal aspect is stressed more, and when 'transcendental' is used, the precondition aspect is stressed.

We categorize entities / processes in general using two terms: 'transcendental' and 'transcendent'.

(1) Concepts are conceptual events based on universals that are a blend of the pure and actual aspects: transcendentals which are the general ways of being of actual entities, and elements in extension-motion namely transcendents, which are the conceptually processual, conceptually instantiated parts of reality. Concepts cannot be just any one of them. The actual aspect here is the specific conscious stuff of the event of concrescence ("growing together")[8] of the purely ontological potentials / ways of entities (the 'pure aspect') with elements of spatiotemporally measurable extension-motion actuals in a highly variegated manner. Thus, an actual concept is a extension-motion entity, and so it is also a realization of connotative ideals in consciousness. It can also be called an actual entity, but need not be an actual occasion in the Whiteheadian sense.[9] Specifically, it is a

---

8 Concrescence takes place when the "[…] objective intervention of other entities constitutes the creative character [….]" Whitehead, *Process and Reality*, 220. 'Objective' refers to any form of contribution of elements by actual entities to others.

9 Whitehead defines the plurality of actual entities thus: "'Actual entities' – also termed 'actual occasions' – are the final real things of which the world is made up." Whitehead,

transcendently transcendental actual entity, so called because, by our convention, *transcendentals here concresce* ("grow together") *with transcendents* in a conceptual manner.

(2) There are also actual entities / events, processually constituted by ontological transcendentals / universals and physically extension-motion transcendents, without involvement of conscious instantiation. These are transcendents, since by assumption extension-motion transcendents concresce with ontological transcendentals / universals, and not with conscious universala. Concrescence of transcendents is with transcendentals that are the ways of being of beings in processes. These transcendentals are called ontological universals, since they are primarily in processes, and only then in mind in a conceptually veiled but connotative fashion.

Against the background of such a two-pronged assumption, the present work holds that it is ontological transcendentals / generalities / universals – active as ways of being of processes – that ingress into conceptually (consciously) and/or physically extension-motion actuals. Every process is a transcendent to every other. In general we call the totality of all such actuals as Reality, which is the Absolute Transcendent with respect to every entity. The Divine too is a Transcendent, but It is the consciously absolute transcendent, and not ontologically absolute: the Divine is not the same as Reality. Everything in Reality does not prehend consciously in a developed fashion as conceptual entities within it do, or as the Divine does infinitely deeply. The Divine pole of Reality is infinitely conscious and active in every finite extension-motion region. The Divine is infinitely conscious, but not so absolutely as to become Reality and have it for Its own self:

Process and Reality, 18. Any entity, inclusive of God, is an actual entity. An actual entity may be understood as final actualisation or reality, i.e., *res vera*, "true thing". Whitehead, *Process and Reality*, 22. "[…] [F]rom the standpoint of any one actual entity, the 'given' actual world is a nexus of actual entities, transforming the potentiality of the extensive scheme into a plenum of actual occasions; … in the plenum, motion cannot be significantly attributed to any actual occasion; … the plenum is continuous in respect to the potentiality from which it arises, but each actual entity is atomic; […] the term 'actual occasion' is used synonymously with 'actual entity'; but chiefly when its character of extensiveness has some direct relevance to the discussion, either extensiveness in the form of temporal extensiveness, that is to say 'duration', or extensiveness in the form of spatial extension, or in the more complete signification of spatio-temporal extensiveness." Whitehead, *Process and Reality*, 77. In short, conceptual entities are also spatiotemporally measurable extended.moving actual entities.

Reality includes, and is not the same as, the Divine.[10] The Divine prehends not merely transcendentally in the particularistic sense, but Transcendentally,[11] in the sense of infinite consciousness. This is the best possible instance of conscious involvement of To Be, Reality and Reality-in-general. Though the Divine is not explicitly dealt with in this essay, making a philosophical theology possible is also one of the motives at the back of it.

The Transcendental is the all-subsuming and pre-conditional pure potential, To Be: it belongs ontologically to Reality. The latter ontologically possibilizes both transcendentals and transcendents by involving itself in them at the level of the 'why' of transcendents: only ontologically, because no conceptuality is involved in a Transcendental (which is, in fact, purely ontological), except in its conscious representation. The Divine is conscious of Reality (Itself and the cosmos) in the best manner – in the infinite, not fully ontologically absolute,[12] manner of identity possible, because the Divine is not part of the cosmos or vice versa. Therefore, the Divine is conscious of Reality at the dimension of the ingression of the Transcendental within the Absolute Transcendent, Reality. This Transcendental is To Be. The Divine is aware of Reality (Itself and the cosmos) infinitely consciously from the zone of Reality's To Be. This is absolute knowledge, because the Divine is infinitely intensive in its processual existence, though the Divine is not identical with Reality. The Divine is different from the cosmos, and self-identity in consciousness does not transpire between the Divine and the cosmos. To Be

---

10 This is no pan-en-cosm-ism. Cosmos is included in Reality-in-total. Similarly, the Divine is also included in Reality-in-total. But the ontologically transcendent (not being identical to) but fully infinite presence and activity (immanence) within the ontological stuff of the cosmos makes the Divine conceptually have the cosmos fully within Himself, and not vice versa. The Divine is not a / the Transcendental. Hence, I hold pan-en-theism. The cosmos is held in infinite activity by the infinitely conscious being of the Divine. The Divine and the world are members of Reality-in-total.

11 Here infinite and absolute foreknowledge is part of the Transcendental consciousness of God, which is tinged with the Transcendent nature of Himself and the universe. The infinite extension-motion intensity of God takes care of the absoluteness of foreknowledge.

12 'Infinity' in God here need not be the 'absolute infinity' by which God is one and the same with the world in order to know it fully, since 'absoluteness of identity' should mean 'infinity of infinity of … *ad infinitum*' identification in all parts. Absolute consciousness of the world by being absolutely not in contact with the world implies the property of being a vacuous Aristotelian Unmoved Mover God. The absoluteness of God's foreknowledge can be due to his absolutely intense imbuing all extension-motion regions of Reality-in-total.

already involves all that it possibilizes (i.e. transcendentals and transcendents) simultaneously in a universally unique and several way. Hence, the specificity of actual entities is not at stake in such awareness. Reality is more than the Divine or the universe severally. It has an ontology that runs at the level of its To Be. The Divine's ontology works at the To Be level only consciously. Clearly, our conceptual grasping of Reality involves the ideal of Reality-in-general that generalizes on all that there are, in their generality and particularity, imperfectly. But due to the infinite intensity of the Divine in all extension-motion regions, the Divine in its parts of infinite activity in all infinitesimal extension-motion regions experiences all that transpires in the cosmos. In the Divine the experience at the To Be of Reality realizes experience from the viewpoint of all the three Einaic categories, if all the experience of the Divine is summated unto infinity of infinity of ... *ad infinitum*. Therefore, the Transcendent and the Transcendental ultimates are best experienced in the Divine, if the Divine exists.[13]

I shall argue that our three ontological-cosmological-epistemological categories can be relied on as theoretically creative of specific categories of particularity and local-ontological and scientific systems proper to such categories, since they allow the specific as a real possibility within the nexus of the cosmologically total, the epistemologically general and the ontologically general. These categories are creative of systems only in their togetherness. *Therefore, they are mutually collusive* for their mutual implication and for the implication of the specific.

The history of philosophy has produced many categorial schemes. Most of them have dealt with things directly in the classificational manner. What about continuous mutual implication of the most general of these generalities, which philosophical systems do not purport to do while attempting coherence, adequacy and maximum correspondence-level truth-probability and applicability? Classificational categories do not imply mutually. Any scheme of categories which maximally imply each other by their very nature are called collusive. Simple and direct generalities of classificational categories can only be transcendentals and/or transcendents of higher kinds, since they together include and imply some particulars. These are possible only with the actual discovery and application of the inductively synthetic concept of the purely transcendental foundation, the *most abstractly verbal and* universally *relevant* "To Be", with its nature of collusion with the other highest categories: Reality-in-total and Reality-in-general.

---

13 The process God does not know the world in its future freedoms fully. A concept of the continuously creating spiritual-bodied Divine fully knowing the world in all its aspects due to his infinite activity in every finite extension-motion physical region but is not identical with the world is in preparation.

Other categories of the ontological tradition like substance, actual entity, essence, space, time, relation, causality, structure, object, number etc. are only subordinate to these. The traditional view of categories entails setting clear boundaries within the range of application of each category and providing a set of defining characteristics necessary and sufficient for membership in the class of categories. This can no longer be the meaning in the context of Einaic system-building. Through the centuries a more lenient position has been advocated. It allows ambiguous boundaries and different bases for assessing category membership. By means of their Transcendent and/or Transcendental universality, our maximal categories tend to reduce the traditional ones into overly specificistic predications. I shall adduce arguments to show that particularisms of the latter type are of relative irrelevance for General Ontology.

We are familiar with the concept of Being in Heidegger. My study proposes a different, abstractly ontologically nomic (Greek, *nomos*, "rule") and processual concept of To Be. The rationale of the universally nominal and verbal concept of To Be here is that, although the showing-itself of To Be is in and through the Being-thinking and Being-Enowning *Dasein* and all other entities as in later Heidegger, its deepest conceptual status shows it as the deepest ontological universal, and as such it is not to be restricted to the anthropological or particular-bounded viewpoints. We may speak of Being that shows forth from the 'to be' of Being-thinking humans and of entities in terms of *Alétheia*, "Unconcealment (from covering, oblivion, sleep)", and *Ereignis*, "En-owning" or "E-venting". For this reason, it is necessary for completeness here to bring in also the ideal of the To Be of Reality. To Be, as the Transcendental and universal *par excellence*, is the widest context rendering reality and thought possible, even when its process in humans (in particular) and entities (in general or total?) are *Alétheia* and *Ereignis*. Just as one uses the so-called universals like set, object, cause, asymptotic approach, infinity, process, generality, totality etc. to connote (never to merely denote) those that belong to those classes by their common features, one may also use the highest ontologically denotative and denotable (see below) concept of Reality and connotatively denotative (see below) concept of Reality-in-general to study Reality categorially through the To Be of Reality.

The objectual orientation pole – abstract (pertaining to universals) or concrete (pertaining to particulars / natural kinds, and tokens) – that is denoted by a denotative term is called a denotable. Anything connotatively denotative is a conscious activity, and connotatively denotable is a conceptual entity. Consciousness connotes the many entities / events / processes through the involvement of universals, and at the same time tends to denote the processual or transcendental entities as such. Actual concepts and epistemological formations from them in the mind are to be treated here by referring also to their universal-boundedness from processes from outside and from within.

Transcendentals (universals as conditions) are necessary for ontology, cosmology and epistemology. Moreover, we need something objectually present beyond thinking, beyond the Being-thinking human's Interpretation, for us to be made to objectuate and objectivate. This aspect, Reality, is defined here as something given most generically and by inductive totalization in and beyond particular ontological commitments. To Reality belongs the process of showing-itself of To Be in human and in the cosmos.

Obviously, humans do not have Reality as such in consciousness. Hence, something abstractly / universally / nomically objectual-causal (*objectual* in the sense of the implied affectation of Reality in human consciousness; and *causal* in the sense of processes in Reality being the cause for the ingression of such connotative universals in consciousness) from Reality and reality-in-particular ingresses in thinking to reflect the totally causal behaviour of Reality. In order to represent this realm of re-presentation in consciousness, we *posit* Reality-in-general as the conceptual and Transcendently (in terms of the Transcendent, Reality in total and in parts) Transcendental (active at the level of transcendentals and the highest Transcendental, To Be) category, which is objectually different from the process of showing-itself of To Be in and through Reality. Reality-in-general is the appearing together of Reality and its To Be in consciousness.

4. Methodology of Mutual Collusion

There is an ontologically collusive science-enhancing system in Whitehead's Categories of the Ultimate: Creativity, one and many. Contrarily, Aristotle, Aquinas and Kant have already maintained a thoroughly differentiating emphasis in their categorial schemes. Husserl platonically absolutized essences as pure epistemic universals in absolutely acausally oriented (because he did not connect the universals to their causal-processual foundation in Reality) and consciousness-imbued, specificity-imbued abstraction from the objects of awareness and plunged them in conscious experience alone, in order to make them purely phenomenological and based in "experience". Heidegger has turned phenomenological essences into the Being-historical giving-forth and Enowning (*Ereignis*) of Being, but with detriment to its universal and abstract nature and altogether leaving out the totality of Reality with its ontological and cosmological structures. The particular sciences and analytic ontologies revel in inductively partially generalized and *ad hoc* categories useful only for empirical research, not for collusive ontological reflection. Whitehead does not collude his categories by To Be. Einaic Ontology attempts ever better systematicity by its collusive categories that go even beyond the particularisms of the sciences and the above-said philosophies.

Any categorial scheme in which the members maximally imply each other by their very nature is a collusive scheme. Collusion facilitates categorial synthesis in systems. The present study synthesizes the ontological Transcendent pole Reality, with the purely ontological Transcendental pole To Be, by involving the ideal epistemological pole Reality-in-general. The three categories are projected as Einaic (Greek, *Einai*, "To Be") categories, based on the purely transcendental science of Einaiology (To Be studied via Reality and Reality-in-general) and the most Reality-imbued science of General Ontology (Reality studied via To Be and Reality-in-general). These may be done only together as Einaic Ontology, due to the mutual theoretical dependence (collusive nature) of the categories. This fact too projects this science as second generation ontology, as we learn in what follows.

Einaic categories yield an ever-evolving system, since by nature their senses collude with and imply each other. Proper mutual collusion of these highly Kantian-sounding but non-classificational categories that are at the foundations of knowledge and Reality, justifies derivation of a systemic and ever self-improving ontology beyond contemporary classificational cosmologies, epistemologies and ontologies, and beyond the traditional purely classificational distinction between the two purely transcendent and transcendental poles – the metaphysical Divine and the To Be – of Reality. It remains to be evidenced that the Transcendental and Transcendent modes of maximal categorialization (not classification) by a unique maximal domain of definition that we adopt here, will give clarity to speculative (not to mere common-sense, pragmatically "clear", particularistic) thinking.

The *methodology of mutual collusion* implied within the nature of our categories is describable with more details than earlier in the following manner: Reality is the highest Transcendent. It is the highest token entity by inductive totalization of all possible objects of implicit ontological commitment. But ontological commitment to token entities is through particulars and universals. Therefore, Reality is the Transcendental Transcendent. Similarly, Reality-in-general is conceptually *of* Reality. That is, the quality of the Transcendent is present in it. Further, it is the conceptual occurrence of Reality in its ontological To Be, which is the highest ontological Transcendental. Hence, Reality-in-general, which is qualified by Reality and To Be, is the Transcendently Transcendental category. In short, the Transcendent and the Transcendental aspects of all that are the case are the instruments of collusion. These never appear separately. The following three chapters will deal with these issues in more depth.

Transcendentals are universals. Husserl has worked much on universals and the formal nature of concepts. Therefore, from the very initiation of our inquiry into the highest, broadest and deepest of categories of all thought, it is good to digress a bit to Husserl's work and differentiate his concept of abstract objects from ours. Pure essences in Husserl are conceptual universals. They do not have extra-mental existence. He speaks much about the ontological implications of the contents and formality of concepts. "'*Abstract*' contents are *non-independent* contents, '*concrete*' contents are *independent*. We think of this difference as objectively determined, perhaps, as follows: concrete contents are by their nature such as to be capable of existing in and for themselves, whereas abstract contents are only possible in or attached to concrete contents."[14] Extra-mental existence of concrete objects is the implication of the above, and is part of an *a priori* formal ontology in Husserl:

> This objective distinction between abstract and concrete is the more general, since immanent contents are only a special class of objects (which naturally does not mean of *things*). The difference in question could therefore more suitably be called a difference between abstract and concrete *objects* or parts of objects. If I here still continue to talk of 'contents', I do so as not to give persistent offence to most of my readers. The distinction has arisen in the field of psychology, where intuitive illustration naturally grasps at sensuous examples. Here the interpretation of the word 'object' to mean a thing, is so dominant, that to call a colour or form an object might be felt disturbing or even confusing. We must, however, bear in mind that *talk about contents is not here at all restricted to the sphere of contents of consciousness in any real (reellen) sense, but embraces all individual objects and parts of objects as well*. We do not even restrict ourselves to the sphere of objects which are intuitive to us. The restriction has, rather, an ontological significance: objects are certainly possible, that in fact lie beyond the phenomena accessible to any human consciousness. The distinction, in short, concerns individual objects as such with unrestricted generality, its true place is in the framework of an *a priori* formal ontology.[15]

Husserl defines also abstraction:

> Taking the objective (ontological) concept of 'abstract contents' as our foundation, we may mean by 'abstraction' the *act* through which it is not prised loose, but none the less made the peculiar object of an intuitive presentation directed upon it. It appears in and with the relevant concretum, from which it is abstracted, but it is specially meant, and

---

14 Husserl, *Logical Investigations I*, 428.
15 Husserl, *Logical Investigations I*, 428–49.

further not merely meant (as in 'indirect' merely symbolic presentation) but also intuitively given *as* what it is meant as being.[16]

However, the distinction he brings up within abstraction disqualifies his abstract contents from being equated with what we understand as the sense of the abstract / universal. In what follows we glimpse him making the distinction:

> If we pay attention to one of the side-surfaces of a cube which appears before us, this surface is the 'abstract content' of our intuitive presentation. But the genuinely *experienced* content, which corresponds to this phenomenal side-surface, itself differs from the latter: it only forms the basis of an 'interpretation', through which, while it is sensed, a side of the cube differing from itself manages to appear. The sensed content is not in this case the object of our intuitive presentation: it first becomes such an object in psychological or phenomenological 'reflection'. Descriptive analysis none the less teaches us that it is not merely contained in the total concrete appearance of the cube, but that it is in a certain fashion set in relief, accented as against all other contents, which in *their* presentation of the side-surface in question do not function representatively. It naturally remains so emphasized, when it becomes *itself the object* of a presentative intuition peculiarly directed to *it*, except that then (in reflection) this intuition is added. This setting-in-relief of the content, which itself is *no act*, but a descriptive peculiarity of the phenomenal side of the acts in which the content becomes the bearer if its own intention, could likewise be called 'abstraction'. We should thus have laid down a totally new concept of abstraction.[17]

In addition, the following is the characteristic of abstraction, which allows us to understand 'universal' in Husserl differently (ours being the *a priori* ontological condition – transcendental – for there being anything and for there being any content in thought):

> If we assume that abstraction is a peculiar act or (in general) a descriptively peculiar experience, responsible for setting the abstract content in relief from its concrete background, or if one sees in the manner of such setting in relief the essence of the abstract content as such, one comes to yet another concept of the abstract. Its difference as against the concrete, is *not sought in the intrinsic nature of its contents*, but *in the manner in which they are given*. A content is said to be abstract, in so far *as it is abstracted*, concrete in so far as it is *not* abstracted.[18]

The highly *a priori* and ontologically committed nature of the abstract and its content is not held in this renewed definition of abstract content. This is an ontological defect in Husserl's consciousness-imbued definition of abstractness of

---

16 Husserl, *Logical Investigations I*, 429.
17 Husserl, *Logical Investigations I*, 429.
18 Husserl, *Logical Investigations I*, 429–30.

abstracta as dependent solely on the act of abstraction. If abstractness is so dependent on abstraction and not at all on conscious connotation via abstracta, nor on the presentation of To Be to consciousness in terms of Reality-in-general and the ontological universals of causal processes, there cannot be ontological universals our kind in Husserl: whether they are the representative *a priori* qualia and ways of causal processes is not clear in him. An Einaic-level set of categories and such an ontology are thus blocked away from the world of Husserlian phenomenology. I propose universals and Reality-in-general as imbued by Reality in its To Be. This makes us admit To Be as the final criterion of construction of universals and categories through mutual collusion of categories.

We have an ontologically general definition of abstractness of anything abstract: Abstractness is that quality which necessarily *is of* and *is implied* in the related objectual processes *and* in the universal/s that thus represent aspects of concreta / tokens in their patterns. Objectual implication implies committing to ontological independence of things as (causal) processes by means of conceptual universals that are by nature veiled reflections of ontological universals / ways in processes. If extra-mental existence of objects and the dependence of their generalities / ways on all of a type are a somewhat *a priori* affair, the question of psychological contents of abstract objects (universals of various grades) also has a somewhat *a priori* aspect. The *a priori* aspect is not presupposed in the actual consciousness about a thing, but given in its *a priori* presupposition which is a Quinean-like but broader ontological commitment. Hence, we should have some extra-mentally ontological foundation for conscious abstractions and their presentation in terms or words as universals. The ontological foundation is given in the *a priori* presupposition, and not in conscious universals as in Husserl. Even his concrete contents are irreducibly consciousness-dependent and not couched in ontological commitment to there being representative processes extra-mentally.

Similarly, eternal objects in Whitehead are ontological potentials / universals working as instruments of novelty in actual entities[19] through the agency of (the

---

19 "In such a philosophy [of organism] the actualities constituting the process of the world are conceived as exemplifying the ingression (or 'participation') of other things which constitute the potentialities of definiteness for any actual existence. The things which are temporal arise by their participation in the things which are eternal [eternal objects / pure potentials]. The two sets are mediated by a thing which combines the actuality of what is temporal with the timelessness of what is potential. This final entity is the divine element in the world, by which the barren inefficient disjunction of abstract potentialities obtains primordially the efficient conjunction of ideal

category of) Creativity in things, which is active in and through the Divine (Creativity being not an equivalent of the Divine). Here the base of such abstractions in the To Be of Reality for their justification is not stressed. Ontological universals can possibilize everything ontological only if they are based in (1) their ontological commitment to the specificity of *something* without reckoning their specific qualities represented in the universals, and (2) their ultimate possiblizing orientation to To Be that reflects the fact of Reality.

Objects apriorily universal in various grades contribute the need for ontologically maximal categories at their limits. They are not all categories, since apriority is spectral in grade (which theory needs much developing especially to avoid the analytic philosophical misuse of the term) in complementarity with aposteriority in the same spectrum. Those that are most applicably universal of all variously apriorily universal objects qualify for the name 'categories'. In short, one such final category should automatically imply and facilitate the others. Such mutual implication and facilitation is the quality of the highest mutual collusion of final categories. There will always be criteria of collusion, and the last of the line of criteria is the To Be (that-ness / thus-ness) of Reality. Hence, universals are ways of processes extended for theoretical concreteness of the collusion of our categories.

Einaic Ontology is the science that studies anything from the point of view of the re-universalised and collusive categories of To Be, Reality (that collusively

---

realization. This ideal realization of potentialities in a primordial actual entity constitutes the metaphysical stability whereby the actual process exemplifies general principles of metaphysics, and attains the ends proper to specific types of emergent order. By reason of the actuality of this primordial valuation of pure potentials, each eternal object has a definite, effective relevance to each concrescent process. Apart from such orderings, there would be a complete disjunction of eternal objects unrealised in the temporal world. Novelty would be meaningless, and inconceivable." Whitehead, *Process and Reality*, 39–40. Square brackets mine.

Whitehead's concept of eternal objects pours much light into the concept of universals / essences in Einaic Ontology. Insofar as eternal objects can be active purely conceptually, they are epistemologically connotative universals; and insofar as they are the original commonalities in things / events / processes, given in ontological commitment as ways of processes, they are ontological universals. Whitehead does not make a clear distinction between epistemologically connotative and ontological universals, nor does he discuss the connection between the ontological commitment that yields ontological universals in epistemologically connotative universals. Yet, it goes to the credit of his genius that he brought the so-far static Platonic ideas into dynamic interaction in prehension within the whole of Reality.

qualifies actual, specific connexity by the Einaic orientation of the ontologically particular), and Reality-in-general (that collusively qualifies every conscious essence by the Einaic orientation of the conceptually specific). Mutual collusion of our categories is possible under the supra-classificational, ontological, connective nature of the universal To Be. Mutual collusion by the Einaic categories yields greater synthesis of the whole system of thought by allowing also the best imaginable place and maximum clarity to the particular, since the particular is active always within Reality. Thus, collusion is the methodology of the Einaic categories.

The place philosophy has given to reality-in-particular – be it cognitive, logical or ontological – remains to be scrutinized for genuineness by asking if anything greater than the particular is better implied by the fact of generalities of particulars. The preference for the particular and the practical has to be provided for in the context of the universal and the total, without adumbrating the importance due to the particular. The whole exercise in the present work is to supply the Einaic ontological *sine qua non*, i.e. an adequate and collusive concept of To Be and its correlates (universals), to future general ontology and scientific ontology without detriment to the importance of the actual. The two thousand and four hundred years of western philosophical thought show the problem in the foundations of thought to be issuing from a lack of blend of Reality and thought from their universal and particular aspects, down to token entities. Insofar as the universal-bounded Reality and thought cannot very much be separated into exclusive facts or particulars, we need to found thought on categories that blend the universal, the particular and the token, to the ultimate satisfaction of *derivability* of maximum truth-probability.

It seems insufficient to recur to centring Reality merely on the Interpretive *Existenz*, "standing out", of *Dasein* (as in earlier Heidegger), because it is an inversion of the ontological importance of the To Be of Reality *in terms of its conceptual occurrence* and realization in and through human *Dasein*. Particulars and universals are genuinely themselves if (1) a mode of having them mutually tinged (collusive) allows a categorial framework that penetrates through any one category of that system into the realm of the other categories – without unnecessary purely non-ontological foundation of the occurrence of the collusive condition "To Be" on *Dasein* – and (2) a mode can be had that makes systems after systems possible through ever more systematically ontological collusions of the senses of the same categories.

The mutually penetrating but categorically different nature of the categories cancels spurious identity of indiscernibles between the categories in the ontological scheme, but allows for the practical application of the categories in propositions. Our three categories have the Universal Law of Causality as the basic principle of formation of ontological universals. This requires extension-motion differences to remain differentiating elements of universals. Therefore, only the acausal connotative formulations of universals each type is identical to another of the same type. The ontological universals of each minute way of processes remain different – at the most partially identical and partially different from similar ones. But the principle of identity of indiscernibles is:

> "[…] any of a family of principles, important members of which include the following: (1) If objects *a* and *b* have *all* properties in common, then *a* and *b* are identical. (2) If objects *a* and *b* have all their *qualitative* properties in common, then *a* and *b* are identical. (3) If objects *a* and *b* have all their *non-relational qualitative* properties in common, then *a* and *b* are identical."[20]

Evidently, (1) yields identity. It is based on identity of all properties. (2) is based on identity of only qualitative properties. Existence of hidden qualitative properties differentiate otherwise indiscernible entities. (3) is based on non-relational qualitative properties. It is a contingent truth, "[…] since it appears possible to conceive of two distinct red balls of the same size, shade of color, and composition. Some have argued that elementary scientific particles, such as electrons, are counterexamples to even the contingent truth of (3)."[21]

The Heraklitian identity of indiscernibles is famous: 'One does not step into the same river twice'. Quine argues closer to my understanding of identity of indiscernibles, in terms of Hume, thus:

> […] [I]f identity is taken strictly as the relation that every entity bears to itself only, he is at a loss to see what is relational about it, and how it differs from the mere attribute of existing. Now the root of this trouble is confusion of sign and object. What makes identity a relation, and '=' a relative term, is that '=' goes between distinct occurrences of singular terms, same or distinct, and not that it relates distinct objects.[22]

This is enough admission of the fact that not only such minimal realities but also our ideal non-classificational categories are identical only to themselves. Even when our categories represent processual realities in their domains of definition proper, they resist further maximization. The ideal of Reality persists in

---

20 *The Cambridge Dictionary of Philosophy*, second edition, s.v. "Identity of Indiscernibles".
21 *The Cambridge Dictionary of Philosophy*, second edition, s.v. "Identity of Indiscernibles".
22 Quine, *Word and Object*, 116–17.

extension-motion as itself. Reality-in-general persists in human consciousness as the epistemological ideal itself, and To Be persists in Reality as the deepest ontological ideal itself. The first is a unique and total token, the second an epistemlogically universal connotative exemplified in conscious concepts, and the third the ontological way of being of (based in the process of) Reality. Moreover, any and every specific entity, idea or way of being within these three is identical only to itself, and not to any other, given the uniqueness of universals that go to form them. Similarly, our three categories are identical severally to themselves, and not each to another. It is these that remain differentiable subjects and predicates in statements. Therefore, the collusive quality of our categories makes them different by themselves.

We cannot rule out from any category the aspect of absolutisation and relativization. Thus, it is important to produce for ontological consumption a mutually qualifying, mutually explaining set of categories that absolutize only in terms of each other and yet have space for improved theoretical expression. Such a scheme can characterize and justify the cause of inventing specific categories for regional ontological enterprises. This speaks for the relevance of the topic under study.

The reason why the ultimate Transcendent category is Reality-in-total, and not actual entity or reality-in-particular, is that nothing less than the To Be of Reality is capable of justifying any category. *Reality*, which is the highest ontological realization of To Be and the unique domain of definition of To Be and Reality-in-general, *must be a category*, and it shall be the highest Transcendent category subsuming all actual entities. Reality-in-general is the highest epistemic realization of the concept of To Be and of Reality. Therefore, it must be the highest epistemic Transcendental category, with itself as the unique domain of definition of To Be and Reality. To Be, for convenience called a category, is in fact not one because it does not categorize anything other than itself, but that is the case with the other categories of ours too. To Be is the deepest ontologico-epistemological *sine qua non* of Reality and Reality-in-general. Thus, the three equally well characterize and play with each other, and no other does as much. This state of affairs can be encountered neither in any scientific ontology, nor in any other metaphysics. The categories of physical sciences may clearly be subsumed under the universality and applicability of our categories. Hence, the project of constructing and justifying the Einaic categories as *the most general formal* categories of ontology is relevant for philosophy, especially scientific ontology, today.

A comment is in place regarding *maximization of categories*. To be sure, reality-in-particular – be it transcendents or transcendentals – co-imply the maximized Transcendent and Transcendentals. As a body of categories, no other classificational category can collude with each other so well, because in the very process of mutual collusion such categories will automatically involve and necessitate the three maximal categories at least as part of the theoretical and ontological-commitment level realities. If we use these maximized categories as primitive notions, the ontology that results will have the double merit of (1) maximum collusion, with its effect: greater systematicity, and (2) theoretical implication and necessitation of reality-in-particular, with its result: possibilization of regional ontologies and the sciences via the same old local categories of the sciences at work in reality-in-particular. Maximization of the Transcendent Reality is by inductive totalization; and maximization of the Transcendentals Reality-in-general and To Be is by inductive generalization, both of which presuppose the axiom of infinity.

5. Aim 1: Scientific-Ontological Categories and System-Building

The history of philosophy has produced for system-building many categorial schemes that deal with data directly in the classificational manner. In contrast, the result of our sort of continuous mutual implication and collusion of the most general and total of ideals as categories is greater coherence, adequacy, relevance and applicability in system-building. The basic quality of ordinary science-based (merely empirical) ontological generalities is clear: they can at the most be transcendentals and/or transcendents yielded by reality-in-particular and thought around them. Proper mutual collusion of our Kantian-sounding but maximized Transcendental and Transcendent categories at the foundations of all scientific knowledge of Reality justifies derivation of an ontological system beyond contemporary cosmologies, epistemologies and particularist ontologies, and beyond the traditional purely classificational distinction between the two purely transcendent and transcendental poles (the metaphysical Divine and Being) of Reality.

Since the characteristic of mutual collusion runs into the very nature of every category of the scheme, and since each of the fundamental categories – Transcendental or Transcendent – needs each other, we can ensure greater systematicity in the produce of scientific thought and expect ever greater truth-probabilities in the ontological *systems* that spring up. This possibility overcomes the epistemological point made by the original versions of the analytic, phenomenological, hermeneutic, neo-realistic and postmodern trends in the 20th century, i.e. their antagonism towards overly foundationalist absolutist metaphysics that have reified everything in the rush for certainty and order.

Relevant notions from the works of a few representative thinkers and three major disciplines (philosophy of science / cosmos, epistemology and ontology) appear in this study, to the extent that they contribute to the question under discussion, i.e. the need to re-generalize and re-totalise ontological categories for system-building. Other authors will be treated when absolutely necessary for the mainstream discussion. I swiftly move past some of the masters of categorial thought by using their contributions to work out a system of more fundamental categories. First of all, therefore, this study is *analytical* of the categorial contributions of these thinkers *in general* and the scientific and ontological disciplines. It is *critical* of them *with a view to discovering what are amiss* in their categorial presuppositions, *and to try and remedy these lacunae by building a system of system-building categories* that looks more elegant, positively consequential and permits from within its structure *continuous systemic betterment* of this very categorial scheme and system in all possible ways. Since the aim is mutual collusion of the deepest and the highest of all transcendentals and transcendents to produce better systems, this study colludes the categories and draws the truth-probabilistic results of system-building.

The ontologies of many thinkers of cosmological, analytical, phenomenological, hermeneutic, logical, epistemological, pragmatic and ethical persuasions qualify for a critical categorial examination from the point of view of the Einaic categories. While working out a systemically better scheme, Einaic Ontology facilitates analysis and evaluation of the categorial foundations of our present knowledge of Reality. Various metaphysics and ontologies as well as contemporary cosmological theories of the origin and evolution of the universe could have been brought in to create a system. These are diffucult here due to the limited space a book on categories has. Different contemporary philosophers of physics and mathematics, however, theorize on the conceptual foundations of such theories. I find no system-builders in these sciences except Whitehead and Peirce. I deem it enough to draw relevant material from these disciplines and thinkers to show the way for a contemporary system rather than directly draw from cosmology, physics and mathematics (or from the writings of the stalwarts in these fields) and build one such system at such an early stage of development of the categories. This fact puts limits to the study, and brevity of the work on the categories demands the same.

6. Aim 2: Synthesis of Nominal 'Being' and Verbal 'To Be'

The way of proceedure of the chapters shows the secondary aim of this project: demonstrating that the deepest verbal universal To Be and its most universal nominal form Being are in fact the same. Martin Heidegger has been quite vociferous about the primacy of the verbal meaning of Being and the supposed viciousness of the nomic-nominal universal meaning of it employed in traditional metaphysics. But his Being has worked in lieu of a universal: Whenever he speaks of the question of Being he unwarrantedly presupposes (1) that his non-universalistic and aletheial Being is the verbal aspect of all beings, and it resists any universalization; and (2) that the beingness (Greek *Ousía*, German *Seiendheit*) of beings, which he presumes to be the is-level universal of all beings, is somehow the only focus of study in metaphysics of all sorts.

As we derive the concept of Reality in Chapter 1, the question of substance will be understood to be different from that of individual substances. Chapter 2 argues for the epistemological reality of universals and Reality-in-general as the epistemological *conditio sine qua non* of all discourse. This is done by recourse to laws of nature and their nomic (rule-type) expressions. Chapter 2 shows that nomic expressions of laws of nature are, in fact, nominal expressions in connotative acausal formulations of universals. It is the combined work of Chapters 2 and 3 to show that the causal-processual ways / universals behind them are ontologically the ways of being of beings, and that To Be is the highest verbal-nominal universal Way of Reality. A way is always processual-verbal. It is also universally present in beings by ontological instantiation, since ways, as nominal in their expression in language, are themselves representative of ontological universals. This interpretation – the combined effort in Chapters 2 and 3 – I believe, will go a long way in solving the metaphysics-ontology dichotomy that Heidegger epitomized in his life's work.

The preparatory Einaic analyses of the concept of To Be in the history of metaphysics are presented in a scattered manner in Chapters 1 and 2, and it culminates in Chapter 3, because an adequate and clear view of the concept of To Be has been most impotant from the very start. Einaic Ontology generates a scientifically adequate and metaphysically fundamental concept of To Be. Heidegger has been the greatest 20th century protagonist of a metaphysics based on return to the concept of Being. It is good to show how I view his concept of Being, how I differentiate the Einaic concept of To Be from his concept of Being and how the nomic-nominal and causal-processual To Be is a better synthesis. The Appendix presents a short sketch of what I find to be the deficiency in Heidegger's concept of Being, by assuming that the reader is familiar with the difference between the earlier (concept of Being based on that of *Dasein*) and later (concept of *Dasein* based on Being) Heideggerian notions of Being. I show how and why the integrated nomic-nominal and processual-verbal concept of To Be is more ontologically, scientifically and anthropologically relevant. The Appendix indicates what I find are defects in Heidegger's notion of Being, emphasizes my mode of understanding the paradox in his concept of Being and helps the reader understand and appreciate the Einaic synthesis of the nominal, nomic and processual-verbal in the concept of To Be in Chapter 3.

Chapter 1. Ontological Categorial Transcendent of Cosmology

Introduction

It is archaeological today to recollect that the concept of substance used to be considered as something that needs nothing but itself to exist. Two millennia of scientific and metaphysical reification of concepts has reinforced such definitions. The present chapter re-interprets the concept of substance or matter as ultimately continuous in the values of physical measurements and universal constants in the totality of all real worlds – be the 'real' actual or pertaining to the actual. In this manner it prepares the way of subverting the more than two millennia-old over-enthusiasm for substance-absolutism and -particularism (begun at least from Plato and Aristotle) and concludes the ultimate Transcendent category of all ontology: Reality-in-total. It synthesizes beyond the concepts of particularized reality and substance in some select thinkers, in contemporary physics and in the philosophy of science, by presupposing that classificational categories always yield particularism. It begins with classes in general, unifies the phenomenal-noumenal foundation of categories in Kant and derives a realistic non-classificational category (Reality-in-total) from the foundational requirements of the concept of substance in modern science. I take for granted Kant's concepts of phenomena and noumena as discussed in the asterisked footnote, A249 (omitted in B):

> Appearances, so far as they are thought as objects according to the unity of the categories, are called *phaenomena*. But if I postulate things which are mere objects of understanding, and which, nevertheless, can be given as such to an intuition, although not to one that is sensible – given therefore *coram intuitu intellectuali* – such things would be entitled *noumena (intelligibilia)*.[23]

I synthesize phenomena and noumena in preparation for a non-reifying but continuously existent substance. In Kant the categories have the function of processing the data of aesthetic appearances into the phenomena of understanding. A phenomena-noumena synthesized into the infinite-infinitesimal Reality is epistemically prior to sensibility. Their unity presupposes relating the maximized

---

23 Kant, *Immanuel Kant's Critique of Pure Reason*, 265–66. Note that: (1) *Coram intuitu intellectuali* means 'before (in front of / according to) intellectual intuition.' (2) All references to the *Critique of Pure Reason* are to N. K. Smith's translation.

thing-in-itself aspect intrinsically to the phenomenal aspect. The maximized epistemological aspect of phenomen related to noumena is Reality-in-general. The basic thrust of Kantian categories is this: "[...] the pure category does not suffice for a synthetic principle [...] the principles of pure understanding are only of empirical, never of transcendental employment, and [...] outside the field of possible experience there can be no synthetic *a priori* principles."[24] The present work takes inspiration from this and extends categories from the transcendental to the Transcendental and Transcendent realms.

To digress a little: C. S. Peirce defines category and juxtaposes it with the universe of metaphysical subjects: "[...] [A] Universe and a Category are not at all the same thing; a Universe being a receptacle or class of Subjects, and a Category being a mode of Predication, or class of Predicates."[25] Insofar as the categories developed in this work are least classificational, the predicational quality of our categories has no classificational sense. The primary sense of 'category' in this essay is of 'ideal / maximal predication'. This helps to overcome ontological particularism without theoretical detriment to tokens and particulars. Two vices visible so far in the 19th and 20th century history of metaphysics are (1) the tendency not to take 'To Be' as the highest predicable and (2) the theoretical non-availability, in ontological particularisms, of the ultimate substance of which it may properly be predicated. Reality can be where the purely ontological To Be inhabits. Reality is not a predicable, but To Be is the highest of predicables.

This chapter makes a realist interpretation of the concept of 'stuff' / 'substance'. This is different from Lars-Göran Johansson's physical-phenomenalist instrumentalism,[26] but with the concept of a pronouncedly infinitesimally and infinitely continuous causality. So far substance is considered merely as a phenomenalistic and instrumentalistic concept in QM. I show the need of ontological commitment to the reality of the wavicle within the ontological concept of Reality. My causal interpretation of nonlocality makes superluminal velocities possible. The so-called 'nonlocality' is in fact of extension-motion (measured spatiotemporal) nature causal and 'local' if superluminal velocities are real. Then, in accordance with the concepts of extension-motion, locality and real superluminal velocities, I reinterpret the criterial character of the velocity of

---

24 Kant, *Immanuel Kant's Critique of Pure Reason*, 265 (A248).
25 Peirce, *Collected Papers of Charles Sanders Peirce*, 4:545.
26 For example, "[...] [T]he indivisibility of the energy exchange implies that it is impossible to describe further details. Hence it seems that a deeper physical explanation is not to be had. This is the limit where physical explanation must stop." Johansson, "Realism and Wave-Particle Duality", 334–335.

light. This completes making the infinitely and infinitesimally objectual-causal phenomena-noumena into the category of the mutual continuous Reality.

Quine uses the term 'objectual' in the following context: "We give content to the ontological issue when we regiment the language of science strictly within the framework of the logic of truth functions and objectual quantification."[27] Evidently, he presupposes only a particularistic ontology. The context of ontological economy, the need of constant wagering of the Ockhamian razor-war, and the objective referentiality of strict truth functions and quantification render his use of the term mostly a language- and logic-based affair. This is further made clear by the following statement: "[…] substitutional quantification, far from being ontologically innocent, is simply ontologically inscrutable except through some stated translation into this objectual idiom."[28]

Quine means by 'objectual' the commitment to there being some actual or abstract entity (I add: causally processual or representative of causally produced qualities) behind all quantification. Objectual quantification for him is referential quantification, if every argument of the variable of a particular quantification is the case. This gives only specific tokens through universal-bound particulars. I take 'objectual' to all possible infinite and infinitesimal realms of occurrence of causality, beyond particularistic quantification, through commitment to the totality of what are mediate to perception and cognition but are continuous with phenomena that are immediate to perception and cognition. This is the main aim of my causal realist interpretation of QM and relativistic micro-, macro-levels of possible causal dimensions. The Kantian prelude is meant for quelling certain theoretical hurdles on the way.

Integration of the mutually continuous, infinitely and infinitesimally objectual-causal phenomena and noumena into the category of Reality-in-total paves the way for basing concretistic categories in the conceptually *a priori* / Transcendental but actually Transcendent category from a cosmological and trans-cosmological viewpoint. The first in the line is to redeem cosmological categories from pure idealism, and the second studying the maximal categorial orientations in quantum mechanical and relativistic theories.

---

27 Quine; *The Roots of Reference*, 136.
28 Quine, *The Roots of Reference*, 136.

1.1 Redeeming Cosmological Categories from Idealism

1.1.1 Entities and Predicates as Particularist Categories

Frege, Russell and Whitehead clarify the concept of 'number' using Boolean algebra,[29] a part of set theory. They have held that a class of objects is *not identifiable with* the corresponding actual collection. The collection merely *corresponds* to the class. By 'class' is meant "[…] a collection or aggregate of distinguishable elements, each of which is called a member of the class."[30] A collection is the aggregate of the member objects (i.e., objects identical[31] for their specific condition/s, not the heap of the actual sub-parts) of a class. It is a mistake to identify a class of classes with the heap of corresponding heaps (sub-parts).[32]

For example, the attribute (a general term for predicable properties, natural kinds and relations) of having rational animality (identity condition of a class) is not the same as the attribute of being featherless bipeds (identity condition of a less essential layer of attributes of another class), but the former is the necessary attribute of the class of rational animals, regarding the concept 'human'. Converting classes conceptually to attributes does not do away with having to consider classes as entities[33] and, in the end, as partial ways of being. All human talk essentially does this: mixing groups of entities (particulars) with attributes / *qualia*. This mixes the mathematical and scientific need of entity-talk and the ontological necessity of attribute-talk into being ingredients of *all* sorts of entity-talk. Here begins the transition from mechanistic-materialistic cosmology that does away with attributes, to organismic ontology that mixes them both, as Whitehead's theory of extension points out:

---

29 Boole, *An Investigation of the Thought on Which Are Founded the Mathematical Theories of Logic and Probabilities* (especially the first few chapters), creates an algebra of logic and probability by deriving the concept of number from the logical value and significance of the symbols 0 and 1, based purely on the laws of thought.
30 Nagel and Newman, *Gödel's Proof*, 16.
31 This identity is not that of identity of indiscernibles by the criterion of absolute identity, i.e., "If objects *a* and *b* have *all* properties in common, then *a* and *b* are identical." *The Cambridge Dictionary of Philosophy*, second edition, s.v. "Identity of Indiscernibles".
32 For example, the class of a type of dog-parts (say "a heap" or "bundle" of dog-heads) of the dogs in the kennel too corresponds to the class of all dogs in the kennel. But the class of all dogs in the kennel have only dogs, not dog-parts, as members. Chisholm, *A Realistic Theory of Categories*, 45–46. This allows a many-one relation from 'son' to 'father', for 'son' is a class by itself. His body-parts are not considered in this class.
33 Chisholm, *A Realistic Theory of Categories*, 47.

The Cartesian subjectivism in its application to physical science became Newton's assumption of individually existent physical bodies, with merely external relationships. We diverge from Descartes by holding that what he has described as primary *attributes* of physical bodies are really the forms of internal relationships *between* actual occasions, and *within* actual occasions. Such a change of thought is the shift from materialism to organism, as the basic idea of physical science.[34]

Whitehead holds that, as organismic action and fluency replace 'static stuff', it is also conditioned by 'quantum' requirements. These are the reflections of "[...] vacuous material existence with passive endurance, with primary individual attributes, and with accidental adventures [...]" which have vanished from mathematical physics.[35] Hence, ontological attributes are no more static qualities in modern physics and ontology.

Reducing classes into attributes need not mean annihilating classes but showing the necessity of theorizing of *the ontological interdependence of attributes*, as will soon be clear. Hence we begin from 'talk' and end up in ontology. Chisholm translates statements about classes into statements about attributes by saying that "x is a member of the class of Fs" is equivalent to "x exemplifies the attribute of being F"; and that the statement "the class of Fs includes the class of Gs" is the same as the statement "everything that is G is F". He proposes nine axioms for this reduction. The nine axioms he sets forth are:

(S1) If A and B are classes, then there is the class, A + B, that is the *sum* of A and B – namely, the class of those things that are either members of A or members of B. (S2) If A and B are classes, then there is the class A × B that is the *product* of A and B – namely, those things that are members of A and also members of B. (S3) There is a class that is

---

[34] Whitehead, *Process and Reality*, 309. A note on actual occasions in Whitehead: Actual occasions, events, actual entities and extensive continuum may be characterized thus: An event is a nexus of actual occasions, "[...] inter-related in some determinate fashion in one extensive continuum." (For example, a moving molecule is not an actual occasion, but a nexus of actual occasions.) Actual occasion is the thing within the plenum, the actual world is built up of actual occasions. Whatever there are – considered by abstraction in the sense of existence – are actual occasions. They are single-member types, the limiting type of an event. Whitehead, *Process and Reality*, 73. Actual entity is the atomized (temporalized) quantum of extension, the final real (because "[...] an actual entity never moves: it is where it is and what it is") thing making up the world or reality. Whitehead, *Process and Reality*, 73, 72, 18. "The extensive continuum is that general relational element in experience whereby the actual entities experienced, and that unit experience itself, are united in the solidarity of one common world." Whitehead, *Process and Reality*, 72.

[35] Whitehead, *Process and Reality*, 309.

the null class – the class 0 that is such that, for every class A, A is identical to the sum of A and 0 (i.e., A + 0). (S4) There is a class that is *the universal class* – the class U that is such that, for every class A, A is identical to the product of A and U (i.e., A × U). (S5) If there is a universal class and a null class, then for every class A, there is the *negative* of A – the class -A that is such that (a) the sum of A and -A is identical to the universal class and (b) the product of A and -A is identical to the null class. (S6) (A + B) is identical to (B + A). (S7) (A ×B) is identical to (B × A). (S8) [A + (B × C) is identical to [A + B] × [A + C)]. (S9) [A × (B + C)] is identical to (A × B) + (A × C)].³⁶

He gives Russell's formula to reduce Boolean algebra to principles about attributes and mentions Carnap's extension of the same, the latter making a slight alteration in the implication relation. Russell's original formula was essentially: "The class of x's such that x is F is so and so" may be replaced by "There is an attribute P, which is such that (1) P and being-F are exemplified by the same things and (2) P is so and so." Carnap showed that the results are more plausible if the existential quantifier were replaced by a universal quantifier. In this case the Russellian formula: "The class of x's such that x is F is so and so" can be replaced by "For every attribute P, if P and being-F are exemplified by the same things, then P is so and so".³⁷ Chisholm has thus presented the possible manner of reduction of entity-talk to attribute-talk. In three steps he reduces his nine axioms into truths about attributes: "[…] (1) replace *class* throughout by *attribute*; (2) replace *member* throughout by *instance*; and (3) replace *is identical to* throughout by *has the same instances as*." He concludes: "Wherever a Boolean formula contains a locution of the form 'The class of Fs is so and so', that locution is replaced by one of the form 'For every attribute P, if P is exemplified by the same things as is the

---

36 Chisholm, *A Realistic Theory of Categories*, 48.
37 Chisholm, *A Realistic Theory of Categories*, 49. A Universal Quantifier is represented by the notation xA, meaning, A holds for *all* values of x, usually within a domain of values implicit in the context or indicated by succeeding notational convention. Here one considers all members of a given set. An Existential Quantifier is represented by $\$x$, meaning A holds for *some* arguments of x. Here the consideration is of some members of a larger set. A technical detail: A Boolean formula is any formula in the set B, using the binary operators, namely, + and *, the unary operator, namely, -, and elements of the set, namely, 0 and 1. I do not accept the logical and empiricist thoughts that issue from this as ontologically ultimate, since here only the extremes of truth and falsity are at work, and not the various possibilities between, due to the algorithmic nature of such schemes. I only want to divert the entity-realism of Russell and Carnap to the ontological advantage of speaking of processes. The Boolean formula does not serve this since it has nothing to do with quality-talk due to its exclusion of such talk through the algorithmic structure.

attribute of being F, then P is so and so."[38] This gives no transgression beyond the nine axioms of traditional set theory. Although set theory reifies things, as does also language, reduction of classes into attributes preserves the possibility of including also processual attributes as categorial entities. That is, along with the many physical categories like substance, type etc. there can also be scientific-attributive categories like space, time, causality etc.

The concept of a *thing* became that of an *event* in modern physics. *Extension* as the primary *attribute* of physical bodies became the primary *relationship* of *extensive connection* of physical occasions / fields,[39] but the ontological superiority of extension over measuremental space and mass is forgotten. Space (length / where), time (duration / when), cause (action by itself and other entities in bodily connection / how) and mass (what / how much) – the nominal particulars that are either attributive or entifying[40] (Latin, *ens*, "being") categories, common to Newtonian and modern physics – may also be reduced into mere attributes: spatiality, temporality, causality and extensive connection, all of which are relations[41] *and* attributes. Even if all entities are reduced into attributes, there still remains one question: is not this facility only a logical one? Either this is merely a logical possibility; or else, attributes are so essential aspects of entities that they in fact describe the processual aspect of entities in a non-entifying manner. I am of the latter opinion.

By the afore-mentioned reduction by Chisholm it is clear that all classificational categories are simultaneously physically particularistic and qualitative. This means that it is easier to reduce particularistic categories and entities into qualities than to reduce highest totalities and generalities into particularistic

---

38 Chisholm, *A Realistic Theory of Categories*, 49.
39 Whitehead, *Process and Reality*, 288.
40 The term 'entify' is from Stenlund, *Language and Philosophical Problems*, 148.
41 According to Brentano, a relation is supported by the subject (the "fundament"), which is known *in modo recto*, "in the right mode". But it is referred to its ordered pair (the "*terminus*"), which is known *in modo obliquo*, "in the concomitant mode". [Latin, *obliquus*, "set at an angle", is from *ob-*, "towards, against, in the way of" and Gk. *likuos*, "at an angle". *Oxford Latin Dictionary* (Oxford: Clarendon Press, 1968). In 'A is taller than B', A is known *in modo recto* and B is known *in modo obliquo*. In an Aristotelian manner, Brentano holds also that both A and B should exist for there being a relation. The truth of it may be seen as we admit that the relation is only a possible relation, if B does not exist. Brentano, *Psychology from an Empirical Standpoint*, 271–73, and Velarde-Mayol, *On Brentano*, 74–76. Here Brentano defines intentional relation by recommending that, in such a relation, only the fundament is real (in the sense of existing).

categories. As our discussion progresses we shall further learn that physically particularistic and qualitative entities are not too differentiable from each other. In fact, they exist together. A particular is a particular by reason of its connection with other particulars and due to the mutual qualitative relations between members of the respective class of particulars. Qualitative relations are nothing but the ways of being of beings of the class.

Reducibility of (1) the quanta of action (the actual entities at a given extension-motion region in measured spacetime) and (2) the primary relationship of extensive connection into the terminology of attributes, will allow synthesis of space-time quanta into a continuum of attributively qualified objectual processes,[42] which we will in the end call Reality. Reduction of particulars to heaps or bundles of properties, as done by Chisholm, has meaning only if the whole activity results in nomic concepts that are nominally or qualitatively acausal (ideally processual, without mention of the causal aspect of processes) and not actually processual-causal. Qualities or attributes are not causal, but can be of something causal. In short, such reduction is of particulars into non-concrete attributes.

The concept of relation is applied in this essay differently. Relation formulated as a quality is always acausal (ideal). It involves universals / essences in events, presented by mind as connotative universals. Behind the acausal denotation is the causal event/s. Events transpire as extension-motion processes. Hence, relations are ways of being of events in extension-motion. Such ways of being are what we call attributes / *qualia*. As we reduce entities / particulars into *qualia*, we realize not only that entities and *qualia* are interchangeable, but also that they are bound up with each other.

This needs casting off the particularistic shades of phenomenalism, pure empiricism and nominalism from the concept of Reality, by treating off pure conceptual phenomenalism, objectivism and nominalism in a cosmology- and ontology-compatible way. Such discussion in the next sub-section is epistemological, but it is essential for scientific discussions. Using Quine's concept of ontological commitment and our re-generalization of it in accordance with the proposed new categories, Chapters 1 and 3 will later show that the quanta of action

---

42 The concept of the 'objectual' has already been introduced. It takes further discussions to set forth the concept more clearly. The processual aspect of the same is to be clarified well enough only as we approach the end of this work, since it presupposes more purely epistemological and ontological conceptualising.

are all processes / events and never objects[43] defined forever in extension-motion. The objectual-causal aspect of the ontology that results will be discussed first in the present chapter. Hence, our discussion of the reducibility of the quanta of action (token entities in QM) and their primary relationship of extensive connection, into the terminology of attributes will allow construction of the General Ontology presupposed behind the categorial concept of Reality.

The problems we should face before actually formulating possible ultimate scientific-realist categories are of getting the phenomena-noumena distinction solved and interpreting the QM requirement of fundamental entities and the characteristic of extensive connection within and without individual substances / masses, within the purview of causality and extension-motion. Spatiality and temporality are the measurmental aspects of the categories of Extension and Motion. Hence, while using QM to bring in causal continuity in the world, we should include Extension and Motion as the classficational physical-ontological categories of science and philosophy. This facilitates accepting the physical reality as physically real, and not as mere epistemological or mathematical spacetime continuum. Thus we have a standpoint that creates the first (Reality-in-total) of a set of second generation scientific-ontological categories – which, essentially, is that of a totalised substance. Solving the Kantian bifurcated substance problem can serve this purpose well.

1.1.2 Kant and Modern Physics on Phenomenal Categories

1.1.2.1 Understanding 'Phenomena' and Sensing 'Noumena'

I do not analyse here the role of forms of sensibility and categories of understanding for Kant's formulation of particularism. Instead, I question their basis on the phenomena-noumena distinction, in order for me to advance the ontological category which results from their amalgamation. Kant accepted the centuries-old near-absolute distinction between appearance and thing-in-itself, while working out also the forms of sensibility and the categories of

---

43 "Ever since Heraclitus, there have been revolutionaries who have told us that the world consists of processes, and that things are things only in appearance: in reality there are processes. This shows how critical thought can challenge and transcend a framework even if it is rooted not only in our conventional language but in our genetics – in what may be called human nature itself. Yet even this revolution does not produce a theory incommensurable with its predecessor; the very task of the revolution was to explain the old category of thing-hood by a theory of greater depth." Popper, *The Myth of the Framework*, 59.

understanding. He followed the purely empirical Humean "constant conjunction" tradition of explaining causality. Thus, in Kant the alleged phenomenal character of causality and other categories of understanding, along with the forms of sensibility, have anticipated categorial phenomenalism. He marshalled and infused the implications of conceptual idealism into his forms of sensibility and categories of understanding, by distinguishing between things-in-themselves or noumena (physical things and divine beings in pure reason as intellectually inaccessible posits)[44] and things-as-experienced-by-us or phenomena (as objects of sensibility and understanding).[45] Due to the alleged phenomenal character of all that transpires in experience, he held causality as just a category of understanding and

> […] condemned the practice of drawing any conclusions concerning the cause of an appearance from the appearance itself, as unallowable – in accordance with his conception of the idea of causality and its *purely interphenomenal* validity; and this conception, on the other hand, already anticipates that *differentiation*, as if the 'thing-in-itself' were not only inferred but actually *given*.[46]

But the Moon-in-itself (Moon$_i$), about which we know nothing, and the Moon-as-experienced-by-us (Moon$_e$), which is itself the very experience of Moon, are not identical. If they were, we could have, in perception and cognition, just the unhyphenated object, the "Moon" without the empirical subscript. Moon$_i$ is not located in space and time ('forms of sensibility'), but Moon$_e$ is located. Notice how unfortunately the empirical spacetime is not distinguished from the physical-ontological extension-motion. Moon$_i$ as such probably does not cause or help cause moon-experiences in humans, causality being a category of understanding that

---

44 Noumena are two types: (1) *negatively a limiting notion*, a thing understood as object of sensuous intuition, abstracting absolutely from the forms of space and time; and (2) *positively a posit of the intellect*, a thing understood "in itself" as a supposed intellectual (i.e. understanding-level) intuition into something unclearly existent or non-existent. The former indeterminate abstraction yields no knowledge, but limits knowledge. The latter is "knowledge" in unwarranted metaphysical construction of "thoughtful intuition", since the categories of understanding do not facilitate objects from without the twelve categories. Kant, *Kant's Critique of Pure Reason*, 266–67 (B306–7).
45 'Phenomenon', from Greek *phainein*, "to appear", is thing-as-it-appears. Thing-in-itself or noumenon is an object of dialectically intellectual *noésis*, "intuition", from Greek *noéin*, "know / intuit", not aided by phenomenal intuition by use of the forms of sensibility, i.e. space and time, and the twelve categories of understanding. This section questions not merely the results of Kant's phenomena-noumena distinction, but the very definitions of these foundational concepts of his philosophy.
46 Nietzsche, "The Thing-in-itself and Appearance, and the Metaphysical Need", 80.

applies only in the phenomenal world.⁴⁷ Whether it does in or from the extramental noumenal world is not knowable. Hence, $Moon_i$ in itself was most probably an idle metaphysical posit, since it is nowhere, at no time, and does nothing. This makes one a critical idealist by dispensing with $Moon_i$ by being critical of our process of experience. The other alternatives are to call Kant an empirical realist or a transcendental idealist,⁴⁸ both of which, again, involve the same $Moon_i$, given the indirect need to posit noumena along with phenomena.

Like Berkeley, Kant hypothesized one 'phenomenal world': humans have the same immutable set of basic propensities (forms and categories) to structure and classify incoming stimuli (in sensibility) and thoughts (in understanding). Today philosophical thought is outgrowing this assumption by asking what the categories are for humans to attain better the ideals of truth-probability, not merely by asking how exactly humans know. Today's scientific, realistic, conceptual, instrumentalistic idealisms are Kantian idealism appropriated by conceptual, linguistic and particularistic schemes, some of them in the name of humanly possible realistic science.⁴⁹ Concepts and expressions vary. There is not one unique 'phenomenal world' or world-as-conceived-by-humans. The world-as-conceived-by-an-Aristotelean differs radically from the world-as-conceived-by-a-Newtonian. Or, drop human chauvinism and consider also animals: the chimpanzee's world (experience) is not Einstein's or the honeybee's. This is Kantian phenomenalism pushed to the limits of his absolute differentiations.

If the critical-idealist talk of worlds-as-experienced shows the partial diversity of experience in world-talk, then hyphenated entities like Moon-as-experienced-by-humans are *ersatz* (substitute, simulated) entities. They, like the Kantian Moon-in-itself and others, are just the Moon, the Sun etc. if *any* objectivity is possible in experience. Clever acuities like "The Moon-as-conceived-of-by-Aristoteleans (a phenomenon) was perfectly spherical" are just fancy formulations

---

47 Musgrave, "Realism, Truth and Objectivity", 35.
48 Kant insists that a 'transcendental' falls within the purview of experience. It is a condition for the possibility of experience of sensible objects or phenomena (*Sinnewesen*), not a 'transcendent' rational object of the empirically unwarranted exercise of intellectual intuition, called noumenon. He claims that intellectual intuitions too are already subject to sensible (spatio-temporal) intuition. Kant, *Immaneul Kant's Critique of Pure Reason*, 264 (B303). This is transcendentalism. The present chapter, attempts to overcome its idealist implications.
49 The most suitable example is the failure of inserting superempirical values in the constructive empiricism of Bas C. van Fraassen. A fruitful discussion of the same is made in Churchland, "Ontological Status of Observables", 41ff.

of "Aristoteleans thought that the Moon was perfectly spherical"[50] etc. This shows that sensibility and understanding categorially cognize actual extra-consciousness processes with something like these categories functioning in the mind. Consciousness's conscious or less conscious applications of categories have both extra-conceptual and conceptual aspects in the objectual processes in its sensation-cognition. We do sense actual processes or their likes and understand their processes categorially. Even mere possible-world statements can be filtered by the mind categorially, but the origin of the basic concepts that formed the statement are real-world-tinged. Hence, we need categories that taste the aspects of phenomena and noumena together, lest their separation bring further theoretical inadvertencies.

Crass realism unsemantically formulates claims of the existence of mind-independent physical entities without providing epistemologically tenable ways of representation of truth-probabilities. Contrarily, Michael Denitt points out, in the following syllogism, the very realistic reasoning that people often use to refute realism: "(1) If the realist's independent reality exists, then our thoughts / theories must mirror, picture, or represent that reality. (2) Our thoughts / theories cannot mirror, picture, or represent the realist's independent reality. (3) So, the realist's independent realist does not exist."[51] (1) is false. So, (3) is not derivable from (2) basing on (1). (1) is untenable because it confuses the realist problem with the semantic and epistemological problem of the possibility of exact correspondence, not due to inexistence of real processes.

Hence, the problematic aspect of the realism-antirealism debate should hinge not on whether we can 'see how' exactly anything is, but on that we can 'see that' it is in some way. This naturally relegates the question of 'see as' from direct relevance to the basic question of 'see that'. It is now easy to see that the more phenomenalist-sounding 'see as' is a negligible issue in realism-antirealism debates between 'see' and 'see that', which concerns knowledge 'that', and not knowledge 'how'. Its corollary follows: the realist semantics of ontological *categorial* schemes at the integration of phenomena and noumena need not bother much about 'see as'.

Thus, if the difference between realism and anti-realism lies in accepting or not accepting that an external world exists independently of mind, then a realism like Hilary Putnam's 'internal realism'[52] is actually a hidden version of

---

50 Musgrave, "Realism, Truth and Objectivity", 35–36.
51 Renzong, "How to Know What Rises up Is the Moon?", 62.
52 Putnam holds: "Objects do not exist independently of conceptual schemes. We cut the world up into objects when we introduce one or another scheme of description."

anti-realism, and does not continue to hold fast to a realist position. C. Brown points out that both Kant and Putnam hold that the world we know is empirically real, *and* also that it is mind-dependent.[53] Any realism that holds either or both of empirical reality and mind-dependence have either phenomenal or noumenal semantics, and also categories based on the natural "way" that sensation and cognition are seen to follow when analysing the process by circumscriptive concretism. Such realisms neglect the apparently negligible elements that play from beyond the immediately circumscriptive concreteness of elements from the physical world involved in cognition. Hence, we should have a cosmic-realist (integral of the phenomena-noumena reality) semantics and epismology and a categorial system that facilitates the empirical aspect within the realist kernel of the givenness of Reality.

From the perspective of the cosmic-realist semantic requirements of bridging the phenomenal-noumenal distinction in Kant (and even Putnam), we need the category of Reality. From the cosmic aspect of empirical sensation and circumscriptive understanding, we need a complete overhaul of the foundations of categories of the specific and the merely empirical, and create a few experiencing phenomena and noumena mutually continuous. Only then can phenomena be understood, i.e., through sensing their ontological, noumenal-realist aspect by ontological commitment and understanding the ontological categorial presuppositions of the specific categories common to sensibility and understanding.

1.1.2.2 Conflating Seeing-that, Seeing and Seeing-as

To bridge the difficulties in the distinction between phenomena and noumena (especially the difference between phenomenal and noumenal semantics), so as to derive a fundamental ideal category from their synthesis (instead of particulars like substances), we first notice that saying the sensible depends partly, not fully, also on concepts and words. This problem will be dealt with before approaching the category: the phenomenal overlapping of the otherness of the object on knowledge and the subjectivity of the concept and word should only be seen in one perspective where Kant's perspective is empirical-phenomenal.

In purely phenomenal knowing it is trivially true that a being lacking the computer-concept (and the word 'computer') cannot speak of the computer.

Putnam, *Realism, Truth and History*, cited in Renzong, "How to know What Rises up Is the Moon?", 62.

53 Brown, "Internal Realism? Transcendental Idealism?", cited in Renzong, "How to Know What Rises up Is the Moon?", 62.

A being that cannot *say that P (or whatever)* cannot *see as P*, either. One might say that a being lacking the computer-concept (and the word 'computer' or its equivalent) cannot see it here. But a cat may see it as is clear by its not bumping onto it when the mouse hides under it. Though our knowledge of it is phenomenal, Kant will have no difficulty in assigning physical actuality to the mouse's hiding. Hence there is something amiss while admitting the physical and its ontology. There seems to be a fear of the ontological somehow becoming metaphysical, since things-in-themselves and noumena are treated on the same footing as purely imaginary posits.

Seeing that there is the computer is not the same as seeing it as such, but they are connected by phenomena differently, this difference seemingly playing the hog by completely taking over the show. That is, phenomenalist arguments wrongly conflate *seeing-that* with *seeing*. Between these, there is also *seeing-as*: the cat sees the prey as food and the computer as non-food, though it lacks the prey-concept and the computer-concept.[54] So, seeing-as is insignificant, while discussing seeing and seeing-that. Hence, it need not be conflated with the other "sights", which differentiates the existence of the computer and the rat from seeing-that and seeing-as. Oversight in this regard has produced the post-Husserlian understanding of the phenomenology of human existence (*ek-sistence*, "standing out"), human existence's importance over creatures' just being (inability to "stand out") and human freedom as something totally dissociated from that of creatures etc. and taken inanely for philosophising.

The conflation of seeing-as and seeing presupposes that beings with different 'conceptual schemes' (languages, theories, systems) see the presupposedly same world as different *worlds*, in the absence of support from a basic seeing-that-something. For Kant this is nonsense and inconsistent for his own "seeing-as"-sort concept of phenomenal knowledge, since for him all have the same forms of sensibility and categories of understanding for synthesizing sense-intuitions and judgments, respectively, under the same phenomenal world.[55] The Aristotelean and the Copernican must see (judgmentally, understand) the same sunrise – for us, by ontological presupposition – if they "see" in the sense as Kant means by it

---

54 Musgrave, "Realism, Truth and Objectivity", 36–37. It is important here to recall that higher-order consciousness is distinguished from primary consciousness by the former's ability to go over into consciousness upon consciousness (upon consciousness ... *ad libitum*). A fairly revealing cognitive scientific treatment of higher-order consciousness is had in chapters 9 onwards in Edelman and Tononi, *A Universe of Consciousness*, 102–110.

55 Kant, *Immanuel Kant's Critique of Pure Reason*, 111–15 (A77–83/B103–109).

within the framework of 'forms' and 'categories', phenomenally and not noumenally.

For Kant, speculative-rational seeing of *entities* is trans-phenomenally unjustifiable noumenal knowledge. But *knowledge* of the moral-noumenal postulates and acquisition of moral virtue are justifiable by action. If knowledge is one sort of action, with its *moral* standard set at ever-widening and continuous acts of phenomenal knowing, then speculative reason is itself possibly not merely noumenal, but works at the junction between the phenomenal and the noumenal. So, phenomenal seeing by forms and categories is the empirical basis of speculative reason. Then phenomenal seeing-as and seeing-that have an element of speculative, ontological, seeing. Ontologically, identity of the object as a process is truer than the object of seeing-as. The Aristotelean would say, "I see that the earth is still rotating on its axis", and the Copernican differently; but they *see* (understand) something as relatively identical (*something* noumenal), but, at the same time, as relatively similar (phenomenally).

Kant has forgotten to provide for what he might call mere seeing, as in the case of the cat avoiding a bash on the computer (or box) without possessing a computer-concept. If the so-called actual existence of the computer is noumenal, 'seeing as' will not even happen, since it supposes the existence of something – existence being according to Kant a commonality (a category) in seeing-as, not a generality in seeing or seeing-that. The presence of some commonality in seeing-as justifies us in assuming that the commonality to be found is real in its somethingness. Thus, both the Aristotelian and the Copernican will have *a commonly shared seeing-that: the seeing of the apparent motion of the sun. This is noumenal in Kant.* The noumenal can be presupposed not merely phenomenal-categorially but ontologically, by sharing in some way their ontological categories with the mind and the phenomena. So, *the absolute distinction of categorial schemes of the phenomenal and noumenal worlds is nonsense*. Note that Kant does not posit a categoreal scheme for noumena, but the scheme of the phenomenal extends from that of the noumenal, if instances of seeing-as can have any generality among themselves.

Hence the profundity "The limits of my language [phenomenal categories] are the limits of my world [seeing-as]" is blatantly false where the world is in fact more than seeing-as, unless 'world' is the same as 'language', 'meaning system', 'interpretation', 'semantically instrumental possible worlds', etc. The triviality that my language limits what I can say of the world should be true. But the anthropological profundity, "The world of the Kalahari bushman contains no computers", is not a hidden platitude that, Kalahari language or concepts being what they are, the bushman cannot see or say at any time that what *we* call computer is here. It is absurd to say that computers do not interact causally (and in other categorial ways) – i.e., physical-ontologically with the sensibility of the Kalahari bushman – but only through phenomenal understanding.[56] Vacuity of relation between categories of beings / processes and the mind is unwarranted. Hence we say, the categories of both sensibility and understanding have something to do with things-in-themselves (which are not the same as things-considered-in-themselves) and, that too, however meagre the depths of them are, they as such are brought into knowledge by humans' sensibility and understanding. That there are things-in-themselves is, thus, a *sine qua non* of all seeing-that, seeing and seeing-as, i.e., of cognition.[57]

Unfortunately, purely formal entities (e.g., the epistemological space and time and the physical-ontological extension and motion) without the deepest or highest exemplification, and type-terms of physical actuals, all taken as ultimate categories of cognition, have vitiated the tradition of categories: because these were taken as categories merely of sensibility or things. While conflating seeing-that, seeing and seeing-as into the concepts of perception and understanding, Kant's transcendental deduction did not work out actual ways of forming concepts from the forms or categories of judgment (at the level of understanding), since there is nothing more general in them than their actual workings in the propensities of the unity of apperception, not connecting anything with judgment from the condition for the possibility of the conjunction between beings and concepts. This is Kantian phenomenalist idealism or conceptual idealism. Kant's conflation of seeing-that, seeing and seeing-as does not presuppose their otherwise objectual presupposition of the direct or indirect there-being of some substance.

1.1.2.3 Phenomenalistic Idealization of Forms and Categories

Alan Musgrave discusses the Gem Argument of David Stove, which he claims has converted many a philosopher (like Berkeley and Kant) into idealists (and solipsists) in effect. He discusses here an invalid argument that represents the way in which many from Berkeley till date became idealists. The general form

---

56 Musgrave, "Realism, Truth and Objectivity", 37. I have somewhat manoeuvred my argument using Musgrave's argument, which uses the example of the Kalahari bushman.
57 I have left out the status of cognition as an act / event / process, because generalizing on act / event / process as more general than knowing is not immediately needed for the purpose of this work.

of the Gem Argument is: (1) We can X things (understand) only if C (sensing things-as-such), the necessary condition to X things, is met. (2) This cannot be met. (3) So, we cannot X things-as-they-are-in-themselves. For X substitute 'know', 'perceive', 'think of', 'talk of', 'refer to' etc. or, Berkeley's 'have in mind', Kant's 'bring under the categories of the understanding', Arthur Fine's 'interact with' etc. For C, substitute 'sensing', 'perceiving', etc. in an absolute manner. As a result, Gem Arguments take different implicit and explicit forms.[58]

The assumption that we cannot meet the C-condition (infallibly 'having', 'sensing', 'interacting with', or 'knowing things-as-such') presupposes the mad-dog realist presumption that we have absolutely direct access to events. That is, the Gem Argument shows the need of solving the realist-idealist debate by integrating phenomena and noumena into one continuous stuff, thus saving categories from phenomenalistic idealization. This would facilitate construction of the realistic (thingly) categories of conceptual events by actual beings / events, even in the case of philosophically treating the concept of error. This requires categories that vow to be neither mad-dog realist nor idealist.

Another fallout of Kantian-like phenomenalism is the quasi-realist metaphysics that makes theoretical things, natural kinds, particulars etc. in science (electrons, blackholes, genes ...) as products or projections of mental, epistemic or linguistic 'talk' / 'discourse'. By commonsense, then, there were no such things before such talk, and these theoretical entities only seem to belong to a phenomenal world of entities. Science teaches that they are spatially and temporally 'located', not produced. Therefore we need a way of saving the basic intuition of spatio-temporal thinghood / eventhood in science (not necessarily its phenomenalism, particularism or pragmatism) from quasi-realism,[59] by an ontology that allows all that are to be what they are, but simultaneously allows the element of talk, projection, production, seeming, phenomena and also error to be involved in the process. If quasi-realism holds, then science does not, and this ends up in Berkeleian conceptual idealism.[60] So, something that saves science should be created as the realist foundational category/ies of scientific ontology.

---

58 Musgrave, "Realism, Truth and Objectivity", 40.
59 A widely discussed view is that realism admitting existence of abstract and concrete events and objects in their interplay is best suited to explain the success of science. The existence of facts as testable explains the way knowledge progresses. The explanatory role of theory seems to carry with it an ontological commitment that is problematic to the quasi-realist, e.g., Blackburn, *Essays in Quasi Realism*, 18.
60 Musgrave, "Realism, Truth and Objectivity", 41–42.

Musgrave says that as the exemplar of epistemic truth-theories Carl Matheson's ideal limit theory also suffers from severe relativism and idealism. Ideal limit theory holds that "[...] truth is to be equated with the limit of ideal scientific practice: what is true is just perfect science pursued to its conclusion."[61] But ideal scientific practice might converge on two incompatible limits, say, an ideal particle theory and an ideal field theory. An ideal limit Martian science will differ from our science pursued to its ideal limit. This means that there is an ideal *limits* theory. Then incompatible ideal limit theories are all true. Two incompatible theories cannot both be true of the world. Perhaps, each is true of its own *ersatz* world-as-it-is-according-to-my-community's-ideal-theory. Thus, idealism allows absolute, phenomenalistic and instrumentalistic relativism without anything concretely existing to talk of and contradicting the global realism presupposed by all ontologies of science.[62]

Brian Ellis's 'internal realist' ideal limit theory holds that truth is a limit (not limiting) notion of reasonable belief. Internal realism is independent of what we think or say. This ends up in *ersatz* or hyphenated entities, which are nothing but the phenomenal names for entities and processes, with respect to our mind:

> The way the world is[,] is relative to the sorts of beings we are. That is one of the consequences of internal realism. It does not make it wrong for us to believe what we do about the world. Nor would it necessarily be wrong for us to believe something different if we were different sorts of beings. Nor is there any third standpoint from which the belief systems of two different sorts of beings could be compared and evaluated, or, if there is, then it enjoys no privileged position. So, according to the internal realist, there is no way that the world is absolutely, [there are] only ways in which it is relative to various kinds of beings, none of which can claim absolute priority.[63]

Here the phenomenal entities are not the hyphenated the-way-the-world-is (the noumenal reality) and the-way-the-world-is-relative-to-us-humans (the phenomenal reality). One does not find a way of rationally intuiting the fact of the world as it is, as ideally and totally what it is in itself, without any ontological dependence on the mind. Thus, Ellis realizes the trouble and gives up realism: "There is not and cannot be any absolute truth, and therefore there cannot be any way that the world is independently of how we, or some other creature, would evaluate its beliefs about it."[64] The commonsense but universal knowledge that

---

61 Matheson, "Is the Naturalist Really Naturally a Realist?", cited in Musgrave, "Realism, Truth and Objectivity", 29.
62 Musgrave, "Realism, Truth and Objectivity", 42.
63 Ellis, "What Science Aims to Do", 71. Square brackets mine.
64 Ellis, "What Science Aims to Do", 72.

the world existed long before us, is thus contradicted.[65] Hence, far from the ideal limit/s, we need a realism in which the general actuality of the whole of reality somehow engenders everything specific and still remains ontologically different from reality-in-particular.

1.1.3 A Category from Phenomena-Noumena Continuity

Kant's reply to Herz about Herz's response after reading the copy (sent to him by Kant) of his *Inaugural Dissertation* puts succinctly the attitude with which the phenomena-noumena distinction is to be understood. To the objection of Herz, Lambert, Sulzer and Mendelssohn that "[…] because change (at least in our representations) is real, time must be so too […] Kant's response was: I no more deny that alterations are something real than I deny that bodies are something real, though by that I merely mean that something real corresponds to the appearance."[66] This is the same attitude as what we find in the once-famous analytic-philosophical stance that we do not need to analyse the metaphysical structure of existence and attributes of processes, since much of it depends on the semantics and logic of the language. Kant's refusal to assign any 'reality' for 'change' is comparable in aftermath to his denial of 'reality' to 'bodies', if 'to be real' is taken in the sense of being related to something existing (not merely phenomenally actual) corresponding to the appearance / phenomenon. This is the sense in which he understood the phenomena-noumena distinction. His concept of 'being real' applies strictly to his phenomenally unapproachable noumena and things-in-themselves, and never to phenomena. Against the background of the foregoing sections we say that phenomenally noumena may be unapproachable but, if 'to be real' includes also 'not to be fully within phenomenal impressions', then 'to be real' means also 'to have something extra-phenomenal extra-mental in the vicinity'.

Critical scientific realism[67] attempts to circumvent the Kantian impasse of phenomenalist solipsism and to commit to there existing an extra-mental world.

---

65 Musgrave, "Realism, Truth and Objectivity", 42–43.
66 Kant, *Theoretical Philosophy, 1755–1770*, lxxiii. He does not mean that *some thing* corresponds to phenomena / appearance (though phenomena and appearance differ slightly in meaning). He admits the possibility of there being noumena. Only to the extent noumena are possible are things-in-themselves and changes in bodies possible.
67 "By 'scientific realism' philosophers mean the doctrine that the methods of science are capable of providing (partial or approximate) knowledge of unobservable ('theoretical') entities, such as atoms or electromagnetic fields, in addition to knowledge about the behavior of observable phenomena and, of course, that the properties of these and

It holds the minimal ontological assumption of an actual world independent of human minds, concepts, beliefs and interests. Contrary to Kant (and "critical idealists"), it has a physical, extension-motion (but called spatiotemporal) structure obeying natural causal laws. According to Niiniluoto it is a lawful flux of causal processes.[68]

> The suspicion that this is philosophically problematic arises from the following fallacy: the equivalence (1) x exists at t iff the sentence 'x exists' is true at t implies that the left side is false at those moments of time t, when a concept or a sentence referring to x does not yet exist. But the correct form of (1) is, (2) x exists at t iff the sentence 'x exists at t' is true. The meta-linguistic sentence on the right side [that is, the metalinguistic statement: 'x exists' in (1)] is temporally indefinite, and if it is true now it is true at any time. Even if the sentence 'Dinosaurs exist' was not yet invented at t_0 = 100 million B.C., it is nevertheless counterfactually true that 'Dinosaurs exist' would have been a true sentence at t_0.[69]

By way of analysing Kant, Niiniluoto argues: (1) Kant's 'phenomena' could denote "[…] expressions of our *partial knowledge* of things as they are 'in themselves' in mind-independent reality." (2) Our knowledge of reality, laden with conceptual frameworks, is partially true. (3) Ever-better frameworks and empirical testing of scientific theories formulated in them yield a rational method of approaching ever-deeper partial descriptions of the world. (4) The "unknowability" of things-themselves (which is perceptually slightly more distant for the mind than its phenomena / "showings") – the noumenal object of our knowledge – is thus rejected. (5) As in Hintikka, the object of perception is an object, not a phenomenon or sense datum between the physical object and me. (6) The logic of perception allows *seeing-as* [recollect also Wittgenstein] statements like 'A sees b as an F' and admits causal influence from a (some) being(s) to sense organs, by which 'this is an F' is the content of perceptual experience that does not disclose the whole object. (7) Even "[…] veridical perception reveals only a part of its object, which is *inexhaustible* by any finite number of observations." (8) By critical scientific realism, "[…] scientific knowledge with its empirical and theoretical ingredients – obtained by systematic observation, controlled experiments, and the testing of theoretical hypothesis – is an attempt to give truthlike description of mind-independent reality."[70]

---

    other entities studied by scientists are largely theory-independent)." Boyd, "How to be a Moral Realist", 151.
68 Niiniluoto, "Queries about Internal Realism", 49.
69 Niiniluoto, "Queries about Internal Realism", 49–50. Square brackets mine.
70 Niiniluoto, *Critical Scientific Realism*, 91–92.

All these show that it is possible to think of a phenomenal-noumenal continuity and call it Reality. One might favour a double-aspect interpretation of Kant's two-worlds theory. Absolute differentiation of the aspects without medium is artificial, since an absolute difference implies an infinite gulf. The double-aspect theory has to somehow show what the real connection or ground of connection between the two aspects is. The double-aspect theory is thus nothing but a continuity theory. The totality has the merit of being the ontologically absolutely committable, really existing, conceptually (linguistically, semantically, epistemically ...) never fully approachable Reality. This is not the totality of two worlds. It is the loose but fully causal togetherness of whatever in reality exists. The other part, if separated, is Reality-in-general, the part of conceptual ideals and whatever reflects of Reality in concepts and whatever belongs of Reality to the conceptual showing-itself world and things-for-us world. This continuous concept of Reality or THE WORLD is the first category foundational even of concretist epistemic categories. I attempt this by a phenomena-noumena synthesis for ontological discourse.

We move beyond Kant to Reality, via the reality of attributes and universal-boundedness of things and totalities of things. We need to derive the actual connection of ontological attributes and universals with the connotative attributes and universals in the epistemic faculty. Identifying something involves language and activities like thinking, perceiving, using language etc. But this does not mean that we can remain just in a realm between the existent and the phenomenal. I quote Rescher, where the author dodges between realism and idealistic phenomenalism:

> But already with Leibniz, the thesis is advanced that the barrier between appearance and reality should not be so drawn that reality is construed in strictly extra-phenomenal terms (reality = the sphere of what is extra-phenomenally the case); rather and more plainly, reality simply answers to *the truth* (reality = the sphere of what is 'really' [i.e., in fact] the case). On this view, reality is not an unattainable something that lies wholly outside the phenomenal realm, thanks to its *ex hypothesi* extra-mentalistic character. Rather 'reality' comes to be specified – nay, *defined* – *on the side of the dispositional appearances*, as comprising that sector of appearance that is *authentic*, factual, correct, or what have you. We come to the view adumbrated above that *reality* is to be understood not 'externally', in terms of mind-independence, but 'internally', in terms of what really and truly is the case from a standard, albeit mind-involving perspective. (Accordingly, we do not abandon a correspondence theory of truth, but merely endow it with a suitable interpretation.) The contention is that reality-as-we-think-of-it (= *our* reality) is the

only reality we can deal with, and that this is not mind-independent, but construed in mind-involving terms.[71]

Rescher falls short of scientific realism, which holds that, for sure most of what we call Reality is mind-independent, and the *activity* of sensing, cognizing, conceiving, explanaining, imagining etc., which is also really part of Reality, is mind-dependent, and that mind is part of Reality. In this sense, the mind-dependent "world" will have to be explained in terms of the mind's activity which is part of Reality – a reality that Rescher seems to miss out. Einaic Ontology goes all the way out to real-ize the phenomenal objects involved in cognitive activity, through the "ways of being" interpretation of ontological qualities / universals and the inherence of ontological qualities in connotative activity. This is the fundamental quality of consciousness. I quote Niiniluoto discussing Rescher, where Niiniluoto seems to share the above view:

> Rescher [...] asserts that to be *identifiable* is mind-involving as well, since the realm of possibility is mind-dependent, but this is less convincing. If an object x has *some* mind-involving properties (e.g., can be identified, seen, thought, etc., by human beings), x itself may nevertheless exist in a mind-independent way. The possibility of the identification of a physical thing, like a dinosaur or a chair, is indeed based on its mind-independent properties (location in space and time, causal continuity, qualities).[72]

Niiniluoto argues that the case with UFO's is similar. They are not self-identifying flying objects, but are potentially identifiable (IFO's) by us as extended parts or chunks of the world flux, by reason of *its* continuity and similarity with the world flux.[73] The distinction between UFO's and IFO's is an argument against Kant's things-in-themselves and things-for-us. The distinction is only in interpretation: an IFO is not a veil hiding a UFO from us, but a *"partial truth"* about a UFO. A UFO is not a negatively definable correspondence to an IFO, as a property-less bare particular, as a *no-thing*. It is a complex composition of properties. The IFO gives a partial description of it, of its properties expressible in a language L. The "existence" and "properties" of IFO's depend on the "reality" / "existence" of UFO's. Knowledge about IFO's gives us truth-like[74] information

---

71 Rescher, *Conceptual Idealism*, 168–69.
72 Niiniluoto, "Queries about Internal Realism", 50. The very way of seeing anything involves also that of seeing-as. This does not preclude seeing-that. Seeing-that allows mind-independence to things. Reality is the highest mind-independent existence.
73 Niiniluoto, "Queries about Internal Realism", 50.
74 "The 'possible worlds approach' to truth-likeness, advanced by Graham Oddie ([*Likeness to Truth,*] 1986) and Ilkka Niiniluoto ([*Truthlikeness,*] 1987), characterises truth-likeness in terms of the distance between a possible world and the actual world [....]

about UFO's.[75] This shows that phenomena and things-for-us are dependent on and interspersed with the actual world or Reality. the same may be shown, using the concept of internal realism,[76] more convincingly in the following manner by involving language, conceptual systems and the world:

A theory is characterised in terms of the basic states, or traits, it ascribes to the world. To each basic trait, there corresponds an atomic formula which says that an individual possesses this trait. Let us call them *atomic states*. Possible worlds correspond to all conceivable distributions of truth-values to atomic states. So, if there are n atomic states (corresponding to n basic traits), there are 2^n possible worlds W_i ($i = 2,..., 2_n$) ... A theory T which is false of the actual world W_A may, nonetheless, be *truth-like* in the sense that the possible world W_i it describes may agree on some atomic states with the 'target theory', i.e., the theory which describes truly the actual world. This partial agreement is employed to explicate the notion of truth-likeness." Psillos, *Scientific Realism*, 264–65. Niiniluoto holds also that "measures of truth-likeness are 'contextual'." Psillos, *Scientific Realism*, 269. One of the strongest forms of possible world theories is by David K. Lewis. David M. Armstrong makes a point against this theory: "David Lewis's ontology, as is well known, is exhausted by a *pluriverse* (the totality of being) consisting of all the possible worlds. [This concept of the pluriverse is different from the concept of the 'pluriversal' ("many for...") of postmodern (anti-) epistemology.] He defines a 'proposition' as the class of all the worlds for which the particular proposition is true. (Notice that his system does not allow for a null world, and hence he cannot form the class of the null worlds. As a result he cannot admit impossible propositions, a weakness, I would argue, because we can and do believe and assert impossibilities.) Suppose that one thinks of each of these worlds as the truthmaker for that proposition in that world. Let the proposition be that <cats catch mice>. It will be true in our actual world, and many other worlds. This gives us the class of worlds that Lewis identifies with the proposition that cats catch mice. But it is clear that this bunch of worlds is not a minimal truthmaker for the proposition, though it may be *a* truthmaker. (The mereological sum of episodes in our world involving the generality of cats catching mice when given the opportunity seems to be nearer to what is needed for a minimal truthmaker.)" Armstrong, *Truth and Truthmakers*, 20. Square brackets mine.

75 Niiniluoto, "Queries about Internal Realism", 51.
76 The internal realist theses are: "(IR1) What objects does the world consist of? is a question that it only makes sense to ask within a theory or description. (IR2) There is more than one 'true' description of the world. (IR3) Truth is some sort of idealized rational acceptability." Here are their exact negations in the metaphysical realist theses: "(MR1) The world consists of some fixed totality of mind-independent objects. (MR2) There is exactly one true and complete description of 'the way the world is'. (MR3) Truth involves some sort of correspondence relation between words and external things or sets of things." Niiniluoto, "Queries about Internal Realism", 46.

Thus every interpreted language or conceptual system L, whose terms have a meaning through [...] conventions [...] determines a structure W_L, consisting of objects, properties, and relations, and exhibiting the world as it appears relative to the expressive power of L. Such structures W_L are fragments or "versions" of THE WORLD. This view can be regarded as a formulation of the internalist principle IR1, since it allows the reality to be structured in many ways. It also denies the sort of metaphysical realism which assumes the existence of an ideal "Peirceish" language L such that THE WORLD = W_L.[77]

There is no ideal language W_L. So THE WORLD = W_L (the world reflected as such in L in "my world") is not the case. The concept of THE WORLD = W_L is non-epistemic. If a community has language L, the ideal W_L is not what they *believe* about the world, but "the way the world *is*" relative to L. If a member of this community has false beliefs (expressible in L), his "life-world" differs from the ideal W_L. This does not lead to relativism. No one has THE WORLD = W_L. All structures close to the ideal W_L are fragments or versions of the same WORLD. L is not a mere ideal. Hence, L and THE WORLD are compatible, but not the same. This saves us from reifying metaphysics, which mistakes (not identifies) what is said in L with what is actual in THE WORLD.

Rescher seems to mistake L for what is sayable about THE WORLD in L. Then categories become purely subjective to L and conceptual, and never ontological. For us there are both ontological universals and conceptual universals. If categories were merely subjective, then THE WORLD would have no absolute and inherent categorial structure (because it is the way the world is relative to L) and could not "choose" the structure W_L.[78] THE WORLD may have inherent, purely ontological categorial structure – the way in which it is what it is – but would not merely have the categorial structure it may have directly *with respect to L*, if we can admit that THE WORLD is an ontological, *a priori*, ideal presupposition behind the structure W_L and every perception.

In short, the "existence" and "properties" of IFO's depend on the "reality" / "existence" of UFO's, and knowledge about IFO's (which Kant would call 'thing-as-perceived / for-me') gives us truth-like information about UFO's (which Kant would call 'thing-in-itself'). Nevertheless, to the extent that we are considered less justified by merely empirical verification in admitting anything other than a solipsistic 'I' deprived even of its body, IFO's exist merely by ontological commitments, which naturally point to ontological universals alone. If the mind is connected to a body and the body to a world by reason of a trans-solipsistic way of knowing, we are justified in synthesizing phenomena (the world-as-I-perceive)

---

77 Niiniluoto, "Queries about Internal Realism", 51.
78 Niiniluoto, "Queries about Internal Realism", 51.

with noumena (the-world-in-itself) with respect to ontological universals – which are in THE WORLD – and speak of those universals as connotative *abstracta* in L.

Metaphorical talk about "choices" in the game of exploring reality is interpretable as in decision- and game-theoretical semantics. Choice of versions of language L (never the ideal language L) is my first move in choice,

> [...] followed by "Nature's choice" of the structure W_L. The game continues by my attempt to study the secrets of W_L. And all true information about W_L, *viz.* about a fragment of THE WORLD, also tells something about THE WORLD. As each W_L is a structure for language L, we can directly apply Tarski's model-theoretical definition of truth for sentences of L. For each L, we can define the class of truths in L: $Tr_L = Th(W_L)$.[79]

Here, $Th(W_L)$ is a theory in the language L of THE WORLD. The union of all classes Tr_L over "all possible languages" L is not well-defined. That allows defining an objective correspondence-level notion of truth about mind-independent reality: "[...] a sentence h in L is *true in THE WORLD* if it is true in W_L",[80] where the conception of truth is a version of the correspondence theory, but not metaphysically suspectible.[81] Correspondence in truth is such when, corresponding to a belief, there is a fact, which connects a semantic notion with at least a token entity. A token entity is had fully rationally only in ontological commitment. Hence, the correspondence level truth that goes beyond Tarskian truth is possible in a probabilistically ever better truth-approaching ontological commitment correlating semantic notions with ever-higher and finally the highest token entity possible. It makes Tarskian particularist and correspondence-level truths possible as a special case by providing a framework for truth claims. This highest token entity is nothing but the totality of whatever exists mostly in much independence and some dependence on the knowing mind and its L.

The total-token entity in ontological commitment is theoretically and practically from beyond seeing-as and seeing, i.e. from seeing-that. Seeing-that in its maximized form is not the Heideggerian, highly phenomenalistic and phenomenological 'as-is-given-in-*Dasein*'. Though seeing-that is given in consciousness, it

---

79 Niiniluoto, "Queries about Internal Realism", 51–52. It should be noted that Tarski was a defender of a version of the correspondence theory. What he has worked out is that part of discourse, which is evidently conformable with correspondence. His model-theoretical definition of truth for sentences of L too defends this correspondence-compatible aspect of discourse, based on such semantics. Truths of higher than the particularist correspondence levels are not imbibed in his semantic theory of truth.
80 Niiniluoto, "Queries about Internal Realism", 52.
81 Niiniluoto, "Queries about Internal Realism", 52.

has an a priori ontological commitment validating it as from beyond consciousness, in ontological independence from consciousness. This is Reality, and it is at least the totality of all once-upon-a-time possible worlds that are now realized. Put differently, it is the totality of all realised worlds, with the objectual-causal roots of universals of all actual processual worlds within itself and the objectual-causal roots of all to-be-realized possible worlds based in the Divine. Thus we have at least THE WORLD, one of Kant's things-in-themselves, as readily available for phenomenal world-talk in language. It is possible to rationally posit a real world of things-in-themselves, a truth-probable world-talk and truth as an ever-growing province of relatively better-justified (by ontological induction / inductive totalization) beliefs about actual and possible states of affairs and, in general, about the whole Reality (that includes at least the world, and at most God, worlds and souls possessing abstract "ways of being"). This is possible only if things exist in totalizable totalities. *Ontologically, totalities with their own internal and external relations are what there are in THE WORLD.*

A synthesis of the Kantian phenomena with the noumenal "world" is thus necessitated by sensibility's and understanding's *being at least of* the world (if not of God and souls), through the forms of sensibility and categories of understanding. Presupposes are ontological universals in the relations of things in totalities, and from them and only thus should they allow the forms of sensibility and the categories of understanding to work in mind and in L through the showings-themselves of things. The concept of the world as the thing-in-itself cannot be postulated by (merely intellectual / understanding-level) inference of any kind (even via the category of causality), because these are possible only empirically, i.e., through begging the question in understanding, based on extension-motion (for Kant, spatial and temporal) sense intuitions.[82] So, ontological commitment to totalities with another layer of ontological commitment to their ontological universals is a must for any discourse in L.

Even epistemological establishing of apriority in categories presupposes the world as itself and as the continuity of noumena and phenomena, by an ontological commitment that totalises beyond solipsistic phenomenalism. Hence, the thought-relation of things-as-perceived does not compel us to suppose that the thing apriorily given in the ontological presupposition of there being anything, however theoretically far removed it is from direct perception, is merely thought-related. Universals are also in thought, but *as properly connotative (consciously noting-together) universals*. Ontological universals lie at the level of the

---

82 Kant, *Immanuel Kant's Critique of Pure Reason*, 81–82 [B58].

presupposition of there being anything, there being relations, there being common attributes, there being totalities of entities / processes, and there being connotative universals. *The ontological universal-boundedness of totalities of things is no proof that things are merely mind-dependent.* That is, phenomena must finally be showings from the categorially prior and actual things-in-themselves in their totalities and their final totality that is Reality – not from reality-in-particular – since every ontological universal implicitly refers to a broader universal and a broader totality.

Such a final totality and minor totalities are possible only if the universality of *a priori* facts is present ontologically in totalities of substance. Hence, it is to be taken for granted that Reality is *more* apriorily (but Transcendently, i.e. in the manner of ideality that is imbued with the Transcendent, Reality) categorial than sub-categories that are based on Reality. The concept 'Reality-in-total' is a singleton universal, to go by the set theory of universals. It is also the objectual-causal, Transcendent category. It is not on par with Reality-in-general, the epistemologically ontological category to be derived in Chapter 2. Reality-in-general is the connotative universal of universal of … *ad libitum*. Reality is the maximal token entity. The said apriority is of the very essence of the notion of Reality. That is, the concept of Reality is the most universal-imbued compared to realities-in-particular. *Such a concept of Reality is a renewal in the concept of matter or substance.* Substance as Reality is all that there actually exists – this being the most adequate definition of substance – and not what needs itself (actual entity) and not anything else (actual entities) to exist – this latter definition being only of particular substances. Reality is continuous within and between the levels of phenomena and noumena. This concept of "continuity" in substance will be further elaborated upon after the following study of the categorial status of Reality from the point of view of the physical sciences, because the fundamental physical sciences of QM, STR sand GTR have a different way of blocking the concept of continuity. The fundamental ideas of the STR suffice to represent the relevant facts of the science of the cosmos. Moreover, in such a short work it is impossible to deal with as many important aspects as possible of these sciences. I deem it sufficient to discuss some of their most relevant ontological categorial foundations in order to arrive at and fully justify the concept of the continuity within Reality.

I discuss two important blocks to continuity here: (1) the theory of "impossibility" of measurement of the microscopic level of phenomena in QM has given rise to perplexing views of the concept of substance; and (2) the theory of "impossibility" of superluminal velocities has created a limitation of measurement of velocities in the cosmic sciences. These suffice for us to instil the permissible and necessary amounts of continuity in our concept of substance, 'Reality'. Thus, the following sections think categorially of the extent and meaning of continuity in the fundamental aspects of physical science.

1.2 Microscopic Categorial Features of Quantum Mechanics

1.2.1 Apriority in Scientific Categories and Realism

Kant's forms and categories are epistemic categories (of sensibility and understanding)[83] and are *post factum* with respect to the existence of things and without explicitly admitting the superior apriority of ontological commitment to there being anything in their individuality or totality. There are the forms and categories, because there are phenomena-based sensibility and understanding. This is so also in any epistemology, cosmology or ontology that takes the cognitive act as categorially conditional or *a priori* to there being 'anything' whatever.

The sense in which 'there are forms and categories' is applied to discourse and to factuality is purely epistemological in Kant. This is epistemic apriority. To do scientific ontology we need to transcend this cognitive deadlock in categories by frame-working the scientific, epistemological and ontological necessities, by constructing universally and particularly applicable most total and universal collusive categories. This is possible only if the phenomena-noumena continuity is taken seriously and culminated by a category of the purely ontologically total order proper. In order to see if the concept of Reality can have any categorial significance we question the nexus of the empirical categories of physical sciences in the light of the World / Reality as the phenomena-noumena continuity. Space, time, mass, causality etc. in the sciences and in Kantian and other categories will then be *post factum* for the *a priori* fact of Reality. I do not intend here to study the afore-said scientific categories specifically but in their general characteristics.

Modern physical categories are based on Aristotelian and Newtonian categories. They hinge on three hypotheses of the scientific realism discussed in Copernicus's *De Revolutionibus*: (1) epistemological: disengaging thought from absolute truths of the disjunction "truth or probability", (2) ontological: dismantling theoretical descriptions into computing fictions non-descriptive of real processes, and (3) axiological: orienting science to presenting correct bases for

---

83 Kant, *Immanuel Kant's Critique of Pure Reason*, 82 (A42).

calculation (not persuade one that it is so), by constructing models of 'reality'.[84] I make my own conceptual appropriation of these hypotheses:

(1) *The Epistemological Hypothesis:* The scientific and cosmological aspect of the crux of this hypothesis is clarified in a 20th century example. Bohr is cast as anti-realist and Einstein as realist in their debate on the 'reality' of quantum mechanical probability distributions. Bohr argued for abandonment of visualization of atomic processes and contentment with an abstract 'symbolic' formalism (based in mathematical instrumentalism[85]) for predicting the statistical distribution for a range of experimental outcomes. Einstein is supposed to have argued for the crass realist demand for describing the quantum mechanical object in terms of its properties irrespective of the apparatus of observation.[86] This duel is representative of the Copernican attempt to disengage science from the "truth / probability" stance in theories. No extent of the debate has yielded a great resolution. The solution seems to rest on the truth / probability issue. Therefore we should make the micro- and macrocosmic ontologically mutually continuous upon each other based on the theory-laden lower truth-probabilities attracted by the empirical and pragmatic concepts formed from categories of reality-in-particular. I shall argue that the concept of maximally theory-laden Reality yields the highest truth-probabilities that have to do with ever better approximations

---

84 Folse, "The Bohr-Einstein Debate and the Philosophers' Debate over Realism versus Anti-realism", 289–90.
85 Scientific instrumentalism may be explained thus, so to derive the concept of mathematical instrumentalism: "[…] [*I*]*nstrumentalism* about the theoretical entities of science tells us that a scientific theory is just a convenient way of talking about observations. According to this view, the theory is no more than an instrument (just as James said all theories were) which enables us to predict one lot of observations on the basis of another lot of observations. In spite of the fact that scientific theory makes free use of what look like names for classes of things that are not observable – names such as 'atom', 'electron', 'proton' – those are just fictions. In reality there are no such things. All we really have is observations and a more or less complicated scheme for calculating what observations we can expect to make, given observations already made. The theory is a calculus: put observables in and it puts out further observables." Kirk, *Relativism and Reality*, 91. Mathematical instrumentalism is the form of instrumentalism in which experimental, theoretical and mathematical methods and results remain the only possible representations and presentations of actual facts so that the question of actual facts count least for science and mathematics. Theoretical justifications unto ontological commitments beyond mathematical descriptions are unwarranted.
86 Folse, "The Bohr-Einstein Debate and the Philosophers' Debate over Realism versus Anti-realism", 290.

to correspondence between the object over there and the work of theory and experiment. Thus follows the relevance of the truth-probabilistic interpretation of the epistemological hypothesis: truth is probability of correspondence, insofar as we can observe only from specific epistemic levels. If Reality is phenomena-noumena continuity, Bohr's probabilism pertains not to probabilism in nature but that in our approach to truth.

(2) *The Ontological Hypothesis:* This Copernican hypothesis dismantles the very ontological need to posit actual objects. The cosmologico-ontological part of the QM physics behind this problem rests on recognizing that the epistemology of mutual objectivation of entities is permitted by the fact that there always are ever-more infinitesimal layers of subatomic wavicles of actual non-zero mass. Einstein and Bohr disagreed due to lack of such an ontology and its epistemology of physical 'reality'. Then the systems (in one or two microscopic levels) and their properties, whose behaviour QM describes, will not be final in the array of epistemic levels of observation. Bohr appears anti-realist, ruling out description of atomic systems apart from the circumstances of observation by mesoscopic[87] apparatuses. He insists that the interpretation of experimental phenomena or measurements of the values of observed properties of observed objects, as pictures (stories) of (a) trajectories of particles moving through space or (b) wave disturbances propagating across a field. Bohr denies that these pictures are pictures of observer-independent microsystems with "real" behaviour in space and time, i.e., for us, in measured extension and motion. But Einstein "[…] wants to accord an 'element of physical reality' to just those *unobserved* properties to which Bohr denies reality."[88] In the light of our discussion of Kant we know that the ontological solution of this cosmological problem rests on (a) making the macroscopic, mesoscopic, microscopic etc. phenomenal-noumenally continuous and (b) treating the respective epistemologies as genuine with respect to appropriate levels. With these our mesoscopic viewpoint can be treated as truth-probabilistic with respect to its own and other levels. That is, truth-probabilities can be only epistemological and not ontological. Therefore, phenomena-noumena continuity in Reality is a realism which can still assure epistemological truth-probabilism.

---

87 The level of objects and concept-formation in direct observation I call mesoscopic, in contrast to the microscopic and the macroscopic. Though the particles involved in the apparatus are microscopic, they are objectified or observed or theorized about by the experimenter from the mesoscopic layer, in terms of imaginations of the object layer.
88 Folse, "The Bohr-Einstein Debate and the Philosophers' Debate over Realism versus Anti-realism", 290.

(3) *The Axiological Hypothesis:* This less discussed hypothesis says that the correct aim of science is merely "[…] to present a correct basis for calculation" and not to "[…] persuade anyone it is so" through models of reality.[89] If Copernicus's epistemological and ontological hypotheses are well understood and resolves, the scientific instrumentalist axiology of mere calculation of probabilistic causality devoid of (a) action of physical-ontological causality everywhere in Reality and (b) truth-theoretic probabilities and persuasion evaporates. This hypothesis (scientific instrumentalism) is not our immediate concern at the start of this section, but we engage in an ontological and epistemological dispute with it by Chapters 1 and 2, wherever we analyse scientific and mathematical instrumentalism and perspectival absolutism.

In short, it must be admitted that the probabilism discussed in the QM world is irreducibly ontological in Reality, and epistemological in the formation of truths in that epistemic reference-activity is ontologically universal-laden. The Bohr-Einstein debate shows how, from the mesoscopic level, they were stuck to the micro- and macroscopic demands and tried to elucidate mesoscopic categorial expectations from the micro- and the macroscopic levels, without regard to the probabilism demanded by our microscopic status. Attitudes at the interface of realism and idealism were present in, and were thrust on the thought of Einstein and Bohr, and they were unmindful of the continuous infinitesimality involved in the layers of the prehensive[90] mutuality of "objects" and the continuity between the macro-, meso-, micro-, nano-, and other levels of "observable" objects / processes, despite the fact that any and every stipulation of the layers or sub-layers cannot be put in connection with the human epistemic layer without adequate categorial and methodological precautions to account for our mesoscopic attitudes and expectations. An epistemology without such precautions is the result of concretism and phenomenalism in the categories of thinking.

This is a problem of realism based on inadequate scientific-realist assumptions that do not allow a general ontology beyond mesoscopic expectations. Where should adequate assumptions be based? Not merely on particulars but on Reality as the total-token entity and its purely ontological and epistemological categories. Realism holds electrons, genes etc. as real. Anti-realism throws

---

89 Folse, "The Bohr-Einstein Debate and the Philosophers' Debate over Realism versus Anti-realism", 290.
90 This epistemologically cosmological aspect of Reality-in-total may be called at least partially as prehensive, as in Whitehead's theory of prehension, which presupposes quantum theory in a field-theoretic, relativistic, cosmological and biological setting. For a succinct discussion of this, see Whitehead, *Process and Reality*, 238–39.

off theories as more or less dispensable, useful and easily replaceable fictions. The difference and the decision lie in whether theories refer or not, within the backdrop of the tools of science (experiment, observation, theory). One takes reality-in-particular within an already theory-laden procedure of fixing reference. This process of fixing conceptual tools is taken as phenomenalistic, as the process is epistemological. As a result, the wider implication of ontological reference to Reality, of trans-epistemological theory-ladenness and universal-ladenness, is forgotten. Reference is primarily an ontological concept based ultimately on Reality-in-total, not merely epistemological or semantic or instrumentalistic. This requires questioning the adequacy of epistemological apriority in scientific categories. I do this in what follows in this sub-section.

Russell postulated "acquaintance" as a direct referential relation between mind and entities.[91] But the verb 'refer' is equivocal. Rorty says that the term 'refer' can mean either (a) a factual relation between an expression and some other portion of reality (irrespective of anybody's holding it), or (b) a purely 'intentional' relation between an expression and a nonexistent object. Thus, (a) is 'reference' and (b) is 'talking about'. We cannot 'refer to' Sherlock Holmes but can 'talk about' him.[92] Talking about is a notion of common sense. 'Reference' is a term of philosophical art. It makes no difference whether we 'refer' or 'genuinely refer', if no one knows whether the object expressed in a term actually exists or not.[93] But

---

[91] Russell, *The Problems of Philosophy*, 22ff, 79, 97. Griffin summarizes Russell's concept of 'acquaintance' in this work thus: "[…] Russell held that empirical knowledge is based on direct acquaintance with sense-data and that matter itself, of which we have only knowledge by description, is postulated as the best explanation of sense-data. He soon became dissatisfied with this idea and, inspired by his logicist constructions of mathematical concepts, proposed instead that matter be logically constructed not of sense-data, thereby obviating dubious inferences to material objects as the causes of sensations. The actual data of sense, however, are too fragmentary for the construction of items with the expected properties of matter (such as permanence). To solve this problem Russell was led to postulate unsensed sensibilia in addition to sense-data. It is important to realize that for Russell, unlike most sense-datum theorists, sense-data were physical, located in physical space at varying distances from the place at which common sense located the material object. It is thus logically, though not practically, possible for more than one mind to be acquainted with the same sense-datum at the same time. Sense-data are merely sensibilia rather than sense-data." *Routledge Encyclopaedia of Philosophy*, first (1998) edition, s.v. "Russell, Bertrand Arthur William", by Nicholas Griffin.

[92] Rorty, *Philosophy and the Mirror of Nature*, 289.

[93] Zhengkun, "Truth and Fiction in Scientific Theory", 266–67.

[…] clear use of terms cannot solve the puzzle whether the scientific theoretical terms refer or not. According to Quine's view of '[o]ntological commitment', when a person talks about a thing, he has received an ontological judgment. Any scientific theory takes an ontological position, or possesses an ontological premise that admits or refuses the existence of something. But "ontological commitment" does not tell us what there is.[94]

For Quine explains ontological commitment thus: "… to know what a given remark or doctrine, ours or someone else's, *says* there is; and this much is quite properly a problem involving language. But what there is is another question."[95] That there is is a matter of fact derived from ontological commitment, and what there is is the object derived from ontological commitment. This implies reference with respect to the 'that' to the 'what'. The involvement of language too implies some involvement of concreteness and some involvement of non-concrete ideals / universals.

Hence, "ontological commitment" to ontological universals commits ("sends or goes with") only to "ideal existence", not "material existence", of theoretical entities talked of in scientific, empirical terms, i.e., it does not guarantee that the terms of ontological commitment refer to actuals (reference being always to tokens and to particulars / types) or have homologues in the world. Maybe, a certain theoretical entity is a fiction or a theoretical term refers to fiction before we determine that the entity is materially existent,[96] or, if not the entitiy, then at least something at the source- or reference-range of the terms in statements. This is because, as Quine says, reference is secondary to ontological commitment, and takes place "[…] via the impingement of energy on sensory surfaces as this is encapsulated in observation sentences and observation categoricals formed out of them."[97]

Observation categoricals are categorical judgments about actual individual tokens (not types). Quine also insists that there is nothing "more afoot than meets the eye" in ontological commitment.[98] I think this does not mean that ontological

---

94 Zhengkun, "Truth and Fiction in Scientific Theory", 267.
95 Quine, *From a Logical Point of View*, 15–16. Cognitive discourse is or aims to be *about* the world; a particular utterance is or aims to be about some portion or object of the world. Hylton, "Quine on Reference and Ontology", 122.
96 Zhengkun, "Truth and Fiction in Scientific Theory", 267.
97 Hylton, "Quine on Reference and Ontology", 118. In Kant, categorical judgments are affirmative judgments "comprising two concepts related by a copula, typically an attribute (predicate) asserted of a substance or thing (subject)." *Dictionary of Philosophy*, 1982 edition, s.v. "Categorical (Judgment)".
98 Quine, *Theories and Things*, 175.

commitment is about empirically defined reference terms and their corresponding things or processes. Ontological commitment is of concrete "somethings" or to pure ontological universals. That is, in Quine reference is not the basic relation language has to the world. Since language has to do with conceiving and expressing, ontological commitment functions theoretically and in general at the contact point between objects (processes) on the one hand and on the other hand the function of ontological universals in perception, understanding, thinking, theory etc. The existence referred to here is ideal existence, but of concrete things / processes and their ways of being / process.

> The fact that reference is [...] not fundamental shows itself in the fact that we must begin with a set of true sentences, a body of theory that is true, or at any rate accepted as true. Only when the truths are in place can we raise the question of existence. In this sense, acceptance of sentences is prior to reference, and truth is prior to existence. Those objects that a given body of theory is about are presumably the ones that must exist if that body of theory is to be true. They are, in Quine's words, the *ontological commitment* of that body of theory. How are we to understand this idea? For Quine, the answer is quantification theory, which has first-order logic at its heart.[99]

Notice that the priority of true sentences over existence is not at the order of being but knowing or judging. To make ontological commitment to objects which are concrete and freer of the broader universals than types, we need quantification. Quine takes such ontological commitments as apparent when a theory is cast in the notation of first-order logic, because the statements are existentially quantified. Then the theory contains or implies an object of which the corresponding sentence is true. This object – an argument for the vaguely

---

99 Hylton, "Quine on Reference and Ontology", 122. It is interesting to note that Frege too has acknowledged the truth-boundedness of reference (*Bedeutung*). To quote Carl summarizing Frege's concept of reference: "The truth value of a sentence, as one of the components of his former notion of a judgeable content, was considered to be its reference. Frege claims: 'The reference is thus shown at every point to be the essential thing for science [...]. It is by engaging in the quest for truth that we adopt the 'attitude of scientific investigation' [Frege, *Posthumous Writings*, cited in Carl, *Frege's Theory of Sense and Reference*, 116] [...] and the quest for truth leads us to an interest in the reference of our expressions. As we shall see, the connections pointed out by Frege between reference, judgment and knowledge suggest the view that the reference of sentences constitutes the kernel of his theory of reference and provides the general framework for the account of the reference of various other kinds of expressions. The reference to expressions other than sentences will have to be explained in terms of the contribution these expressions make to the reference of the sentences within which they occur." Carl, *Frege's Theory of Sense and Reference*, 116. Square brackets mine.

ontologically committed variable – must exist if the theory is to be true. (Note that a universally quantified open sentence implies the existential quantification of the same open sentence. Our focus on existential quantification here is thus only for the purpose of clarity.) This gives a sufficient condition of ontological commitment. He holds also that this is the *only* way for a theory to be so committed, as is clear in his

> [...] slogan 'To be is to be the value of a variable'. [...] More accurately, 'a theory is committed to those and only those entities to which the bound variables of a theory must be capable of referring in order that the affirmations made in the theory be true. Quine also accepts idioms equivalent to quantification theory as indicating the same commitment, even if those idioms do not use variables [...].[100]

Thus, names (which are universal-laden), as linguistic carriers of ontological commitment, are not on par with quantified variables (which represents a particular, i.e. reality-in-particular) given in a theoretically accepted truth-statement. We should therefore argue that Reality is more *a priori* than reality-in-particular (represented by quantified variables), since quantified variables end up ontologically committing ever wider if the object of quantification is further ontologically universal-laden. Such ontologizing requires more than specific ideal existence (and ideal forms of categories), i.e., ever broader tokens to yield the universals committed to in the ontological commitment to ever broader tokens and particulars. Hence, we argue in the following manner.

This chapter began with the possibility of reduction of classes to attributes / *qualia*, which are also universals of a kind. One of the objections to ontological commitment is cognate to this:

> [...] the predicate of a true sentence must correspond to some entity – a property, or a "universal," as it is often called. Thus it is held that one who asserts that the rose is red is committed not only to the existence of the rose but also to there being a property, redness. The word 'existence' is not always used here; some held that universals have a different sort of ontological status from that of objects and mark it with a different word, such as 'being' or 'subsistence'.[101]

The treatment of existence in ontological commitment can be maximized using the universal qualities co-committed to and the causal-processual origins of such ontological universals. This ends up only in Reality as total-token. Here comes up an important question to be answered: How can anything be expressed in terms of Reality and its *qualia* proper? The present chapter answers this question too,

---

100 Hylton, "Quine on Reference and Ontology", 124.
101 Hylton, "Quine on Reference and Ontology", 126.

through *the justification for deriving the two categories that follow from the category of Reality-in-total*: If an entity or a type / particular in discourse is reducible to and always expressed in terms of *qualia*, then *Reality too is reducible* to and expressible in terms of *qualia*. Qualia too have the nature of categories provided the ones considered are ultimate by co-extensiveness with Reality. Any other category is only insufficiently capable of generalization, which is by co-extensiveness of a token (or a type) with the universal/s proper. Only Reality concedes the most general of universal-ladenness, and so also of the highest of ontological commitment to universals into which Reality is reducible. An argument of a quantified variable at the level of commitment to Reality has the highest concreteness of term-reference. Then, to determine the ontological reference of a scientific term within the purview of a system that has Reality-in-total as a category becomes non-problematic, if the latter is co-implied in them all as the maximal case token reducible to its most universal universals committed to therewith. Hence, the conclusive argument begun before a paragraph ends with the conclusion that *reduction of Reality into all its ontologically committed ontological universals / qualia pissibilizes ontological commitment to the total token Reality via the ever broader ontological commitment to ontological universals yielded by ontological commitment to particular somethings.*

Van Fraassen's anti-realistic constructive empiricism claims the ontological reference of a scientific term by 'observability' in the strictest particularistic sense: "Science aims to give us theories which are empirically adequate; and acceptance of a theory involves a belief only that it is empirically adequate […] [A] theory is empirically adequate exactly if what it says about the observable things and events in this world is true."[102] But we should fix the meaning of 'empirical' differently if everything empirical is theory-laden. If theory is never satiated without being pushed to the side of ontological commitment and *qualia* (eternal objects in Whitehead[103] are parallels with some difference) in ontological and scientific

---

102 Van Fraassen, *The Scientific Image*, cited in Zhengkun, "Truth and Fiction in Scientific Theory", 267.
103 "Eternal objects have the same dual reference. An eternal object considered in reference to the publicity of things is a 'universal'; namely, in its own nature it refers to the general public facts of the world without any disclosure of the empirical details of its own implication in them. Its own nature as an entity requires ingression – positive or negative – in every detailed actuality; but its nature does not disclose the private details of any actuality. An eternal object considered in reference to the privacy of things is a 'quality' or 'characteristic'; namely, in its own nature, as exemplified in any actuality. It refers itself publicly; but it is enjoyed privately." For us, an

enterprise, then 'empirical' is never absolutely demarcated from the General Ontological. That is, we can express anything in terms of Reality and the *qualia* proper to the particular at discussion if we explain it using all the causal relation of the processes that have gone into it. Clearly, this General Ontological notion of 'empirical' is from the possibilities given in General Ontology, along with ways of demarcating it from 'non-empirical'. This needs access to thinking through the highest token entity (Reality) through particulars, which by its nature involves even the less relevant universals. Thus, talk about empirically traceable particulars is possible within General Ontology, which treats Reality. But van Fraassen seems to equate observability with direct observability through naked organs of sense. Detecting through an apparatus is, for him, different from observation.

In order to know how anything can be expressed in terms of Reality and its *qualia* proper, we consider a criticism of van Fraassen's constructive empiricism of observables and unobservables, by keeping in mind that van Fraassen favoured direct observability as a condition for scientific realism. The following is a set of objections by Churchland to his concept of observability:

> Consider some of the different reasons why entities or processes may go unobserved by us. First, they may go unobserved because, relative to our natural sensory apparatus, they fail to enjoy an appropriate spatial or temporal *position* [....] Second, they may go unobserved because, relative to our natural sensory apparatus, they fail to enjoy the appropriate spatial or temporal *dimensions*. They may be too small or too brief or too large or too protracted. Third, they may fail to enjoy the appropriate *energy*, being too feeble or too powerful to permit useful discrimination. Fourth and fifth, they may fail to have an appropriate *wavelength* or [...] *mass*. Sixth, they may fail to 'feel' the relevant fundamental forces our sensory apparatus exploits, as with our inability to observe the background neutrino flux, despite the fact that its energy density exceeds that of light itself.[104]

---

ontological universal does not refer; it is a way of process of the many. Whitehead, *Process and Reality*, 290.

104   Churchland, "The Ontological Status of Observables", 39. The defect of van Fraassen's concept of the 'empirical' may be understood in the following manner. At work, scientists do not accept this limited view of van Fraassen. Dudley Shapere argues, citing the solar neutrino experiment: "[...] [a]lthough the central core of the sun lies buried under 400,000 miles of dense, hot, opaque material, astrophysicists nevertheless universally speak of the experiment as providing 'direct observation' of that central core." Shapere, *Reason and the Search for Knowledge*, 342. If we can explicitly infer the effect produced by a theoretical entity (atom, electron, neutrino, etc.) under experimental conditions (i.e. not directly), use the effect to expand knowledge of other entities, make artifacts in practice (e.g. the electron-microscope), then we posit the theoretical entity and endow the term with reference. Quantum physicists

This argument comes close to, but short of, positing different layers of prehension, like the macro-, meso-, micro-, nano-, ultra-quantal- and other layers of entities *ad libitum*, all of which have their own observable effects, which need not directly impinge on the organs of sense or the apparatus at the mesoscopic level. The very concept of experiment will have to be extended, so as to make his observability-criterion also theory-laden. Only a series of changes in scientific revolutions can guarantee the relative tenability of a theoretical entity (e.g. 'atom', 'electron'). So, observability does not suffice, nor empirically empty ontological commitment. Both these are theory-laden, i.e., laden by the need to further explicate by use of further ontological commitments, observations etc. of ever-widening circles of entities, processes and the relevant universals and abstract entities.

This indicates that van Fraassen's mesoscopic epistemological apriority bears no bet in justification of ontological categories that potentially base scientific categories. We need to posit the most general condition for the possibility of ontological commitment and observability, applicable to all sorts of layers of measurement of entities / particles, in order to conclude anything out of the mess of the question of phenomena-noumena continuity. Phenomena involve observation, imply ontological commitment (to some processes being spoken of or not, with foundation on real universals at the causal origin and result-range of processes), and point further to the 'that' and 'what' of seeing ever widening realms of entities. If phenomena and noumena are continuous, there is theoretical apriority in their continuum Reality. This is not a transcendental (involving some pure universals) apriority, but a Transcendent apriority involving actual entities in their totality and the universal To Be. General Ontology is thus at the roots of particular observables. *Entities and purely ontological Transcendent apriority are also involved. This is no epistemological apriority but General Ontological.*

This completes our shedding of conceptual phenomenalism and inadequate objectivism from ontology and allows treating both entities and attributes on an ontological footing through an ontological category from phenomena-noumena continuity. The reality of the cosmic and epistemic micro-meso-macroscopic continuity of Reality is now clear, as the infinite-infinitesimal, objectual-causal and cosmological categorial contribution of QM and STR to future truth-probabilistic scientific ontologies.

---

        suspect 'hidden variables' models due to lack of observable effects in measurement. 'Hidden variables' may be fictitious under existing conditions. Zhengkun, "Truth and Fiction in Scientific Theory", 268.

1.2.2 Wave-Particle Duality and the Concept of Reality

It is classical to consider the dual nature of the quantum object ("quanton") in quantum events as a violation of realism, depending on the experiment we perform. We examine this from the foundation laid for scientific realism in the previous pages, and suggest a viable synthesis of the conflicting realist and idealist interpretations, a synthetic view of the seemingly dualistic nature of matter, and thus an ontologically coherent and adequate concept of Reality. Even if experimental conditions miraculously tell us when the objects are waves or particles, still the question of how theoretical conversion from one to the other state of micro-existence happens remains unanswered. The important features of QM stumbling the realist are: (1) wave-particle duality and (2) the settlement of the EPR paradox thought experiment by admitting the non-locality concept. Without achieving a satisfactory view of these the category of Reality in the inductive totalization of meso-, micro-, nano- and other levels of objects / events, as supervened by universals will remain questionable in the face of the anti-realism of wave-particle dualism and the resultant probabilism of scientific and mathematical instrumentalism and the discontinuity that these imply in Reality. A QM realism and its ontologically tenable epistemological probabilism will thus result, without resorting to the anti-realism of instrumentalism.[105]

Though Newton had taken all possible micro-particles to be corpuscles, he had also considered the photon, as energy quanton, to be perfectly wave-shaped, i.e. without extension. Thomas Young's double-slit experiment (1803) with visible light showed that light propagates like waves. He measured the distance between (1) the interference fringes, (2) the photographic film and the double-slit and (3) the two slits – and calculated the wavelength of the particles using the well-known formula $n\lambda = d \sin \alpha$, and concluded that light travels as waves. The calculation "presupposes" that each object passed through both the slits in

---

105 "[...] Bohr ... insisted that quantum theory refers only to 'observations obtained under experimental conditions described by simple [classical] physical concepts'." Bohr, *Atomic Physics and Human Knowledge*, cited in Bunge, *Treatise on Basic Philosophy, Volume 7: Formal and Physical Sciences, Part I*, 169. Square brackets mine. Further, Heisenberg points to the same in the following: "[...] [t]he statement that any light quantum must have gone *either* through the first *or* through the second hole is problematic and leads to contradictions. This example shows clearly that the concept of the probability function does not allow a description of what happens between two observations. Any attempt to find such a description would lead to contradictions; this must mean that the term 'happens' is restricted to the observation." Heisenberg, *Physics and Philosophy*, 52.

the screen (because particle motion by one particle should always be through a single slit, and *waves are taken as just paths, not objects*). That is, at calculating the wavelength when it passes through the screen with the double-slit, we are made to assume that each object is at two places simultaneously. This, according to him, shows the *object* to possess wave character.[106] The double-slit experiment allows choice between photons, electrons, etc. (in general, "quantons"[107]), as the test particle. An intensity distribution results on a photographic film, yielding the wave character of the *object* that passes through the double-slit. But each particle is in a definite position when hitting the photographic film. Each has made a spot on the film, the position being measurable with an accuracy of the mean diameter of the grains in the film emulsion. This accuracy is smaller than the distance between the two slits in the screen. Thus, measurement at interaction shows the object to be a particle.

If the object was really a particle at hitting the film, the simultaneous intensity distribution of hits of particles (indicating waves over the target region) cannot be explained.[108] The quick-minded conclusion is that of an alternation of wave- and particle properties with respect to objective circumstances and apparatuses.

---

106 Johansson, "Realism and Wave-Particle Duality", 329–30.
107 Bunge uses the term 'quanton' to avoid the classicism and mathematical instrumentalism involved in the concepts of waves and particles. He gives his reasons in the context of discussing the Schrödinger- and Heisenberg equations: "The state function (or vector) ψ is a complex valued function of the space and time coordinates. Its precise form is determined, up to constants, by the precise form of the [H]amiltonian $\hat{H}$ as well as by the initial and boundary conditions. In the simplest case, that of a single quanton free from external forces, ψ looks like a classical plane wave; in the case of an attractive central force (such as that exerted by an atomic nucleus on an electron), ψ looks like a classical spherical wave. These purely *formal* analogies gave rise to the misnomers 'wave function' and 'wave mechanics', just as the use of the [H]amiltonian formalism associated with the Schrödinger equation suggested the misnomer 'particles' for the ultimate specific referents of quantum mechanics. Bohr [...] held that the undulatory and the corpuscular views were mutually complementary, whence we had to keep them both and play dialectical games with them. On the other hand Heisenberg [...] admitted that they are only 'mental pictures' and 'are both incomplete and have only the validity of analogies which are accurate only in limiting cases'. Since they are indeed just analogies, and since they cannot be both correct, we shall adopt neither of them. We hold instead that the central referents of quantum theory are *sui generis* entities deserving a name of their own: quantons." Bunge, *Treatise on Basic Philosophy, Volume 7, Part I: Formal and Physical Sciences*, 171.
108 Johansson, "Realism and Wave-Particle Duality", 329–30.

One forgets here that these can be properties primarily with respect to our measuring. Realism and empiricism are both at stake here. Observer-independence and observability of the object in the wave- and particle states are therefore to be interpreted.[109] It is not enough to merely say that the wave shape is an unwarranted conclusion from the probability distribution on the target plane. We must know why it denotes waves and/or particles.

The central problem of the war between Bohr (and others), who were representatives of the Copenhagen Interpretation, and Einstein (and others), who attacked them in the name of realism, is that of an unwarranted muddle of realism and classicism in physics:

> In our view Einstein and his coworkers were right in asserting the realistic thesis that the world exists without our assistance. But they were wrong in supposing that the world is composed exclusive of things all the properties of which have sharp values (i.e. eigenvalues) at all times. This hypothesis is not *realist* but *classicist*. It amounts to claiming that the world is composed of classons, and therefore must be describable by classical (or neoclassical) theories.[110]

Realism may be defined as the attitude to justified true belief in terms of ontologically and epistemologically compatible logical results of any or the entirety of theories about what there are and what is the case, through *a priori* commitment to mediately experienced processes, their universals and empirically cognitive and/or immediately perceptual approach to them. Here the epistemological

---

109 This may be understood in terms of the solution to the problem of Schrödinger's cat: "The quantum state involves a linear superposition of a reflected and transmitted photon. The transmitted component triggers a device that kills a cat, so according to U-evolution the cat exists in a superposition of life and death [...] [T]his is resolved because particles in the cat will almost instantaneously suffer hits, the first of which would localize the cat's state as *either* dead *or* alive." Penrose, *Shadows of the Mind*, 334. This is realism concerning the existence of the particle in the form of a wave, and realism concerning the state of a particle while it is being expressed statistically by the "Schrödinger time-evolution of the wavefunction of a particle, initially localized closely at one point, subsequently spreads out in all directions." Penrose, *Shadows of the Mind*, 332.

110 Bunge, *Treatise on Basic Philosophy, Volume 7, Part I: Formal and Physical Sciences*, 175. Cushing puts Pascual Jordan's classicism differently, quoting him on individual photons passing through a polarizer: "denial of the classical concept of causality is not to be understood as a temporary imperfection of our knowledge, but is inherent in the nature of the thing – again showing how incorrect our previous, classical concepts were." Jordan, *Physics of the 20th Century*, cited in Cushing, *Quantum Mechanics*, 131–32.

aspect of apriority is that of commitment to truth-probabilities from facts of physically and perceptually mediate experience. The ontological aspect of it is commitment to facts of physical processes within physically and perceptually mediate experience. Empiricistic apriority is based on ontological commitment to phenomena in their immediacy in the perceptual act and in their givenness; and realistic apriority is based on commitment to things-in-themselves, from the mediacy in the perceptual act and in their givenness. The former yields empiricism, empiricist idealism, solipsism, relativism etc.; the latter results in realism of various kinds.

Truth in realism may be defined as justified true belief in terms of ontologically and epistemologically compatible ("compatible" to the needs) logical results of any or the entirety of theories about what there are and what is the case, through *a priori* commitment to processes and empirically cognitive approach to them. Kant's phenomena and noumena are mutually continuous. Similarly, commitment to the immediate and the mediate in perception and cognition is such that they are taken as mutually continuous. This alone facilitates theoretical acceptance of processes. *Classicism in physics* is the cognitive attitude to the processes of the physical world by which, without justification by mediate or immediate ontological commitment, one takes for granted that processes possess the exact values of measurement that the scientific mind attaches to them from mesoscopic cognition. *This is the epistemological variety of determinism* that defeats the infinite and infinitesimal determinations implied in ontologically committed objectual causality.

The problem of wave-particle duality can be resolved from a realistic understanding of the confusion between epistemological determinism and ontological causalism. In our study of wave-particle duality in QM it will be shown that epistemological determinism (e.g. that the eigenvalue ["own (sharp) value"] determined or determinable for a wavicle is always and everywhere what the wavicle actually possesses) has vitiated the concept of causality at the infinite and infinitesimal levels of physical objects / events. One suggested twirling convergence of realism and instrumentalism is by Cushing:

> In the end, we may be left with an essential underdetermination in our most fundamental physical theory and this issues in an observational equivalence between indeterminism and determinism in the basic structure of our world. [...] One possible conclusion is that, as a pragmatic matter, we can simply choose, from among the consistent, empirically adequate theories on offer at any time, that one which allows us best to 'understand' the phenomena of nature, while not confusing this practical virtue with any argument for the 'truth' or faithfulness of representation of the story thus chosen. Successful theories can prove to be poor guides in providing deep ontological lessons

about the nature of physical reality. Such an enterprise would appear to be consistent with Quine's dictum....[111]

Cushing suggests an observational equivalence between indeterminism and determinism and argues that, pragmatically, theories that work are more preferable theories that seem to offer ontological lessons. To vindicate this position he takes recourse to Quine's dictum: "... the world intrudes first as a surface irritation and remains thereafter as a constraint on our imaginations (in constructing scientific theories)."[112] The confusion here may be between representation, reference and (Quinean) ontological commitment. Reference need not be to an object, but to a fact about an object. *Reference is not representation; it can be representation in cases of clear correspondence to objects.* Materialist positivism represents objects by reference. *Ontological commitment is* not merely to what terms represent, nor merely to what is given in reference (by propositional meaning), but to what is mediately presupposed in reference and accepted by *a priori* admittance, i.e. *to objects / events / processes as such with their causal possibilizing agents in extension-motion in Reality.*

Hence, the instrumentalistic indeterminism of the Copenhagen interpretation and the realistic causal determinism of Einstein (the latter partially implies both ontological and epistemological determinism at one go) need not mathematically imply any ontology of objects / events / processes for them to be of significant consequence in results and predictions. But attempts to justify the reality or unreality of wave- or particle nature of wavicles through causal explanations of their shape, structure, paths and influences should allow advances in the very physics and mathematics of the problem and lead QM into micro-, nano- and other more infinitesimal realms of physics in integration with near-infinitesimal, nano-, micro-, meso- and macrophysics. Such purely physical and mathematical consequences of ontology beyond the present achievements in QM make such ontologies worth the trouble.

This is the ideal background to discuss the real nature of the wavicle and to know whether any influence by the apparatus does the trick of making the wavicle alternate between exhibiting the wave- and particle natures, if it possesses any other possible shape/s of motion and interaction, and if ontological commitment to quantons with an ontologically justified nature in place of wavicles with dual nature is of any significance. Fixing the real nature of the wavicle as wavicle

---

111   Cushing, *Quantum Mechanics*, 214–15.
112   Cushing, *Quantum Mechanics*, 215. Here Cushing renders his recollection of Quine's statement during a public lecture at Wittenberg University in late April, 1992.

need not mean fixing their eigenvalues with their experimental wave- or particle nature. Instead, it means fixing why and to what extent these can be fixed basing on the possibilities yielded by the infinite and infinitesimal causal nature presupposed by the notion of purely realistic quantons.

The Copenhagen interpretation of the Uncertainty Principle says that the probabilistic predictions of quantons remain unaltered despite removal of all possible perturbations by apparatuses and our choice of apparatuses.[113] This interpretation insists that wavicles (and so, Reality) are in themselves so close to being infinitesimal that they resist being perceived as waves or particles by supposedly allowing our statistical alternation between the two states for the purpose of observation and prediction.[114]

The eigenvalues equation tells where exactly the ontological search for the nature of the wavicle has reached. An eigenfunction is any of a group of independent functions, which are solutions to a certain differential equation. An eigenvalue is any of a group of values of a parameter (an eigenfunction), for which a certain differential equation has non-zero solution within stipulated conditions.[115] The

---

113 "In fact, our ordinary description of nature, and the idea of exact laws, rests on the assumption that it is possible to observe the phenomena without appreciably influencing them. To co-ordinate a definite cause to a definite effect has sense only when both can be observed without introducing a foreign element disturbing their interaction. The law of causality, because of its very nature, can only be defined for isolated systems [...]." Heisenberg, *The Physical Principles of the Quantum Theory*, 62–63.

114 "Some, such as Schrödinger and Einstein, saw the indeterminacy relations as a statement of man's limited ability to penetrate the details of natural processes, not as a complete description of how nature acts. Others, such as Heisenberg and Bohr, felt that in the indeterminacy relations and in Bohr's associated principle of complementary physical descriptions, humans had discovered the deep truth that nature itself follows only statistical rules. In its fully developed form – the *Copenhagen interpretation* – this latter view represents the public philosophical view of most physicists to this day." Wheaton, *The Tiger and the Shark*, Cambridge: Cambridge University Press, 1992.

115 *The New Shorter Oxford English Dictionary on Historical Principles*, s.v. "Eigen-". To put it differently, "The eigenvalue of a matrix M is a number λ which satisfies the equation $M\lambda = \lambda\psi$, with $\psi \neq 0$. In quantum mechanics, the matrix M will correspond to a particular dynamical variable (such as position, energy or momentum) and λ will correspond to the value obtained by measuring that dynamical variable if the system is in the state described by ψ. Ψ is called an eigenstate of the system." Coughlan and Dodd, *The Idea of Particle Physics*, 222. More accurately, an eigenvalue may be explained in the following manner: "For a linear transformation T on

solution of an eigenvalue is representable as an eigenfunction. "[...] [T]he *eigenvalues equation* for an operator $\hat{A}$ representing an arbitrary dynamical variable ("observable") A [...] reads $\hat{A}u_k = a_k u_k$, where a_k is the k-th eigenvalue (admissible value) of $\hat{A}$, u_k the corresponding eigenfunction (admissible solution), and k a real number."[116] There are two separate issues here: the measurement of energy state and the apparatus of measurement. The question of measurement may be understood in the following manner, within the meso-nature of the apparatus, this latter being different from the issue of the *physical influence* of the apparatus on the observation:

> According to the Copenhagen school, every a_k [...] [in the eigenvalues equation] is one of the values that an *observer* may find when *measuring* the property A with a suitable instrument – of *any* kind. However, it is plain that [...] [the eigenvalues equation] makes no allusion whatever to any observers, instruments, measurement techniques, or measurement operations. The only interpretation [...] [the equation] tolerates is a strict or literal one, namely that a_k is one of the possible values of A – whether or not we happen to measure it. So much for the semantic aspect of the question. The methodological aspect is just as clear: the results of a precision measurement depend not only on the thing measured but also on the measurement method, and they are rarely exact. (They can be accurate only if the eigenvalues are denumerable and widely separated.)[117]

That is, the very results of the measuring process have some probability inculcated by the combined effect of the very micro-nature of the wavicles and the meso-nature of the apparatuses that measure and identify. This probability is not the same as that probably caused by the possible *physical influence* of the apparatus on the possible sub-micro or sub-sub-micro levels of interaction between the wavicles at issue and the wavicles within the apparatus, because these influences are far too negligible for a level of difference (that between the micro- and the sub- or sub-sub-micro levels) that is comparable to that between the micro- and the meso-levels. The fear of possible influences by the apparatus on the nature of

---

a vector space V, an eigenvalue is a scalar λ for which there is a non-zero member v of V for which $T(v) = \lambda v$. The vector v is an eigenvector (or characteristic vector). For a matrix A, the eigenvalues are the roots of the characteristic equation of the matrix (they are also called characteristic roots and latent roots); the number λ being an eigenvalue means there is a non-zero vector $x = (x_1, x_2, ..., x_n)$ for which $Ax = Ax$, where multiplication is matrix multiplication and x is considered to be a one-column matrix." *Encyclopaedic Dictionary of Mathematics*, s.v. "Eigenvalue".

116 Bunge, *Treatise on Basic Philosophy, Volume 7, Part I: Formal and Physical Sciences*, 172.
117 Bunge, *Treatise on Basic Philosophy, Volume 7, Part I: Formal and Physical Sciences*, 172.

the measurement-result regarding the wave- or particle nature could be warded off as really negligible in the following way:

> In general, a precision measurement of a property A will yield a whole set of values. One usually compresses this multiplicity into a formula such as $meas\ A = a'_k \pm \varepsilon_k$, where a'_k is the arithmetic mean of a set of measured values, and ε_k the corresponding relative error, which is characteristic of the measurement method as well as of k and of the size of that set. As a rule a'_k differs from the theoretical value a_k. If both were always identical, as implied by the Copenhagen interpretation, it would be possible to scrap all the research projects devoted to measuring the eigenvalues of all dynamical variables, since these eigenvalues would be given accurately and a priori, hence once and for all, by [the eigenvalues equation]. Fortunately, the workers at Cern, Dubna and Fermilab need not worry: the eigenvalues equation warrants only an objectivistic interpretation since, by hypothesis, the property A represented by the operator $\hat{A}$ is a property of an individual quanton, not of a microsystem including experimental equipment and experimenters.[118]

If there are influences by the apparatus, if they have been reduced to the minimum, if the experimenters make sure experimentally that they do not affect the mean value much and if they possess theoretical ways of warding off such influences in the calculations, then any such influences would be so many times minutely infinitesimal compared to the already infinitesimally minute values we are discussing, as to be capable of affecting the values non-negligibly. Moreover, if such influences do happen and are provided for, they appertain the measurement of the eigenvalues and will not very much affect the nature of wavicles as such: because, whatever the influences, the wavicle will be some sort of a wavicle. In the process of determination of the values such sways are sure to interfere, if, as will be the case, there are sub-micro level and sub-sub-micro level and smaller exchanges between the quantons in question at the experiment. And these have not been accounted for by the current level of experimental and theoretical achievements.

---

118 Bunge, *Treatise on Basic Philosophy, Volume 7, Part I, Formal and Physical Sciences*, 172–73.

For example, the gravitational effect[119] of the whole earth on the particle in question is negligibly minute, because gravitation is a weak force, not strong.[120] Strong forces have a greater influence, but these are micro-level and hence are already being warded off by the apparatus. Even if there were unidentified forces of this nature active on the wavicles in question, their influence need not always alter the expressions of the wavicle into either a pure wave or a pure particle, since these expressions need not be the expressions as such of the wavicles but our measurements of the same. Further, the sub-sub-micro-level influences are so negligibly small with respect to measurements of the sub-micro-level influences, *they do not seriously alter or interchange the wave- or particle natures* of the wavicle at all. The so-called weak force too is of intra-sub-sub- ... micro-level origin. Therefore, there should first be ways of determining which of them really does substantially affect the expression of the wavicles non-negligibly.

Nevertheless, no interactions can completely alter the supposedly actual choice – between wave and particle expression at each given moment – of the nature of wavicles. If there were interactions of the same (micro-) level as the wavicle, those objects that affect the wavicle in question would already have produced a non-negligible effect. These perturbations are measurable at least statistically, and at the very least are detectable in principle at a future time. But we were speaking of the negligible effects which occur from much more near-infinitesimal layers of activity in the wavicles and of determining their deep structure,

---

119 The discipline of Quantum Gravity attempts to unify gravitation with the strong forces of microphysics by quantising gravity. This does not mean that gravitational effect on quantal wavicles in the present world is considerable. Quantum Gravity applies to sufficiently condensed states of matter, as at the realms of the Big Bang universe and black holes. To such astrophysical realms, even Planck-length, -time, -mass, -energy and -temperature are all large scale compared to actual inter-particle scales of length, time etc. of such realms. Coughlan and Dodd, *The Idea of Particle Physics: An Introduction for Scientists*, 179.

120 The distinction between weak and strong forces is made on the basis of their effect on subatomic particles. Weak force is "[...] the weakest of the known kinds of force between particles, which acts only at distances less than about 10^{-15} cm, is very much weaker than the electromagnetic and the strong interactions, and conserves neither strangeness, parity, nor isospin." *The New Shorter Oxford Dictionary on Historical Principles*, s.v. "Weak". Strong force is "[...] the strongest of the known kinds of force between particles, which acts between nucleons and other hadrons when closer than about 10^{-13} cm (so binding protons in a nucleus despite the repulsion due to their charge), and which conserves strangeness, parity, and isospin." *The New Shorter Oxford Dictionary on Historical Principles*, s.v. "Strong".

motion, shape of expression etc., and finally also of the measurable effects of these realities on the totality of the wavicle.

In short, our issue comes to mean the following question: Will there not be sub-micro level strong forces that can cause cumulatively large effects, which may not immediately affect the measurement but will already have determined the structure and motion of the wavicle from within the very wavicle in question? Our answer finally has now been that *the perception of the wave- or particle nature of the wavicle may be seriously altered, perhaps only enhanced, by such perturbations.* If they will both be seriously altered and enhanced, then they are not disjunctive, but cumulative, in the act of their occurrence. Then it should necessarily be in their extension-motion location of occurrence. This is already a causal cumulative contribution from the sub-, sub-sub- ... levels of the same processes. We cannot imagine how they can be a mere non-causal, non-extension-motion influence in their measurable states!

This shows that, if quantons are in some way related to other quantons, they are causally related particles. If quantons are basically structured from within, and these are wavicles structured by others, they are infinitesimal in their real deep-causal structure, all of which we are never capable of measuring by any unique method or device at any given time. Then there is no meaning in supposing that their measured specific state vectors or eigenvalues are the final determinations available of quantons in their entire inwardly infinitesimal structure. Some aspects of measurement that contribute to one specific measurement of a wavicle may be more or less well determined. The very perception of wavicle as wave and/or particle is not altered by the statistical nature of *our* measurement of wavicles. Only the intensity or measure of alteration in the wave- and particle paths will be affected by the perturbations one can think of.

Now to put it bluntly: "Whether nature is only statistically consistent or whether humans are simply unable to penetrate to a consistent core soon became matters of only marginal professional concern to physicists."[121] The internal structure and nature of wavicles do not apparently have anything to do with eigenvalues, since they are themselves statistical, and not absolutely set. These measurements experimentally determine in some way the *internal* structure, motion, shape etc. of quantons, sub-quantons (and sub-sub-quantons, if they too are quantisable in any fashion, whatever) etc., and as generalisable as actual waves and particles at the same time, because any measurement implies the wavicles' being anything other than absolutely perfect straightline motions. That is, the eigenvalue, as a

---

121 Wheaton, *The Tiger and the Shark*, 307.

statistically measured property, does not need to be identified as setting the exact value of the quanton's wave- or particle dimensions.

It is now clear that Einstein's claim in itself of lack of completeness in QM was far-fetched due to the incompetence of his classical realist stand to claify his point. He argued well against the supposed completeness of the existing group of theories of QM. But his argument was in favour of some absolute measurements – which cannot be had. Bohr and co. were for complementarity of wave and particle – not for one and the same moment of the experiment but for alternating between them from moment to moment. This too is not to be attested by any wavicles. The incompleteness of QM should not in the final analysis reflect in the explanation that wavicles are at times waves and at other times particles. It happened falsely to be called the theory of simultaneous complementarity of the wave- and particle natures. It is true that it is of the very nature of actual objects of non-infinite velocity[122] that they are exemplifications neither of the absolutely non-extended concept of waves, nor of the concept of perfectly spherically extended particles surfing on in straight line motion or wave motion. The fact of actual wave-formation implies the non-infinite magnitude of velocity and makes the wavicle simultaneously elongated (non-spherical) and extended (non-vacuous) particles moving in finite speed. Finite but large velocity propagations can assume only such motions.

Every measured value of a state or function is a certain value. So, it does not encapsulate all possible near-infinitesimally internal and infinitely contextual causal influences on the structure, motion, shape, qualities etc. of wavicles. The near-infinitesimally internal effects on the extrinsic nature (structure, motion, shape, qualities etc.) of a wavicle cannot mean an infinite number of motions, shapes, qualities etc. in its actual conduct within a context. Nor can an exact quantity show up in any conceivably unique manner. Any manner conceived is in fact a probabilistic generalization over many statistically conceived "exact" wave paths – not over many uniquely actual quantities or values. The fact that these are experimentally concludable paths does not mean that the wavicle as such in its actual shape, motion etc. is merely conceptual when conceived – as is clear from the ontological nature of connotatives in mind. Just as ontological commitment apriorily commits to objects / events / processes, so does it commit also to their ways of being / conducting (say, one special sort of motion), beyond

---

122 Heisenberg argues differently in his *The Physical Principles of the Quantum Theory*, 62: "It became apparent that ordinary concepts could only be applied to processes in which the velocity of light could be regarded as practically infinite."

their ways of showing in mind. Now we elucidate on what are determinable in ontological commitment to wavicle motion.

It is now better to assume *the wavicle as elongated, extended particles in wave motion*. The non-infinite magnitude of velocity of any wavicle simultaneously makes the wavicle elongated and extended particles in wave shape. That this is a determinable *general* nature of all wavicles is not a probabilistic statement. In fact, what are probabilistic about the nature of wavicles are the more specific details like: (1) the exact length, breadth and depth of the elongated particle shape which are not fully determinable, and (2) the exact nature and frequency of departure of the elongated particle from a probabilistically determined wavelength of exactitude, which are also not fully determinable. This realistic explanation makes room for a realistic interpretation of the general nature of wavicles. Look theoretically into any level of structure of motions in a wavicle, and we realize that not merely the internal structural motions, but also external sub-sub- ... quantal influences determine them in some fashion. This allows us to look beyond the wavicle and connect it to causal influences within it from beyond itself at any given time. This broadening of sources of causality can go on, and it will end only with Reality. This is the best way of extending ontological commitment from particulars (natural kinds) to its broadest setting. This is also the way of broadening the idea of theoretical intake of causal influences in the given nature of a wavicle or a conglomeration of wavicles. Hence, for clarity in the physical science we need ontological commitment to objects / events / processes in every wavicle-experience interpreted statistically as wave or particle. These do not rest there: they are theoretically satiated only within the causally related universals and finally Reality.

If the track of a moving object is uniformly bumpy in two dimensions, motion is two-dimensionally wave-like. Each bogey of the wave expresses its own partial inner unity of extension-motion regions in terms of wavelike particle motion, the particle-formation being intermittent – otherwise it would define uniform smeared-out path which is with infinite velocity. The wave-like oscillation is a pulsation of the packet / particle of energy. To retain both the natures as real, beyond the problems of probability, inadequacy of measurement etc., we need to *consider bottlenecking* (intermittent constraining) as a reality about wavicle motion / wave-like particle motion in finite velocity.

If pulsation takes place by a rough bottlenecking of the energy packet at the nodes and releasing it at the crest, then again bottlenecking at the node and releasing at the trough, it is wave motion. Bottlenecking can only be spatially three-dimensional and temporally one-dimensional, thus adding up to four-dimensional motion. Uniformity of acceleration in a straight line is only a statistical levelling out at the given state at nodes, midway between the crest and trough of every unit wave. With respect to every two adjacent crests and troughs of each wave there cannot be uniformity of acceleration in speed, due to bottlenecking and releasing. (This brings up the question of uniform limit motions, which will be treated under STR below.) The elongated three-dimensional wavicle cannot move in the fourth dimension, if it had been totally absent at nodes. But it does. This makes discreteness a notion of statistical levelling. We have no apparatus to show the results of this levelling with precision. It might be argued that at least within a wave (comprising a crest and a trough) the speed is uniform, only the direction of the vector constantly changes in three dimensions, and therefore uniformity is only on a large scale.

If the speed were uniform, the shape of motion should also have been uniform at all points. Being uniform everywhere would mean that the whole wavicle is absolutely a perfectly spherical stuff from a supposed beginning of the crest to the end of the *same* crest, thereafter a pure non-entity at the node, and then, miraculously, a spherical stuff from the beginning of the tough, and so on. If this latter explanation were adhered to, it would have to be admitted that at nodes wavicles are non-existent in motion, and that energy is simultaneously absolutely a wave and a particle – both contradictions in terms (i.e., in 'extension', 'finite velocity' etc.). That is, wave is only a trail, a path, a form. What propagates in waveform must be non-vacuous (hence, spatially three-dimensional and temporally one-dimensional) energy particle, and energy must comprise extended and elongated particles moving in wave fashion by bottlenecking of non-zero four-dimensional magnitude, then releasing from that state into another non-zero four-dimensional magnitude, then bottlenecking and so on in their spatio-temporal coursing.

Every sufficiently elementary entity in the universe must follow such a path. The trail of a meso-object (say, a train) allows arithmetic and plane geometric consideration of its motion as a straight line. For micro-objects, wave shape is non-negligibly essential in all of their motion, due to their microscopic nature. Hence, quantons and even meso-world objects in motion are best treated as wave-like motion of particles and objects. Experimentally, this is attestable; and theoretically too, because absolutely straight-line motion implies the impossible infinite velocity. Our only difficulty is that a quantal wavicle has to be observed by a quantal energy particle, yielding a measure-pattern that yields the coarseness of the measurement of a meso-object by a meso- or micro-object. Humans have not achieved sufficient technical advancement to observe a micro-wavicle by apparatuses that put to use sub-micro wavicles. This fact disallows acquiring the accuracy that can be yielded by *measurement of an already composed quantal object by another already composed ultra-quantal energy wavicle.*

A quanton's wave path is constituted by bottlenecking at nodes – defining wave-like behaviour at nodes – and bulging at crests and troughs – defining particle-like behaviour at crests and troughs. That is, by scientific categories, the actual, spatio-temporal, physical (gravitational or other forms of causal) and extending influences will be more visible by suitable apparatuses at crests and troughs than at nodes where bottlenecking sends the space-time measure more time-like but still, though meagrely, extended. The only possible explanation for bulging alternatively towards the crests and the troughs will be that, when the particle is released from bottlenecking it spins spirally to the direction of the crest on a possibly slightly off-directional axis and, after reaching the next node, spins to the side of the trough. This defines the sinusoidal motion of waves.

In short, the difference between matter particles and energy particles seems to be that the former is not bottlenecked in motion, and the latter is bottlenecked at nodes. There is a seemingly unsolved question here: At how many dimensions does nodal bottlenecking take place? The reason should also be found as to whether and why energy waves are two- or three- or four-dimensional. I personally hold it as extensionally three-dimensional and four-dimensional in extension-motion. According to Feynman there is some similarity in wave motion between electrons and photons: "Electrons behave in this respect in exactly the same way as photons; they are both screwy, but in exactly the same way."[123] Energy particles travel winding. That is, they are four-dimensional in motion. In that case, our representation of the whole concept of bottlenecking is only a plane-geometrical three-dimensional explication of the phenomenon in four dimensions.

The velocity at the important points – crest, trough and nodes – cannot be uniform from the point of view of sub-micro and sub-sub-micro and other more minute levels. The exact velocity we speak of is only a statistical levelling out by techniques of measurement and calculation. *The* **minute differences of velocity at different wave points** *and the constant change of direction in the sinusoidal motion must be capable of making* **the wavicle change the direction after it enters via one slit** and gets affected by the sub-micro and other smaller motions at the material that makes up the slit and other unrelated wavicles in the vicinity. To the standard of motion set up by the micro-level calculations, it should have hit one

---

123   Feynman, *The Character of Physical Law*, 128.

and the same spot on the photographic (or other) plate. This standard is also a levelling out. *From the point of view of the micro-, nano- and other smaller levels of perturbations it is not accurate, nor does it bar off-directional motion. Thus, a certain wave pattern (not one wavicle, which corresponds to one particle) must seem to jump both the slits (but, in each case of entry and deflection, at a different stage of its motion – say, node, crest, trough, or between any two of them) simultaneously – and produce on the plate a net effect of "a probabilistic wave" that has seemingly entered through both the slits.*

This must be so because, for example, at the spots of bottlenecking (nodes) the density of the particle is near zero, and so translational deflection to the same side (or to a direction between the previous and the very next) at which the vector has been defining an angle (due to the presence of the slit) from the node will be possible, instead of changing the side of definition of angle of sinusoidal wave-like motion in a regular manner as pre-determined by the statistical levelling out implied by the perfectly sinusoidal mathematical form. This displaces the spatio-temporal configuration of the effect of the particle/s (the unit wavicle/s) to the near side of particle hits from a different slit. *This makes us replay the process without allowing for the displacement and* **to statistically "detect" the slit of operation as the second one.**

Further, the fact that the measure of compatibility and the wide comparative difference of status between our measuring apparatuses and the sizes and speeds involved make us hold the scientific instrumentalism of probability distributions. **This need not** at all make us say that **Nature blocks us** from perceiving what is happening, or that the quanton is at times a wave and at times a particle. The blurs appearing in different shapes are a testimony to the fact that, though the quanton does follow *some* path at any time and is more or less a particle, and though it really follows the wave path and so it is also a wave, the exact shape of wave path of the now-elongated and then not-so-elongated particle is determined by intra-quantal and extra-quantal causes that are both prior to its constitution and contemporary to its motion. Such constitution point to ontological ways inherited from past causation. The non-symmetry of time-measure sees to it that causal influences from the respective relative future do not affect the respective past of the motion. The blur and the probability distribution are just natural to the apparatus we use, including our eyes, if it were to see such a motion in an animated manner.

That is to say, there is no absolute jump between particle nature and wave nature in one and the same wavicle at the node – neither a jump due to the presence of the slit nor a jump due to any other causal or non-causal influence.

The effect in the plate changes in accordance with the deflecting hit-station "within" the wavicle motion, upon the wall of the slit or upon a sub-sub- … level causal influence from the wall of the slit, which influence is not measurable within the level of subtlety of the apparatus: the hit is either at the trough-state, or at the crest-state. This permits a non-difference in the reconstruction of the path at the two slits. Moreover, as we saw just now, the wave's motion in the extended fashion is three-dimensional, and the measure of motion is uni-dimensional. The actually instrumentalist principle of complementarity that retains a miraculous duality, concluded from the bizarre double-effect on the plate, is therefore not the result of an absolutely temporally discontinuous causal or partially causal phenomenon. It is causally continuous.

The scientific-instrumentalist principle of complementarity of realism and the mathematical instrumentalism in it may be put in the words of Bohr, as quoted from memory by J. S. Bell: "the opposite of a deep truth is also a deep truth": "truth and clarity are complementary."[124] This principle is, from our point of view, mere instrumentalist accrual of practical wisdom and practically helpless mystical intensity based only on a peripheral understanding of the connection between the empirical and the naturally possible actual.

The wave's spatial motion is three-dimensional. Thus it is still partially continuous in the wide sense for two reasons: (1) The probabilistic corpuscular description of wavicles is a circumscription by relatively arbitrarily concretist spatio-temporal measurement. (2) The probabilistic corpuscles are in fact measured by classically separate non-synthesized categories. The nodal bottlenecking (which is non-zero) yields the relatively higher wave-effect at node and particle-effect at crest and trough. The points between the nodes, crests and troughs are also part of particle behaviour, but too intermediately negligible to define a clear causal effect on ordinary mesoscopic objects engaged in microscopic experiments due to the similarity of velocity of the particle and the photons that try to measure their movement from the detector placed between the double-slit and the plate.

This is the case of experiments that involve interaction at nodes (more of wave nature), crests and troughs (more of particle nature). Not the intensity of light beam but the frequency is of consequence here. The nodal wave state defines only a non-vacuous bottlenecking of the wavicle, and this implies a non-zero mass for the wave-like state of the wavicle at the nodes and the comparatively more "spatially" extended size of the wavicle as ensured for non-nodal points in the wave.

---

124   Bell, *Speakable and Unspeakable in Quantum Mechanics*, 190.

This yields also the spatially three-dimensional nature of wavicles. *These two conclusions are clearly realism of the "see that" yielded by ontological commitment, and not phenomenalism of "see as" yielded by the instrumentalism of the usual statistical interpretation.* Thus, what interact at the quantum level are the near-particle states at crests and troughs; not the nodal, bottlenecked, wave-like states.

What then about the validity of the invariants / constants of quantum measurement at the node, crest and trough? We may argue in the above manner also for relativity of such "universal" constants. The discovery of invariants is, in fact, within the context of statistical levelling in measurements. The specific statistical levelling of invariants (not the invariants themselves) must be traced for their levelling a particular reality to the preconditioning dispensation at the origin of every new quantum world of particle-formation at any high-intensity Big Bang-, blackhole- or perhaps even lesser explosions. The "QM" invariants at the origin of the island universe or universe of island universes need not then be the same as those of another. *This ensures the relativity of absolute quantum jumps in measured values and the existence of continuity between levels of energy or matter in the universe.* We are now enabled to presuppose a continuity of *four-dimensional pulsation in the size of the particle* within any unit length of wave and beyond. Interactions with other known subatomic particles – at the micro-, sub-micro, or even minuter regions of causal action – occur more frequently at crests and troughs, where the particle is in due stability to interact with, than at the more wave-like nodes or at points near nodes. Effective observation of interactions at nodes should involve ultra-quantal wavicles, which fact presupposes layers of ultra-QM,[125] trans-ultra-QM etc. in which perhaps superluminal waves could

---

125 Ultra-Quantum Mechanics needs a relatively more synthetic view of the four interactions of nature: electromagnetic, strong (primarily between quarks, and what is left over, between hadrons), weak (between leptons) and gravitational (between all sorts of bodies, including subatomic particles). A possibility that cannot be shoved aside is that, if the question of discrete quanta within the question of the wave nature of the path is settled in favour of continuity of propagation in the wave motion of one and the same particle, the explanation for the discreteness of quanta at their ejection from respective sources, may also be settled by taking into consideration the possible fact that in other universes (produced by other big bangs of massive black holes) the value of the constant of quantum jumps will differ, thus rendering a universal value of *h* impossible. Still, the various levels of ultra-quantal QMs require levels of values of the quantum constant with respect to different universes, meaning that the possible values of it in the different universes will make a continuous spectrum of values.

also incur three-dimensional bottlenecking of particles at nodes and bulging at crests and troughs.[126]

The continuity we now speak of assures that the motion of a spatially extended crest-positioned or trough-positioned particle (photon, electron, neutrino, or any quanton whatever) is not a vacuous presence elsewhere (including the node) as and when it is at a specifically measurable point in the wave path. *This statistically reduces the wave into a path, wherein the geometrical concept of wave has physical significance only because it is partially permeated by energy particles in non-infinite velocity, which non-infinity is why the motion is in wavicles.* The phenomena of wavicles at different layers of mechanics allow continuity of state values of matter / energy in the universe. *Absolute extension-level or motion-level or extension-motion continuity in one and the same wavicle is a myth, for there can be partial discreteness in wavicle motion and inner constitution of one and the same particle as represented by bottlenecking and bulging.* This indicates the possibility of continuous filling of different possible values of constants of quantum measurement in Reality if an infinite multiverse is the case.

To interpret the "entry" of the same wave in both slits in the double-slit experiment: What propagates is not the crest-trough structure. It is the energy packet in the vector direction, with sinusoidal motion at right angles to the vector direction to the extent of the crest and trough.[127] *Nodes are the points of motion-measuremental (temporal) levelling out of the three-dimensinal extensional (spatial) aspect of the speeding propagation*, even in the case of standing waves that do not seem to move forward at the nodes. The motion of the unit wave is sinusoidal, and there is a definite (but not fully measurable or fully predictable) relationship between the wavelength and the wave period. This relationship controls the speed propagation. If devoid of gravitational and other interferences, a wave propagates at sufficiently uniform motion with respect to the straight-line ideal-average of the waves at the nodes. This too does not guarantee absolute uniformity, since the inner-quantal causalities bring up perturbations that unsettle any absolute uniformity in motion.

The fact that the shorter the wavelength the higher is the energy of the transmitted packet shows that the wave oscillations tend to increase by increase of

---

126 For a detailed study of the possibility of real-valued superluminal and local-causal propagation, see my *Causal Ubiquity in Quantum Physics: A Superluminal and Local-Causal Physical Ontology* (Frankfurt: Peter Lang, 2014).

127 The speed of crest and trough is called phase speed. The speed and direction of transport of energy by waves is called group velocity. In capillary waves, group velocity is one and a half times the phase speed.

energy, and decrease and level out closer to straight line motion by decrease of energy. By ordinary algebra, approach to infinitesimally negligible energy should then result in inverse approach to absolutely straight-line motion, i.e. to infinite wavelength and velocity. *This demonstrates that only zero energy / mass (pure vacuum) can travel in straight line.* Let me put it differently. Amplitude is the perpendicular distance achieved by crest and trough from the equilibrium position of the nodes. Decrease in it implies asymptotic approach to zero energy, i.e. to absence of matter or energy in the respective extension-motion field. Only absence of energy / mass in an extension-motion field qualifies that field to follow Euclidean geometry. In short, the concept of zero rest mass is a problem to be reinterpreted as near-zero mass,[128] where alone the approach is to plane geometry Due to presence of matter / energy, wavelength is a distortion of Euclidean straight-line motion by four-dimensional sinusoidal motion.

Johansson attempts to solve the riddle of wave-particle duality in the state-of-the-art manner, by stipulating that every object is a wave phenomenon during motion and a particle when taking part in irreversible interaction with other objects[129] – a problem that we have been grappling with by differentiating the geometrical notions of particles and waves from the physical notions of the same. According to him the riddle of wave-particle duality "[…] should be understood so as to imply that if no irreversible interaction occurs, the object is moving and thus a wave."[130] As far as we are concerned, this is a clear indication that interaction takes place at crests and troughs which are more particle-like; not at nodes which are less particle-like, and so are interpreted as wave-like. The above criterion of Johansson uses three muddled concepts, which we understand differently from him:

(1) *Waves* obey the superposition principle and have a non-negligible extension in space (extension). No actual wave can be in one point: a facility that forestalls spatio-temporal (extension-motion) absolutisation. For us this means only that *wave is a path* defined not only by the movement of the particle but also *by the very forward elongation and sideward shortening of wave, more truly so at nodes.* That is, superposition is of physical energy, not of mathematical waves.

(2) *Particles* are, mathematically, point instants and do not follow the superposition principle. For us this requires considering their inner constitution and

---

128 As we discuss the criterial nature of the velocity of light and postulate real-valued superluminal velocities, it will be clear that zero rest mass is a postulate of the criterial nature of the velocity of light.
129 For detailed discussion: Johansson, "Realism and Wave-Particle Duality", 330–38.
130 Johansson, "Realism and Wave-Particle Duality", 330.

sideward extension. Mathematically, no object is a physical particle and wave simultaneously: a predicament we are always given up to. Physically a wave is simultaneously an elongated particle too: *physically a wave is a particle in tendency to be a wave, and a particle is a wave in tendency to be a particle.* Hence, in physical reality the wave-particle dichotomy does not exist. But mathematically (geometrically), the idealizations that waves and particles are, do present a dichotomy.

(3) *Irreversibility* (thermodynamic) enters physics in systems of a great number of partly independent objects. A higher than the air-pressure mono-atomic gas in a container is an example. If the container is opened most of the gas leaves it. The absolutely reverse process, though not in logical conflict with the laws of motion, never occurs since time reversal is impossible. This is a statistical effect in molecular kinetics. Irreversibility is thus a macroscopic phenomenon.[131] *The irreversibility that does not occur is that of time.* Physical proesses may be reversed; but only if the time proper is absolutely symmetrically reversed can a physical process be reversed absolutely. In normal physical understanding, irreversible interaction with other objects means interaction with objects big enough to have the irreversibility property, but without time-symmetry, i.e., an interaction in which the probability of a reversal of state change is negligible. This, it is said, is true even of QM.[132] I explain it further.

Measurement on a quantum object as it is possible today is an irreversible interaction. It changes the state of the object. Measurement interactions are not the only irreversible processes. In many experiments only a fraction of the prepared objects is measured. The rest is lost outside experimental control. All these objects soon hit some macroscopic objects, i.e. irrelevant parts of the equipment. These too are irreversible interactions, since the hit part of laboratory is in irreversible state change.[133] 'Measurement' refers to measurement where the wave function collapses, but those cases are the problematic ones for the realist.[134] "This analysis describes the wave-particle duality as a mind-independent property of quantum objects, and so far the analysis does not conflict with any empirical evidence."[135] The duality here is for us no physical paradox: the particle is in wave shape and the wave is in particle tendency. Both are the same, a wavicle, physically. One

---

131 Johansson, "Realism and Wave-Particle Duality", 331.
132 Johansson, "Realism and Wave-Particle Duality", 331.
133 Johansson, "Realism and Wave-Particle Duality", 331–32.
134 Johansson, "Realism and Wave-Particle Duality", 332.
135 Johansson, "Realism and Wave-Particle Duality", 332.

often forgets that the so-called dichotomy is only mathematical, where one considers the purely mathematical concepts of wave and particle.

It is straightforward now to ask the question of the double-slit experiment, keeping in mind that (1) collision takes place at crest-state and trough-state and (2) nodal states interact least with any object of quantal or infra-quantal levels. But a solution of the following traditional kind, as is customary, mixes purely mathematical shapes with actual physical shapes and thus confuses the issue:

> How then can this criterion be used as an explanation of the two-slit experiment? The answer is: every object is a wave from the emission from its source to its collision with some big object. When passing the screen the object passes both slits and does not interact irreversibly with the screen: the object neither exchanges energy nor momentum with the double-slit screen. Behind the screen the wave is separated into two parts whose total momentum is the same as it was before the screen. These two parts interfere with each other and the intensity distribution of the total wave shows an interference pattern.[136]

The questions are not answered why the wave splits. Johansson clarifies:

> After a sufficiently large number of registrations the distribution of these waves among the sensors maps the intensity distribution of the waves approaching the sensor row. But it seems as if the sensor row has been hit by a number of particles. There is nothing in the records telling us that waves were propagated on the water surface.[137]

I explain it differently. Even if collision takes place at crest and trough, nodes are not free of its effects. They too experience slight perturbations. The perturbations deflect the wave at its passage through the slit and the measurement is thus mismatched with the projected mathematical wave-shape. This accounts for the measurement. A complaint could be that the explanation above does not show how the wave collapses. Johansson replies: This complaint "[…] confuses philosophical and physical explanation; asking for a detailed account of the collapse of the wave during interaction with certain objects is to ask for more physics, it is not the philosopher's task. Moreover, the indivisibility of the energy exchange implies that it is impossible to describe further details."[138]

The wave collapse is in fact merely the theoretical effect, not physical effect of the shift of path. The only possible realist conclusion of the duality-problem for our world is therefore that the "[…] two important problems for a realistic interpretation of QM can be solved by assuming that quantum objects are waves which

---

136 Johansson, "Realism and Wave-Particle Duality", 332.
137 Johansson, "Realism and Wave-Particle Duality", 334.
138 Johansson, "Realism and Wave-Particle Duality", 334.

exchange conserved quantities in discrete steps."[139] Roughly, this is the causal-realist physical-ontological explanation I have proposed in the previous pages. The only additional qualification is that the discreteness of steps and the differences in allowed values in QM in general are to be referred to the determinations of the respective world's Big Bang or comparable-intensity states of sub-quantal constitution.

The explanation as to how exactly the so-called wave collapse is empirically recorded in any one experiment – a matter of precision – is not for the physical-ontologist to attempt. Just as the exact value of the quantum constant will vary from universe to universe, so also will an exact resolution of the collapse value/s. To go by the slit-jump explained a while ago, the ontology of collapse is a matter of the mathematical wave and particle being perceived. Even the exact value determined in any world of a multiverse of worlds with different extension-motion conglomerations and constitutions is exact only with respect to the given conditions of observation. This point of view enables us to hold that values absent in this world will be present in others, which can be measured only statistically in all these worlds, and never exactly, at any given instance. *This completes the continuity criterion of values of constants and non-constants in a multiverse. In fact there is no exact value in itself for all eternity, since each physical process to be measured is a nexus of infinite number of infinitesimal causal perturbations, all of which cannot be circumscribed by measurement.*

I join the point of view of Johansson to the extent that the conclusion of the problem of wave-particle duality has its additional explanation in the following two human predicaments. I explain it without taking his solution too much:

(1) Mathematics quantises for possible measurement. What is near-infinitesimal is non-representable except by assigning unit- or fractional values that quantise arguments in functions.[140] One then forgets that quantum statistical values

---

139   Johansson, "Realism and Wave-Particle Duality", 338.
140   Though in a different context, a mathematical realization by an applied mathematician-logician-scientist-metaphysician is relevant: "Years ago, in a communication to the Royal Society in 1906, I pointed out that the simplicity of points was inconsistent with the relational theory of space. At that time, so far as I am aware, the two inconsistent ideas were contentedly adopted by the whole of the scientific and philosophic worlds. To say that the event-particle (p_1, p_2, p_3, p_4) occupies, or happens at, the point (p_1, p_2, p_3) merely means that the event-particle is one of the set of event-particles which is the point. The second consequence of the definition is that if the p-system and the q-system are spatio-temporal systems which are not consentient, the p-points and the q-points are radically distinct entities, so that no

that tend to average could in fact be taken as more realistic measurements under the point of view of measuring rods more accurate than light quanta, say, in an ultra-QM. Under such ultra-quantal measurements of quantal discrete values, what is statistical for quantum measurements will turn out to be discrete quanta, circumscribable by extension-motion measurements proper to ultra-quantum measurements. That is, our meso-world's predicament of quantum statistics based on classical statistics need not be the ideal mathematics that the physical processes of ultra-quantum statistics follows. This probability is inherent in the very nature and predicament of mathematical procedures – not because epistemological probabilities are the same as the ontological probabilities of the wavicles, but because there is no particularist way of circumventing both: neither in isolation, nor together.

Scientific realism should go on attempting to overcome the unfortunate effects of this predicament by relativising physical constants by use of an array of mathematical perspectives.[141] The epistemological probabilism advocated here concering exactmess of anything measured or devised should be ontologically realistic in that it accepts the temporally continuous causal actuality of processes as such outside, with the own exact values of any causal process at any measurable extension-motion, but with inexactness concerning the values we propose. This view favours existence of entities as such, but only as epistemologically somewhat circumscribable for theoretical formulations. Some physicists call this circumscription simultaneously *as physical-ontologically probabilistic and as the only possible description.* Our epistemological truth-probabilism shuns this, because the knowledge is here probabilistic and the processes themselves flow of necessity.

---

    p-point is the same as any q-point. A complete explanation is thus achieved of the paradoxes in spatial measurement involved in the comparison of measurements of spatial distances between event-particles as effected in a p-space and a q-space. The ordinary formulae which we find in the early chapters of textbooks on dynamics only look so obvious because this radical distinction between the different spaces has been ignored." Whitehead, *The Interpretation of Science*, 130–31. The context is general physical, but it is applicable to the mathematical presuppositions behind QM measurement too.

141 De Broglie, Bohm and Bell were mathematical realists in QM, influenced by Einstein. Bell's alternative formulation predicts almost all the effects of the traditional instrumentalist quantum theory. Hence, we are justified in holding the realist view in QM. Moreover, paradoxes like Schrödinger's cat do not arise in the acute sense. D'Espagnat, *Reality and the Physicist*, 166–67. Hidden variables as suggested by Bohm are a way of overcoming phenomenalism, as reality cannot at any time be exhausted by theory.

(2) The epistemology of our physical knowledge is phenomenally (perspectivally) determined to be much distant from what is closer to the near-infinite and the near-infinitesimal. These two quantitative dimensions together naturally apportion to QM determinations in the meso-world less truth-probabilities than that available about the meso-world. Truth-probability is inherent in the very nature of the epistemology of our meso-world pragmatics, not merely of mathematics and statistics. In QM and relativistic field theories most physicists depend on mathematical instrumentalism, a corollary of phenomenalism, that adheres to mere mathematical description without having to objectify, because the epistemology presupposed about the QM is of the meso-world and so does not objectify as in the meso-world. The Einstein-Podolsky-Rosen (EPR) paradox (1935) mistakenly assumed that every property of a physical object has a sharp value at a given moment, *assumed like Bohr and others that the appropriated sharp value is what it can possess, and applied this meso-world approximation uncritically to quantum phenomena to derive the non-realism of physical-ontological probabilism of Bohr.* Bohr formulated his objection in a scientific-instrumentalistic fashion, thus falsely circumventing also the non-classical physical-ontological realism of causation essential to any solution. He is then be considered as holding that to be is to be measured.[142]

This epistemology may be clarified in the following manner:

> [...] [G]iven that quantum mechanics employs the mathematics of the real number continuum and the notions of exact position and exact momentum, then there is no good reason to think that extremely small bodies such as electrons do not have these properties simultaneously; certainly our inability to observe that they do is no compelling reason to think that they lack such a conjunctive property. Moreover, the fact (if it is a fact) that we cannot put the notion of exact simultaneous position and momentum to any practical explanatory or predictive use is not a compelling reason to discard it. Besides, the notion has *some* explanatory content: it explains the otherwise puzzling fact that we can measure either the exact position of an electron or its exact momentum (but not both) whichever and whenever we choose. What better explanation of this could there be than the fact that an electron has both of these properties at all times? Furthermore, in the EPR argument, Einstein brilliantly showed that quantum mechanics itself strongly suggests that an electron has both of these properties at any moment; that at any rate is a very plausible interpretation of the results of the EPR experiment.[143]

---

142 Bunge, *Treatise on Basic Philosophy, Volume. 7, Part I: Formal and Physical Sciences*, 175.
143 Murdoch, *Niels Bohr's Philosophy of Physics*, 233–34.

In short, QM is guilty of applying the meso-world concept of arithmetic, geometry and mathematics to the micro-world. That is, not only Einstein but also QM scientists are guilty of carrying forward Newtonian concepts the geometry of matter. This is what compelled Bohr to espouse a pragmatic epistemology of mathematical instrumentalism, which carried a semblance of solution. Mathematical instrumentalism yields descriptive, constructive and other empiricisms. Unfortunately, this is no realism at all, and the causal reality of the wavicle is at stake if we said that the wavicle is simultaneously both a mathematical wave and a mathematical particle. This phenomenalism that shuns the physical is to be overcome by the physical reality of the wavicle, as a particle in wave tendency.

As we have overcome phenomenalism by making the Kantian phenomena and noumena mutually continuous and augmented Quine's ontological commitment by the final Transcendent *a priori* of realism (Reality), we are in a position to assume realism also about the QM world. Such causal QM objects add up in other realstically and causally possible values of quanta and of Planck's and other "universal" constants – in different causally realized universes – plus other possible types of entities, to Reality.

The two afore-mentioned (numbered) points of support for and difference from the position of Johansson, by reason of the gravity of our differences from him, enable the following conclusive position on invariants. There always are seemingly absolutely pre-established invariant quantities of physics, like the Planck's Constant. That is, any impingement of the electron on a metal does not free photoelectric energy from the electron system of the metallic atoms. Only a light beam of energy equivalent to frequency v multiplied by Planck's Constant h – an energy determined to be more than hv_0, where v_0 is a certain low frequency limit – does it.[144] But this is so only with respect to this world, and the equivalent of Planck's constant for other worlds will differ in quantity. What is pre-established forever is an infinitesimally causally determined nature of infinite number (if there are so many) of causal worlds, not the constants determined by specific finite amounts of worlds. *This makes sufficiently large chunks of nature itself capable of yielding only probabilistic measurements, even without any human interference. This does not mean that nature as such is physical-ontologically probabilistic,* in the sense that, as it were, a sort of epistemological truth-probabilism is built into the physical ontological stuff of the multiverse.

---

144 The Kinetic Energy KE of the electron = $\frac{1}{2} mv^2 = h(v - v_0)$, where m is the mass of the electron and v is its velocity.

Human epistemological truth-probabilities are a class apart. The cumulative effect of epistemologically feasible levels of achievement of truth-probability in measurements and constants – say, as in the case of Planck's constant – may be best understood and ontologically surmounted only by a scientific realism that theoretically assumes into the near-infinitesimal physical stuff all possible objectual-causal effects from every near-infinitesmial and infinitely broader realms of entities. The ultimate entifying category of such thought is Reality.

In addition to Planck's constant there are also other seemingly invariant quantities in mesoscopic and macroscopic physics: cosmological and meso-world constants of (1) gravitation, (2) velocity of light, (3) linear momentum, (4) angular momentum, (5) electric charge etc.; in the QM world, constants of (6) lepton number, (7) baryon number etc.; in some particle interactions, constants of (8) isotopic spin (during strong interactions, but not during interactions involving electroweak force); and (9) $E = mc^2$ (universally if the velocity is assumed to be invariant and fundamental, but there could be superluminal propagations too, where this could hold by replacing c by any superluminal C) etc. are invariants experimentally fixed in this universe.

Empirical situations that work as invariants of this sort are, (1) irreversibility of thermodynamic action, and (2) the assumption of invariance like (a) quantum jumps and (b) the 'parity' invariance law of action-at-a-distance between complementary EPR particles when one of them is causally affected. Against Johansson and other physicists I argue that these are such only with respect to our (island) universe. If these may be traced back to the cumulative effects of all finite (or, doubtfully, infinite) *past* causal perturbations at the Big Bang of any one universe or explosions of blackholes, then it is not only that causality from the past is preserved, but also invariants of all naturally feasible kinds are relativized with respect to the worlds.

This opens up for discourse the realistic possibility of the near-infinitesimality of levels of interaction and cumulative causal determination in matter-energy, dark matter (undiscovered "undiscoverable" matter-energy) and whatever other states of matter there are. (Dark energy being anti-gravitation.) The totality of matter that there is in the multiverse of all island universes is in continuity of values of all domains of macro-world, meso-world, QM-, sub-QM- and other near-infinitesimal levels of measurement. *This fact facilitates the reasonableness of the fact that the causal origin of every value is in some way connected to the causal origin of some other values.* Thus, the immediate causal effects responsible for one phenomenon in one observation-layer may be found in other cognate universes that do not interact with the present universe of that observation-layer.

We just see some effects here, which are not causally accounted for here but elsewhere, perhaps at least accessible extension-motion recesses of a measure-mentally spatio-temporally – and also gravitationally – more circumspect but finite-volume universe of universes. This yields us space to think of other such universes of universes with the same states of affairs regarding physical constants and values.

One such test case crucial for the current work is the concept of action-at-a-distance, which may be considered to be that of non-local action or nonlocality. If there were nonlocal actions, there would have been all sorts of cosmologically unjustifiable miracles taking place for no reason at all, and the principle of causality would have to be questioned even where it is accepted to be the case, since everything physical that exists consists of wavicles through and through. Either causality is thoroughgoing, or else it has no sway at all. Discussion of this problem will clear our understanding of objectual causality in all possible realized and realizable worlds, and thus the category of Reality-in-total can be made theoretically more well-founded.

1.2.3 Nonlocality and the Concept of Reality

Under the previous section I made frequent allusions to the actual but less and less experimentally observable (near-)infinitesimality of internal and external causal influences in the often so-called fundamental wavicles. The actually sub-microscopic have no limits in infinitesimality, but they are less and less experimentally observable. The varying measures of infinitesimality of internal and external causal influences in the so-called fundamental wavicles of any layer of observation must be true, since anything is constituted. So there is reason for there being actual but connected layers of phenomena and observation like those of the macro-, meso-, micro-, nano-, ultra-quantal and other levels for constructing causal QM. If infinitesimality is acceptable at the fundamental level, no wave front is identical in measurement with any other of its kind on any absolute (infinite) scale, because each in all its parts is unique by reason of its unique finite extension-motion state measured in spacetime.

That is, all token processes and members of each type of process are different by specific identity. There are very close measurement affinities between mutually approximating objects of one and the same layer of the quantal, which we tend to measure off by common finite standards of reckoning infinitesimal causal effects behind motion. This points to the necessity of there being causal influence on anti-particles even in experimentally controlled causal action of a given particle. Experiments need not directly involve anti-particles in the present physics' ordinary causal manner that transmits causal influences at the speed of light. Yet there must be particle-influences on anti-particles as in the EPR experiments. The alleged fantastic "action-at-a-vacuous-distance" is realistically possible only if there exist causal influences between the particle and the anti-particle, propagating at superluminal velocities – a result we owe to the history of experiments and counter-experiments of the decades-long history of solutions of the EPR Paradox.

If such an experimental-causal alternation of state is possible between a particle and its experimentally immediately and apparatus-wise related anti-particle, it shows the existence of a wave-like, non-vacuous and causal influence between the two, which physicists committed to the ultimacy of luminal velocity term unscientifically as "instantaneous" without evidence for it. There cannot be energy or matter that propagates in the absolutely straight line posited by Euclid in ideal geometry. Infinitesimality of the train of particles is not observable by arbitrarily setting up a final limit to velocity from within this island universe, too. Hence, we must favour finite superluminal velocities to make the near-infinitesimally possible values of causal effects within wavicles of all layers possible. Such velocities will allow interpretation of the concept of wavicles beyond the sub-nuclear and the quantal. To bring this about, let us attempt an objectual and ontological mode of understanding the so-called nonlocal causality – action-at-a-vacuous-distance – in QM.

Gell-Mann gives a simple explanation of the EPR experiment, as modified by Bohm (hence, called the EPRB experiment). It deals with the decay of a particle into two anti-particles, here two anti-photons:

> If the particle is at rest and has no internal "spin," then the photons travel in opposite directions, have equal energy, and have identical circular polarizations. If one of the photons is left-circularly-polarized (spinning to the left), so is the other; likewise if one is right-circularly-polarized (spinning to the right), so is the other. Furthermore, if one is plane-polarized along a particular axis (that is, has its electric field vibrating along that axis), then the other one is plane-polarized along a definite axis. There are two cases, depending on the character of the spinless particle. In one case the plane polarization axes of the two photons are the same. In the other they are perpendicular. For simplicity let us take the former case, even though in the practical situation (where the decaying particle is a neutral pi meson) the latter case applies.
> […] The setup is assumed to be such that nothing disturbs either photon until it enters a detector. If the circular polarization of one of the photons is measured by the detector, the circular polarization of the other is certain – it is the same. Similarly, if the plane polarization of one of the photons is measured, that of the other photon is certain – again, it is the same as that of the first photon. Einstein's completeness would imply that both

the circular and plane polarization of the second photon could then be assigned definite values.[145]

The measurement problem as implying the completeness axiom for physical theory is expressed with great clarity in the words of Gell-Mann:

> If, by means of a certain measurement, the value of a particular quantity Q could be predicted with certainty, and if, by an alternative, quite different measurement, the value of another quantity R could be predicted with certainty, then, according to the notion of completeness, one should be able to assign exact values simultaneously to both of the quantities Q and R. Einstein and his colleagues succeeded in choosing the quantities to be ones that cannot simultaneously be assigned exact values in quantum mechanics, namely the position and momentum of the same object. Thus a direct contradiction was set up between quantum mechanics and completeness.[146]

Recalling that scientific determinism is perspectival absolutism – a type of absolutism of the current and immediately possible scientific perspective and its measurements of physical quantities – I argue that Einstein stood for both realism and scientific determinism of the concretist variety in the expected result of the EPR thought experiment. He tried showing that "[…] if one believes the wavefunction exhausts all the statements that can be meaningfully asserted about a physical system, then one must also accept that the real physical state of the system depends on what befalls another system with which it has previously interacted, no matter how far apart the two systems may become."[147]

By our understanding of the infinitesimality and infinity of causal influences within and from without the wavicle, the wavefunction does not yield an exhaustive explanation. Peter Holland says that Einstein argues: "[…] [A]dherence to the completeness assumption compels one to adopt 'unnatural theoretical interpretations'."[148] Hence, one must relinquish one of the following assumptions: "(*a*) the description by means of the ψ-function is *complete*" (the 'completeness' assumption) and "(*b*) the real states of spatially separated objects are independent of each other" (the locality / separability criterion), under the concept of locality, i.e., "[t]he real, physical state of one system is not immediately influenced by the kinds of measurements directly made on a second system, which is sufficiently spatially separated from the first."[149] It must be noted here that **the locality**

---

145 Gell-Mann, *The Quark and the Jaguar*, 171.
146 Gell-Mann, *The Quark and the Jaguar*, 168–69.
147 Holland, *The Quantum Theory of Motion*, 458.
148 Holland, *The Quantum Theory of Motion*, 458.
149 Holland, *The Quantum Theory of Motion*, 460.

***condition* means** that, from within the criterion of luminal limit-velocity, each of the anti-particles experiences the action as local and separable from the other, and for the combined system of the two it is experienced as non-local.

If the wavefunction is incomplete, it is possible to hold that the real states of spatially separated objects are independent of each other (under the assumption that the highest possible velocity in the universe is that of light). That is,

> [...] for a ψ-function [...] a measurement on 1 [a first atom or other particle] represents a physical operation which only affects the region of space where f_1 is finite and can have no direct influence on the physical reality in the remote region of space inhabited by atom 2. Thus, the real state of affairs pertaining to atom 2 must be the same whatever action we carry out on 1 (including no measurement at all). Hence, the functions v-, v'- [wavefunction in z-direction and eigenfunction in the z'-direction of atom 1] must be simultaneously attributable to atom 2. But this is impossible, for these states differ by more than a trivial phase factor and represent different *real* states of affairs for 2. Einstein concludes that the coordination of several ψ-functions with what should be a unique physical condition of 2 shows that ψ cannot be interpreted as a complete description of the physical condition of a system.[150]

Einstein believed that it is possible to isolate 1 from 2: physics itself would become an impossible enterprise if such a distant interconnectedness were admitted as a general property of nature, for it would deny the possibility of studying segments of matter in isolation, and physics would lose its empirical basis.[151] If he had attempted to provide functional space at least in the concept for all possible causal effects on wavicles, and conceived these effects as epistemologically penetrable in part, he could have come up with an ontologically committed (as he required realism out of QM) interpretation of the concept of the micro-worlds' localized wavicles, which are non-circumscribable by approximate meso-world appropriations and by the concept of localized sub-microworld wavicles that are non-circumscribable by micro-level approximations. This would have inspired him to see the possibility of solving the question of positively superluminal yet finite distances between the anti-particles of the EPR paradox in a "local" but in extension-motion not fully isolable manner, by postulating "locally" justified superluminal velocities by reason of the merely experimental status of the limit-velocity of light and the need to posit different past levels of finite amounts of near-infinitesimal causal influences within a given wavicle.

Murdoch clarifies the original intentions of the EPR argument and reformulates it into two parts. The first part explains the concept of completeness of

---

150   Holland, *The Quantum Theory of Motion*, 460.
151   Holland, *The Quantum Theory of Motion*, 460.

theory and gives the condition necessary for completeness. Murdoch refers to EPR in *Physical Review 47*: "[…] [E]*very element of the physical reality must have a counterpart in the physical theory*. What they [the authors: Einstein, Podolsky and Rosen] mean by 'counterpart' is that an element of physical reality should be represented in a state description within the theory."[152] This very condition tastes realistic classicism, and needs revision into ontological commitment to processes, instead of a vague counterpart in the physical theory – which musters some superluminal yet finite causal influence between a particle and its anti-particle, and *perhaps* even between particles themselves and anti-particles themselves.

Before expatiating on this requirement towards the end of this section, we study the EPR. According to Murdoch, the first part of the argument is this:

> (a) If a physical theory is complete, then, if x is an element of physical reality, there is a state description within the theory which includes x. (The completeness condition.) (b) There are elements x, y of physical reality that are not both included in any quantum-mechanical state description. (c) Therefore quantum mechanics is not a complete physical theory.[153]

By advising to substitute the concept of prediction with the supposedly ontologically less misleading concept of determination, EPR facilitates understanding of the second part and gives a sufficient condition for the concept of 'physical reality': "If, without in any way disturbing a system, we can predict with certainty (i.e., with probability equal to unity) the value of a physical quantity, then there exists an element of physical reality corresponding to this physical quantity."[154]

This being the case, it is my argument – in digression – that any determination of values (e.g. momentum, position etc.) is a truth-probabilistic determination, not only based on the probabilistic character of our determinations, but also because the very momentum and/or position of a wavicle lend themselves only to probabilistic determinations. This does not mean that nature is in itself probabilistically ontological. The "exact" determination of any one of these quantities *à propos* the theoretically and experimental givenness of particle S_1 of the pair of anti-particles is in fact a meso-world-, or even a micro-world-, sort of levelling out of the infinite number of infinitesimal causal influences within S_1. This does not mean that all causal influences are levelled out in their very givenness. There

---

152 Einstein, Podolsky and Rosen, "Can Quantum-Mechanical Description of Physical Reality Be Considered Complete?", *Physical Review 47*, cited in Murdoch, *Niels Bohr's Philosophy of Physics*, 165.
153 Murdoch, *Niels Bohr's Philosophy of Physics*, 165.
154 Murdoch, *Niels Bohr's Philosophy of Physics*, 166.

can be measurements of great certainty by which at least the fact of a certain level of influence is admitted. That is, eminently clear measurements of certain quantities are the touchstone of there being some causal influence (impingement by or transfer of physical elements) determinable in its ability to strike ontological commitment to certain real (physical) elements of that level of observation.

We go to the second part of EPR and understand it in Murdoch's words:

> (1) We can determine either the exact position or the exact momentum of S_2 at t, but not both. (2) The real physical state of S_2 is the same, whether we determine the exact position or the exact momentum of S_2. (3) Therefore there is at t a single real physical state of S_2 in which position and momentum both have exact values. (4) Operators representing position and momentum are non-commuting. (5) Therefore, there exists a single physical state in which two physical quantities represented by non-commuting operators have exact simultaneous values. (6) The physical state of an object at any time is completely described by a single state vector. (7) Different non-commuting operators have no state vectors in common. (8) Therefore a physical state in which physical quantities represented by non-commuting operators have exact simultaneous values is not describable in terms of a single state vector. (9) But such a physical state exists, viz. the one referred to in premiss (5). (10) Therefore there are elements of physical reality, x, y, which are not included in any quantum-mechanical state description. (Premiss (*b*) of the previous argument.)[155]

This summary of the second part of the argument is straightforward, so we do not discuss it directly. We take for granted the state-of-the-art explanation. Now we move into the EPR argument regarding measurement:

> Referring now to the EPR experiment, the authors argue that since we can determine with certainty either the position or the momentum of S_2 at time t, on the basis of a measurement on S_1, it follows via the criterion of physical reality that the position and momentum which can be determined with certainly for time t must be simultaneous elements of physical reality.[156]

Murdoch opines that this is fallacious. He does this by being forgetful of the fact that what is at issue here is the speed of light as the upper limit of speeds of communication between S_1 and S_2, and not the logical conjunctiveness of the negation of a disjunction, for no one measures with absolute exactitude any measurable quantity concerning a physical phenomenon. He shows the fallacy in EPR to be the following:

> The truth of a disjunction does not entail the truth of the corresponding conjunction. From the fact that we can determine with certainty either the exact position or the exact

---

155 Murdoch, *Niels Bohr's Philosophy of Physics*, 165–66.
156 Murdoch, *Niels Bohr's Philosophy of Physics*, 166.

momentum of S_2 at time t it does not follow by way of the reality criterion that S_2 has an exact position and an exact momentum at t. This argument, however, is not quite what Einstein had in mind. What he intended can be put as follows. Whether we determine at time t the position or the momentum of S_2, the physical state of S_2 at t remains the same, since neither a measurement on the distant S_1 nor the determination concerning S_2 can have any effect on the physical state of S_2. Hence, if we determine the position of S_2 at t, then S_2 must have at t whatever value of the momentum we would have determined had we so chosen; and conversely, if we determine the momentum of S_2 at t, then S_2 must have at t whatever value of the position we would have determined had we chosen to determine the position. From what he says elsewhere, it is clear that this is the argument that Einstein had in mind.[157]

How Gell-Mann counters Einstein's demand for completeness is important:

> But the value of the circular polarization and the plane polarization of a photon cannot be exactly specified at the same time (any more than the position and momentum of a particle can be so specified). Consequently, the requirement of completeness is just as unreasonable in this case, from the point of view of quantum mechanics, as in the case discussed by Einstein and his colleagues. The two measurements, one of circular and the other of plane polarization, are alternatives; they take place on different branches of history and there is no reason for the results of both to be considered together.[158]

This problem has to be reflected upon and conclusions should be reached. These statements are forgetful of the fact that what in fact is at issue in the locality-criterion in EPR is the speed of light as the upper limit of speeds of communication between S_1 and S_2. I argue as follows:

The exchange particles between nucleons are µ-mesons. These constitute the strong force. Beneath them are quarks, which interact via gluons. As of the present scientific knowledge, these take subluminal velocities. By reason of the indefiniteness (not exactly infinity) of the indefinite number of near-infinitesimal properly past causal influences (from the indefinite causal sub-sub- ... layers of the same particles and from their causal external vicinity) on the particles S_1 and S_2, we never have a measurement of absolute exactness. We can ascertain only the most probable dimensions and variances of probable shapes of wavicle motion of S_1 and S_2, which (the dimensions and variances) show up minutely causally at the microscopic or sub-microscopic or sub-sub-microscopic level associated to the wavicles. It is enough that we be able to assign at least the respective dimensions

---

157 Murdoch, *Niels Bohr's Philosophy of Physics*, 166. Towards the end of this quote, he makes reference to Einstein, "Quantenmechanik und Wirklichkeit", *Dialectica*, 2 (1948), 323.
158 Gell-Mann, *The Quark and the Jaguar*, 171.

and variances of motions (and probable measurements in these dimensions and variances) to the wavicles. The causal influences over the two wavicles are quite similar, some quantities of which are opposite in direction. It should also be admitted here that all agree that no physical change of dimension of motion happens without causal influences, since the very physical change is causation in extension-motion. These influences are proper to the immediate causality in question at the micro-level. The presupposed exactness of measurement is also culprit here. The crux of the measurement problem is this:

> What is the actual relationship in quantum mechanics between a measurement that permits the assignment of an exact value to a particle's position at a given time and another measurement that permits its momentum at the same time to be exactly specified? Those measurements take place on two different branches, decoherent with each other (like a branch of history in which one horse wins a given race and another branch in which a different horse wins). Einstein's requirement amounts to saying that the results from the two alternative branches must be accepted *together*. That clearly demands the abandonment of quantum mechanics.[159]

The issue of interpretation here revolves round the question of whether positive-valued propagations could travel from the one to the other particle and vice versa – not merely at the time of causal intervention on the one, but always. These may be part of the undiscovered causalities from within and without the particles. If the two branches measured did not belong to two totally unconnected branches of history, we can accept both together.

Einstein spoke of an isolable 'element of reality', thus giving rise to the possibility of Bohm's hidden variables theory, which attempts to treat undiscovered causalities active from within the inner processual recesses of the particle: "If, without in any way disturbing a system, we can predict with certainty (i.e., with probability equal to unity) the value of a physical quantity, then there exists an element of physical reality corresponding to this physical quantity."[160] This is the viewpoint from which he argued for impossibility of the so-called nonlocality – i.e., the so-called impossibility of local action of causal propagation under a positive superluminal velocity or of a lack of propagation that miraculously brings in or witnesses an action or change in the second particle.

Before we further discuss the issue of nonlocality and the contribution of John Clauser, Alain Aspect and others to it, we should know that the realism of

---

159   Gell-Mann, *The Quark and the Jaguar*, 169.
160   Bunge, *Treatise on Basic Philosophy, Volume 7, Part I: Formal and Physical Sciences*, 206.

locality for Einstein is equivalent to isolability of the concrete. This is *classicism that mixes admitting ontological (continuous near-infinitely and near-infinitesimally causal) determinationism or causalaity along with absolute epistemological determinism*. Holding on to this assumption, EPR propose (1) a necessary criterion of completeness: "Every element of the physical reality must have a counterpart in the physical theory" and (2) a sufficient criterion of reality: "If, without in any way disturbing a system, we can predict with certainty (i.e., with probability equal to unity) the value of a physical quantity, then there exists an element of physical reality corresponding to this physical quantity."[161]

This sufficiency condition is considered to be violated according to the results of later quantum experiments. That is, Einstein's realism in the EPR is an epistemologically absolute deterministic concretism (that we can every causal possible influences in a concrete case) based on (1) exact measurability of changes due to casual influences (which we objected to) and (2) impossibility of superluminal causal influences (which, as we soon see, must be considered to have been proved otherwise). Bohr and others held simultaneously that there is no interaction between the two anti-particles and that the systems are not separated – both on the basis of Einstein's own final limit to the speed of extension-motion propagation. But,

> [...] the quantum potential implies that a certain kind of 'signalling' does, in fact, take place between the sites of distantly separated spin ½ particles in an entangled state, if one of the particles undergoes a local interaction. This transfer of information cannot, however, be extracted by any experiment which obeys the laws of quantum mechanics. The causal interpretation thus provides an explanation of how the correlations come about in each individual process, in a way that is consistent with the statistical noncommunication of information.[162]

If two systems are isolable and the light we see has the highest permissible velocity, the superluminal exchange of causality between the two systems is problematic. The same situation arises when there is a total non-communication, too. Yet, if there is some effect that is beyond the horizon of luminal exchange – be it causally superluminal and local in communication, or non-causally "non-communicative" – it must be reasonable and acceptable. The possibility of nonlocal or non-communicative exchange or a miracle (impossibly superluminal, based on luminal communication) in the EPR experiment implies the need to re-interpret the very ontology of QM, because this alone can account for the realistic case

---

161   Holland, *The Quantum Theory of Motion*, 461.
162   Holland, *The Quantum Theory of Motion*, 476.

of continuous near-infinitesimal recesses of divisibility and the consequent ever more near-infinitesimal wavicle-existence of exchange particles.

Bunge suggests a defective solution: "The original system becomes dismantled only when at least one of its original components gets integrated into another system – e.g. when it is captured or absorbed by another atom."[163] This is in fact an admission of superluminal exchange between S_1 and S_2, because an action-at-a-vacuous-distance can be avoided only if there is some exchange-wavicle between them before one of them is captured or absorbed by another atom, in order for a real physical change to take place in one particle corresponding to the change in the other. This exchange-wavicle can cause an effect in the other particle only if it is positive-valued and superluminal in velocity. Non-recognition of this fact makes Bunge to make the following conclusion about the issue:

> In conclusion, (*a*) when two quantons interact, their state functions become entangled (not factorizable); (*b*) when the two quantons separate widely in space, they continue to form part of the original system although they do not act upon one another, much less at a distance and instantaneously [...] (*c*) spatial separation is no cause for divorce: there is divorce only if there is new marriage; (*d*) non-separability is a consequence of the superposition principle and the Schrödinger equation; (*e*) non-separability is possibly '*the* characteristic trait of quantum mechanics' [...] (*f*) the failure of classical separability or 'locality' (Einstein separability) confirms the systemic world view [...] not however the holistic one, because we do succeed in *conceptually* analyzing the composition and structure of systems; (*g*) in quantum theory there is EPR distant correlation (or EPR effect) but there is no paradox: the paradox arises only if quantum theory is combined wit the classical principle of separability or 'locality'.[164]

We do not admit a miracle, i.e. a spooky action-at-a-vacuous-distance without medium of communication of influence. No physics can accept such a miracle. If it is admitted that the exchange is positive and superluminal, it must also be taken as causal, just to keep it natural and physical – for until then luminal communications have been positive-valued and causal. The facts of continuous near-infinite and near-infinitesimal sub-, sub-sub-, ... -quantal causal influences within wavicles S_1 and S_2, if read together with the need to keep physical (extension-motion level) exchange between the two temporal light cones of the EPR experiment causal, Einstein should have been persuaded of the possibility

---

163 Bunge, *Treatise on Basic Philosophy, Volume 7, Part I: Formal and Physical Sciences*, 215.
164 Bunge, *Treatise on Basic Philosophy, Volume 7, Part I: Formal and Physical Sciences*, 215.

of superluminal velocities, continuous sub-, sub-sub-, ... -quantal causal influences and a multitude of values of the different universal constants in actually realizable possible worlds.

This "local" interpretation, happily, does not violate the discreteness- / discontinuity assumption in QM for this world, but violates the discontinuity assumption in QM for the infinite multiverse. It violates also the speed barrier in STR, which will duly be clarified at discussion of the question of superluminal velocities in the foundations of STR and GTR. *If there is no upper limit for superluminal velocities, there is absolute causal continuity of causal origin of all kinds of particle-values and values of constants in Reality*, though each world considered in isolation remains discrete in the totality of values of constants available therein. In this case it suffices to say that Bohr's statistical instrumentalist interpretation does not do justice to the inner causal processes of particles that may be pinpointed through connection of every entity *with all realized entities in the proper past* of the contemporary world of the entity. Here, discreteness of values in QM breaks down on the large scale extension-motion proper to an infinite multiverse.

In this context notice also that Einstein unconsciously oversteps his classicist concretist determinism and suggests a surprisingly unitary (or, physically somehow monistic) system of physical universe and its physics: "Nature as a whole can only be viewed as an individual system, existing only once, and not as collection of systems."[165] This shows that "[...] *the state of the whole is prior to that of the parts* ([...] the parts are not *physically* determined as aspects of the whole, as they would be in a unified field theory, for instance)."[166] This is monism, not holism. In the holism of Reality that I propose, universes or physical systems are never completely unified, because no communication can travel at infinite velocity. If there is a maximal velocity in a universe or group of universes, others will have other criteria. There can anyway be some causal connections between many neighbouring universes or groups of them. The holism pivots around the highly probable fact that there is continuity of universal constants in the existing and future universes together. As against this, Einstein's monism would have to admit infinite velocities and complete mutual identity of extension-motion regions.

---

165 Holland, *The Quantum Theory of Motion*, 570.
166 Holland, *The Quantum Theory of Motion*, 568. Bohm says: "The relationship between parts of a system described above implies a new quality of *wholeness* of the entire system going beyond anything that can be specified solely in terms of the actual spatial relationships of all the particles." Bohm and Hiley, *The Undivided Universe*, 58.

To concentrate more on the continuity principle, I leave out Bell's contributions to justification or non-justification of the "locality" standpoint.[167] I study the experimental demonstration by Aspect, Clauser etc., of what they call 'non-locality' in nature, in order for me to suggest a causal-continuous phenomenal-noumenal interpretation. Aspect and others[168] have experimentally tested that there are (causal or non-causal?) correlations between particles S_1 and S_2 even when the events of detection of the two photons are for him outside each other's light cones,[169] and that if there are hidden variables they are nonlocal under the assumption of impossibility of superluminal velocities. I argue that, if the light cone of S_1 is transgressed by the communication of the disturbance between S_1 and S_2, then the fixed velocity of light has been violated by a posiive-valued communication that has gone superluminal.

It should be remembered that the possibility of tachyons that E. C. G. Sudarshan and others proposed and the fixing of the velocity of light as a velocity barrier between two universes, are still based mathematically on the questionable way of measurement of motion by the criterion of the very velocity of photons that one aims at measuring. Hence, the mathematical physics behind the concepts of tachyons and photons is equally questionable. The question of superluminal velocities will be discussed for its own sake later in the main section on Relativity. *The present suggested interpretation of the EPR will be complete only after we study the possibility of superluminal velocities in that section on STR.* The continuity between subluminal and superluminal worlds will then follow. This

---

167 Here, instead of studying the involving descriptions of Bell's inequalities, I deem it sufficient to mention that Bell's understanding of realism is a determinist (and concretist) realism, and that this is a presupposition that d'Espagnat takes as a loophole to argue against him: "However, if we examine the proof of Bell's inequality more carefully, the assumption of realism really is one of the premises of a local realistic theory, but this premise is only a *special form* of realism, the deterministic realism, i.e., the existence of a hidden parameter. So that the violation of Bell's inequality can not be regarded as a violation of realism *in general*, e.g. a general statement, such as 'disagreeing with the doctrine that the world is independent of mind'!" Zuoxiu, "On the Einstein, Podolsky and Rosen Paradox and the Relevant Philosophical Problems", 301.
168 Bohm and Hiley, *The Undivided Universe*, 144–45.
169 This violates the Bell's inequality for locality (which shows that the disturbance from 1 is not communicated beyond the light cone of 1). Even the criticism by others of Aspect's experiments (saying, the photon detector's efficiency was not close to unity) may be found to be a contrivance to save the phenomena. Bohm and Hiley, *The Undivided Universe*, 144–45.

is sufficient support for the phenomenal-noumenal continuity via relativization of the macro-, meso-, micro-, sub-quantal and other perspectives based on the ontological tenability of there being Reality as extra-phenomenal (in the sense of a totalized existing thing-in-itself) that can show itself phenomenally.

1.2.4 Ontological Categorial Foundation of Quantum Mechanics

The detection of Ω^- particles supports the existence of K mesons. The mass and charge (i.e. the very existence) of Ω^- particles are determined in an eightfold way. This guarantees the empirical adequacy of the theory. The theory's content is the existence and properties of the particle: not merely the properties alone, but their causal existence in a universal-bounded manner, since the measurement and qualitative determinations depend on other processes within and without, which, together with the immediately available properties, involve ontological universals as active between them in common.

The mathematical-instrumentalist approach of QM to particles does not bother about existence. Similarly, a constructive empiricist too cannot avoid allowing quantities like pressure and volume as observables. *Otherwise theory would have virtually no actual, but only empirical, content.* Once pressure is taken as an observable, then even other quantities (mass, lifetime, charge etc.), detected merely with instruments, become observable, albeit partly, against the backdrop of the near-infinite and near-infinitesimal levels of causal history of anything. Thus, there are good empirical reasons for the existence of entities[170] as processes. By reason of our synthesis of phenomena and noumena, observation is not

---

170 Franklin, "There Are No Antirealists in the Laboratory", 141. Bunge says: "[…] some physicists – notably Heisenberg (1969) – have stated that symmetry precedes existence: that quantons are nothing but embodiments of symmetries. This Platonic delusion stems from the manner the theorist confronts the bewildering array of 'fundamental particles'. Instead of proceeding inductively, or else by trial and error, he imagines that there is a single basic quanton that can be in different mass, charge, spin, isospin, hypercharge, etc., states. He then forgets for a while that hypothetical entity and investigates the algebraic properties of its state space, guided only by extremely general physical principles. In particular, the theorist investigates the group-theoretic structure of that space: he conjectures, say, that the structure is an US(2), US(3), or some other symmetry group […]. But this is no evidence for the power of pure mathematics to mirror the world […]. In conclusion, quantum theory accounts for quantons as things quite different from classons, and therefore it offers a rather counterintuitive ("paradoxical") picture of reality." Bunge, *Treatise on Basic Philosophy, Volume 7, Part I: Formal and Physical Sciences*, 218–19.

merely phenomenal, but also pertaining to real things-in-themselves. Observability need not mean absolute determinability by observation. What remains to be done is to extend the concept of entity to that of Reality, against the background of the continuity of values of quantum and other kinds of measurement and constants. With this in view let us briefly discuss the ontological categorial features of QM.

By tradition, the three basic features of quantum concepts are:[171] (1) discontinuity of quantum processes (replacing continuous trajectory by indivisible units of spatial transition in the given universe); (2) probability or indeterminacy[172] of quantum laws (replacing complete determinism by the concept of causality as a merely statistical trend); (3) wave-particle duality of quantum objects in experimental conditions (replacing analysis of the world into distinct parts with a unique, fixed, 'intrinsic' nature like "wave / particle" by admission of the nature of quantum objects as dependent on external experimental conditions). "That means a quantum phenomenon, even the world, is an indivisible whole in which parts appear as abstractions or approximations, valid only in the classical limit."[173] If it is valid only in the classical view, the quantal particle has no reality at all. Hence, we need a synthesis of the classical and quantum views.

By revamping the classical and quantum views Jiachang and Xinhe give a typically probabilistic and mathematical-instrumentalistic synthesis of QM:

> The quantum feature, namely the indivisibility of the quantum, implies the indivisibility of both the quantum of action and the quantum process. From the former, we cannot

---

171 Jiachang and Xinhe, "Relational Realism or Reform of the View of Physical Reality and Its Logical Manifestation", 362.
172 The indeterminacy / uncertainty principle proposes: "It is impossible to design any apparatus whatsoever to determine through which hole the electron passes that will not at the same time disturb the electron enough to destroy the interference pattern." Feynman, *The Character of Physical Law*, 143. Feynman does also moot the possibility of observation without much disturbing the electron, provided there can be gentle rays of considerably low frequency: "If we want to disturb the electrons only slightly we should not have lowered the *intensity* of the light, we should have lowered its *frequency* (the same as increasing its wavelength). Let us use light of a redder color. We could even use infrared light, or radiowaves (like radar), and 'see' where the electron went with the help of some equipment that can 'see' light of these longer wavelengths. If we use 'gentler' light perhaps we can avoid disturbing the electrons so much." Feynman, Leighton and Sands, *The Feynman Lectures on Physics*, vol. 3, 1-8.
173 Jiachang and Xinhe, "Relational Realism or Reform of the View of Physical Reality and Its Logical Manifestation", 362.

> draw a clear line of demarcation between quantum object and measuring instrument; in principle, we cannot recognize the state of an object as thing in itself, but rather the integral phenomena as the results of interaction. From the latter, we can find a way to a probabilistic feature, i.e., deterministic laws in CP [Classical Physics] which were based on the principle of continuity are invalid in QM, and we can only give probabilistic predictions concerning changes of state.[174]

This deserves analysis from the point of view of integrating the classical understanding of alleged indivisibility of the ultimate particles and the quantum notion of ultimacy of Planck's constant of multiplication, from our new notion of near-infinitesimality of matter-energy at all sub-levels of actualities of Reality. Both ancient and contemporary mathematics make use of infinities, infinities of infinities etc. Even the shift from application of the Euclidean to non-Euclidean geometries in theoretical physics has not done away with infinities of infinities. Some of them are called asymptotic infinities in which two lines, if produced to infinity, would meet at infinity. In this case, at least one of them should curve. It is clear from the statistical and wave-particle QM qualities that every event and measurement, however big or small, depends on many others. Our study of the presuppositions and results of QM shows that, from the point of view of near-infinite and near-infinitesimal causality active in and on any process, all possible values of quanta and physical constants should be available in some or other universe in the multiverse.

The above fact ends in infinity of infinities of *qualia* or *abstracta* for theory to depend on. *Qualia* should be read together with the realist content of empirical and observational posits and the phenomenal-noumenal continuity beyond Kant. This will ontologically commit us to the highest (totalised) inner entifying thrust and Transcendent presupposition of all inquiry, i.e. Reality, as the actually Transcendent (the highest Entity transcendent to any specific actual entity, with capital 'T' to show its highest nature), which is active in thought and processes apriorily. It is given in consciousness apriorily, since it is active in a Transcendental mode of the intellect. It is a Transcendent actual infinity if the multiverse is infinite, it is the totality of totalities, and further on, it is active apriorily (i.e., ideally as condition for the possibility of something) in thinking. It is the final presupposed actual Transcendent category behind actual QM events and measurements. The universals, universals of universals etc., and the conceptual Transcendental Reality-in-general with its relevant connotative universals, all presupposed in measurement (as an expression of thought), are not actuals.

---

174 Jiachang and Xinhe, "Relational Realism or Reform of the View of Physical Reality and Its Logical Manifestation", 362–63. Square brackets mine.

Instead, they are what are to be instantiated in processes or thought about them respectively, and therefore are potentials. A Transcendent too can thus be apriorily (as the condition for the possibility of something) active.

The new concept of the *a priori* entailed here should be clarified. Whatever an *a priori* is must anyway be decided by actuality, since it is to be applied primarily in discourse of and about actuality. If every *a priori* should be founded in reality, it is not a pure vacuous idea. It is the counterpart of something pertaining to what is reality-in-particular or Reality. To that extent the concept of Reality can be treated as a concept with foundation in reality. It is not merely the concept of universals, but of the totality of actualities. As an *a priori* in consciousness, it is only a Transcendental. By reason of the ontological realism that pertains to ontological commitment at the level of Reality, it is also capable of being considered as reality independent of minds. To that extent there is something ontological corresponding to it outside of the phenomenal representations of the mind. It is a Transcendent. Hence, Reality as a category is a Transcendentally considered Transcendent at the foundation of all knowledge, especially of knowledge wrought by application of different scientific categories.

We do not know what possibly constitute Reality beyond the physical universe of actual island universes. At least the specific physical object or physical reality as the object of measurement "exists" in some fundamental sense, but it is not knowable apart from the conditions and theories of measurement and the specific categories and *qualia* / universals we involve for it. We need to inductively totalise the particular. This is given by *a priori* conditionality in thought as Reality. By 'physical reality' Bohr does not denote a pre-measurement, merely theoretical ontological object. The marks on strips of tape, the clicks of counters, the patterns on photographic plates etc. must perhaps be dealt with classically.[175] Or else it may be treated in terms of some concrete things pertaining to the objects at measurement, which need not necessarily be merely classical-bound. This is because we always conflate concepts of the different quantal layers, all of which have elements of the real suffused within their objects. In comparison to this it is interesting to see how Wartofsky misses in his phenomenalist synthesis the point that the QM description too is not the end of the story:

> If, on the other hand, Quantum-mechanics is a fundamental theory *and* complete, then the classical talk is relativised to these quantum-foundations, and the dynamic variables of the classical macroworld are simply the 'phenomena' yielded (as we approach the classical limit), by the essentially statistical or objectively probabilistic 'reality' at the

---

175 Wartofsky, "Three Stages of Constitution", 211.

level of the quantum microworld. Thus, the very notion that the quantum-description is incomplete depends on smuggling classical criteria of completeness into a world – or into an ontological level of the physical world – where they are anomalous. Quantum phenomena, in this sense, are completely described *the way they really are in themselves*, for their *being-in-themselves is the being-for-one-another* of their constituent subsystems, that is, in the interaction.[176]

Wartofsky's system with members that are beings for one another is not real, since the probabilism of QM is for this system really in this system. In the rush to dispense with commonsense categories and to save probabilistic phenomena he throws off reality as such. Not only classical descriptions but even QM descriptions carry commonsense categories – may be probabilistic or not at all. If carefully watched it is easy to recognise that the theory that relativising QM descriptions as incomplete needs purely classical notions is unfounded. The notions of actuality, reality, causality, description, probability etc. needed to render QM descriptions incomplete must somehow be made to differ from both the classical and the QM. We have already discussed how nonlocality with respect to the world of photons is clearly locality under quantons of superluminal velocities. The question of spacelike[177] separation should be encountered by physical-ontological extension-motion interpretation, keeping the superluminal separation between the particles still subject to causal interactions.

For this we need to fuse the particularised classificational categories of Extension, Motion, Causality and Mass into one category of a totalised substance, Reality, which allows all these to spontaneously arise within the respective realms of inquiry, and the causal continuity in Reality makes superluminal causal communication acceptable. If a measurement is taken exclusively for extension (space), the exercise will turn out to be capable of making measurements only as concretist, non-generalisable and non-causal. Then causality becomes merely a statement of the statistical trend in mathematical-instrumentalist explanations,

---

176 Wartofsky, "Three Stages of Constitution", 211.
177 "Two events in space-time are called spacelike separated if each lies outside the other's light cone… In this case, neither can causally influence the other, and measurements made at the two events must commute." Penrose, *Shadows of the Mind*, 295. It is strange that confoundment with the alleged impossibility of superluminal velocities has served for this conclusion. If such velocities are possible, there is in fact no spacelike (extension-like) separation without time (motion) because light-cones would have to be substituted by superluminal cones, which make the reactions transpire in extension-motion, i.e. causally.

physical properties become the probabilistic table of eigenvalues[178] or whatever else, and even the quantum objects described "completely" by wave functions become mere probabilistic descriptions – not of what is the case in Reality, but of the very description from any one elementary point of view in realism.

Just as revealed in our discussion of wave-particle dualism, quantum objects are unique in the sense that they are essentially real, but probabilistic in our present description based on meso-world observation of the micro-world, because *classically devised definitions* of basic scientific categories within limited contexts and without universal ontological (realistic) commitment govern the activity of QM measurements. Causality, like the other categories, is normally formulated within the purview of reality-in-particular. It never goes realistically probabilistic since its surety about causation is not always absolutely determinable as this or that, but such a concept of causation remains instrumentalistically probabilistic in quantum measurements without realism, if the mutual connexity of the proper past of the contemporary world of any particle is not taken for granted apriorily, i.e., when measurement is without the presupposed ontological commitment. This presupposition is what is inductively totalised in Reality.

Hence, in categorial discussions we need a fusion of scientific categories into one that summarizes also the causal past of objects / substances. Thus Reality is the Transcendent ontological categorial foundation of QM. This is no mere recurrence of classical notions, because even QM requires an ontological commitment that goes back to causal roots. There is nothing wrong in going back to the totality of Reality as the objectual background of ingression of causal influences on the object or wavicle in question. We are in fact summating over all the continuous wavicles of all levels of measurement, under the name Reality. This is a notion that surpasses isolated layers of measurement, like the macro-, meso-, micro- etc. The summation of all possible causal roots of any given real and near-infinitesimally and mutually continuous wavicles in QM yields the Transcendental Transcendent Reality that is categorially active behind all phenomena, by reason of its highest Transcendental and Transcendent natures. This very Transcendental Transcendent may be found to serve as the cosmological category behind the

---

178 Eigenvalue is "[t]he specific value of a quantum property associated with a particular *eigenstate*." Eigenstate is "[a] 'pure' quantum state, described by a unique *state vector* (or wave function). Most of the time, most quantum systems are in a *superposition of states*." Gribbin, Q Is for Quantum, 115. "A list of numbers which spells out the set of quantum properties of a quantum entity in a particular quantum state (for example, the *spin* orientation, *orbital* and other properties of an *electron* in an *atom*). It is also known as the state vector." Gribbin, Q Is for Quantum, 175.

relativistic theories too. I inquire into this claim in the following section and apply some clues about QM nonlocality in the foundations of relativistic theories, to convince ourselves of the relevance of the category of Reality.

1.3 Categorial Features of Relativistic Theories

1.3.1 Questions in the Origin of the Special Theory of Relativity

1.3.1.1 Inertial Referential Velocity of Light

STR was proposed by Einstein in order to treat all physical processes in the cosmos as spatially three-dimensional and temporally uni-dimensional events at sufficiently high speeds. The General Theory of Relativity (GTR) takes this concept into the realm of gravitational events and unifies all the four dimensions into the concept of a gravitational field. According to STR, an absolute velocity cannot be measured. It is measurable with respect to the observer. Velocities are ordinarily dependent on the source. But the speed of light, he claims to have showed, is independent of the source that emits it – unfortunately, with respect to the very speed of light that was used to calculate it. This gave him confidence to base measurements and the mathematics of STR upon invariance of the speed of light, which we depend on for sight with naked eyes, telescopes, microscope etc.

The equations of motion of this theory were couched in Lorenz transformations, which is an improvement upon the Galileian transformations. The Galileian Transformations presuppose sameness of the laws of physics in all frames of reference. Based on this Newton had obtained the dynamics of bodies relative to each other.[179] His presupposition of absoluteness pertained to the laws of motion, without a fixed velocity that facilitates a framework of observation of motion with respect to frames of reference. He used the velocity of light unconsciously as the criterion, not its speed, since he thought it to be instantaneous in propagation. Lorentz and Einstein had to follow it unconsciously too. But the detection of its velocity should have changed the scenario.

If we consider two trains in adjacent rails, A stationary and B accelerating, A is the frame of reference to calculate the motion of B. In the first type of frame of reference with stationary / inertial "mass", everything moves with respect to it, and Newton's first law (asserting the continuity of bodies in rest or uniform

---

[179] Books 1 and 2 of *The Principia* deal with the motion of bodies, and Book 3 with the system of the world, which is, again, of bodies at the macroscopic level. The three laws of motion are treated as "axioms or laws" of all possible motion. Newton, *The Principia*, 416–30.

motion in a straight line unless acted on otherwise) that concerns also uniform rest is obeyed. In the second type (about accelerating or decelerating objects), the first law is not obeyed – passengers experience a backward or frontward fling in acceleration or deceleration.[180] If A is itself moving, then B that moves with respect to A moves also with respect to the speed of light – a generalization about motion of bodies in the universe, expressed with respect to the speed of light.

Michelson-Morley- and Kennedy-Thorndike experiments[181] have empirically set the velocity of light, on the way showing also that an absolute space of ether was non-existent or at least unnecessary with respect to the motion of particles. Einstein claims he did not know of the Michelson-Morley finding before he used the velocity of light as the medium for measurement of motion. He received his impetus in part from the Maxwell equations for motion of electromagnetic waves, which had already included the speed of light as an absolute constant.[182]

While inquiring of the nature of light propagation – of measuring the velocity of light and determining it as fixed forever – it went unnoticed that the background 'type' accepted as measuring rod is the very speed of light. We grant that there is no ether as reference background for measuring. Yet it should be remembered that all measurements hereafter are made with respect to another absolute epistemic limit: the velocity of light that allows no measurement beyond itself with respect to the possible superluminal velocities.[183] Galileo was concerned about observational limits. Poincaré and Lorentz too had similar intentions.[184]

---

180 Whitaker, *Einstein, Bohr and the Quantum Dilemma*, 76.
181 Rothman, *Discovering the Natural Laws*, 223ff.
182 Gribbin, *Q Is for Quantum*, 236–37.
183 Relativity theory does not preclude tachyons (coined by the US physicist Gerald Feinberg in 1967 from Greek *tachys*, "fast") with superluminal velocities. It says that tardyons / ittyons (Greek *tardys* and Hebrew *ittys*, "slow") with subluminal velocities do not exceed the velocity of light in vacuum and tachyons would never reach or cross below the velocity of light. Pickover, *Time: A Traveller's Guide*, 150. Although these may be true, it must be admitted that our criterion for all these has always been taken to be the velocity of light.
184 In a paper on the dynamics of electron (1906) Poincaré developed a theory of relativity based on transformations of motion. "The principal difference was that Einstein developed the theory from elementary considerations concerning light signaling, whereas Poincaré's treatment was based on the full theory of electromagnetism and was restricted to phenomena associated with the concept of a universal ether that functioned as the means of transmitted light." *The New Encyclopaedia Britannica: Micropaedia*, s.v. "Poincaré, Henri". Lorentz, influenced by FitzGerald's proposal that bodies approaching the velocity of light experience foreshortening, formulated

Lorentz, Fitzgerald and Einstein finally pointed out the phenomenon of foreshortening and the impending idea of zero-rest mass of bodies at the speed of light, through equations that regard the velocity of light signals from and to the moving body as the fundamental velocity criterial to the calculation. This result is tenable only with respect to the speed of light. The same may be obtained also with respect to any subluminal or superluminal velocity, which too *may be* source-independent and invariant in all directions with respect to this world. The source-independence and invariance of light quanta with respect to the micro-world are results of the arithmetic of zero rest mass and foreshortening, which in turn are the results of measuring the speed of objects by light. When v approaches c it is replaceable by c, and the value in the denominator tends to zero.

The velocity of the wave moving along OM_1 relative to the apparatus is $c - v$. The velocity of the wave motion along $M_1'O'$ relative to the apparatus is $c + v$. $OM_1 = OM_2$. Therefore, the time t_1 required to travel the path $OM_1' + M_1'O'$ is: $t_2 = (L / [c - v]) + (L / [c + v]) = 2Lc / (c^2 - v^2)$. That is, $t_1 = (2L / c) \times (1 / [1 - v^2/c^2])$. The reflected wave from O travels along the hypotenuse OM_2' of the right angled triangle $OM_2 OM_2'$. Then $M_2 M_2' = vt'$, and $OM'_2 = ct'$. That is, $c^2 t'^2 = v^2 t'^2 + L^2$, i.e., $t' = L / \sqrt{(c^2 - v^2)}$. Hence, the time taken by the reflected wave to travel the distance $OM_2' + M_2'O'$ is yielded by $t_2 = 2t' = 2L / \sqrt{(c^2 - v^2)} = [2L (1 / \sqrt{(c^2 - v^2)})] / c$. The difference in time, which would be zero for a stationary observer, now becomes $\Delta t = t_1 = 2L [(1 / (1 - v^2/c^2)) - (1 / \sqrt{(1 - v^2/c^2)})]$.[185]

The denominator of one of the multiplicands of one of the elements, i.e., $1 - v^2/c^2$, enters equations of measurement of relative velocity, momentum, mass etc. and plays the mischief. There *is* mischief, because we measure all these with respect to the velocity of light, which is fixed only experimentally, deduce that **the rest mass of light quantons is zero with respect to c (themselves)**, and then claim that with respect all possible contexts this velocity must be the limit velocity. This is like quoting God's words to prove the existence of God. We measure (and calculate) these properties with respect to light, and say that, *since they are measured with respect to light and found to produce infinite values when v approaches c, light quanta are source-independent with respect to themselves*. One thus claims also that c is the limit and calls it inertial referential velocity. If c were

---

the Lorentz transformations (1904) relating the space and time coordinate of any two systems in motion at uniform and constant velocity relative to each other. These describe the increase of mass, shortening of length and dilating of time at approach of the body to light's velocity. *The New Encyclopaedia Britannica: Micropaedia*, s.v. "Lorentz, Hendrik Antoon".

185 Singh and Bagde, *Elements of Special Relativity*, 10–11.

greater than its actual present value, there would not have been zero rest mass or source-independence at the present actual value. If c were not the limit, the constancy *in vacuo* of the speed of light would have surely been taken to be a meso-world approximation. Then, quantal limits and values (say, Planck's constant) would also be recognized as approximations. In this case there is the more desirable possibility of all sorts of quantal values and limiting values of speeds, at least with respect to each one of the totality of all possible realized and realizable worlds. The base of this totality we call the cosmological-ontological category, Reality-in-total. Without it physical ontology is impossible.

Lorentz transformations presuppose measuring motion by light signals. They are a way of subtracting the effect of distance of the observer's frame from that of the observed:

> Whenever two observers are associated with two distinct inertial frames of reference in relative motion to each other, their determinations of time intervals and of distances between events will disagree systematically, without one being 'right" and the other 'wrong." Nor can it be established that one of them is at rest relative to the ether, the other in motion. In fact, if they compare their respective measuring rods (in the direction of mutual motion), each will find the other's rod foreshortened. The speed of light will be found to equal the same value, c = 186,000 miles per second, relative to every inertial frame of reference and in all directions. The status of Maxwell's ether is thereby cast in doubt, as its state of motion cannot be ascertained by any conceivable experiment. Consequently, the whole notion of an ether as the carrier of electromagnetic phenomena has been eliminated in contemporary physics.[186]

Since the existence of a reference frame called ether is abandoned, since the velocity of light is *sufficiently* constant in all directions *in vacuo* for our apparatuses, and since the speed of discharge of quanta is found to be independent of the source, Einstein and others approved as ultimate the measuring of all motions with respect to the constant speed of light in all inertial frames. This explains the prevalence of relativistic transformations involving the velocity of light, in all relativistic calculations.[187]

One conceptual consequence of these transformations, they being based on the speed of light, is the supposed limiting character of the speed of light.

---

186 *The New Encyclopaedia Britannica: Macropaedia*, s.v. "Relativity".
187 The mathematical equations that relate space- and time measurements of one observer to those of another, moving, observer are known as Lorentz transformations. If the rela-tive motion is measured along the x-axis and if its magnitude is v, these expressions are: $x' = (1 - v^2/c^2)^{-1/2} (x-vt)$, $y' = y$, $z' = z$, $t' = (1 - v^2/c^2)^{-1/2} (t- v\,x/c^2)$. *The New Encyclopae-dia Britannica: Macropaedia*, s.v. "Relativity".

As the speed of one inertial frame of reference relative to another is increased, its rods appear increasingly foreshortened and its clocks more and more slowed down. As this relative speed approaches c, both of these effects increase indefinitely. The relative speed of the two frames cannot exceed c if light and other electromagnetic phenomena are to travel at the speed c in all directions when viewed from either frame of reference. Hence the special theory of relativity forecloses relative speeds of frames of reference greater than c. As an inertial frame of reference can be associated with any material object in uniform nonrotational motion, it follows that no material object can travel at a rate of speed exceeding c.[188]

STR forestalls only movement at a speed greater than that of electromagnetic quanta. To measure superluminal velocities which are thus fettered below by the speed of light, other source-independent inertial superluminal propagations are to be used as reference velocities, if one does not want unnecessary infinities and zeros in theory. The so-called source-independence concept has precipitated the acceptance of the velocity of light *in vacuo* as fundamental to all measurements.

> To transmit a signal means to transmit a momentum and energy (taken to be inseparable in the theory of relativity [...]) which are capable of "switching on" a certain device, e.g. a trigger mechanism [...]. Although there is no privileged frame among all inertial frames, there is one privileged velocity in all of them. Both these circumstances are intrinsically associated with the fact that electromagnetic waves can propagate *in vacuo* [...] [N]o material medium is needed for their propagation.[189]

There *could be* other source-independent propagations which are at the same time superluminal. The assumption of invariance of the velocity of light is based on the experimental evidence of sufficiently uniform motion of light *in vacuo*, *but the experimental evidence is itself based on measurements using light signals as the criterial velocity*. As we use light signals as a criterial velocity we are comparing any other velocity with that of light, and no other velocity can legitimately approach the speed of light. If it does, it will cause *inconsistencies of infinity and zero related to comparison*, exactly like when complex quantities are introduced to show possible superluminal velocities under the yoke of the electromagnetic criterial velocity. Then the resulting zero rest mass is the result of using the speed of light as criterion, and not a result of the so-called limiting character of the speed of light over all possible superluminal velocities. Moreover, the fact that so far no superluminal velocity is directly detected helps us think that our anthropic – visual, apparatus-level – criterion is the highest criterion, since this criterion shows c as insurmountable and source-independent.

---

188 *The New Encyclopaedia Britannica: Macropaedia*, s.v. "Relativity".
189 Ugarov, *Special Theory of Relativity*, 36–7.

Thus, three things are conceptually unaccounted for in STR: that (1) our experiments presuppose presenting things visually and engineering apparatuses for a theoretically final-for-sure visual imagination and relativistic depiction criterioned upon the speed of electromagnetic propagation; (2) EPR nonlocality is indirect evidence for spatio-temporal (extension-motion), "local", superluminal, but causal, signals; and (3) if there were particles of superluminal velocities just as source-independent and just as capable of working as inertial velocity as light is, we could base experiments on any one of the superluminal signals thus mooted. The equations would then posit impossibility of approach of light quanta to the speed of the superluminal velocity at issue, by the same element of equation, namely, $(1 - v^2/c^2)^{-1/2}$ – substituting a superluminal velocity s for the criterial velocity c – as in the equation that bases itself on the velocity of light as inertial velocity. Then, v would not possess zero rest mass even if it approached c (which happens to be intermediate to v and s); not so when it approaches s. Light quanta as the basis of Einstein's relativistic postulates (and many of the impliations of the postulates) would then be unnecessary.

I put the argument in this section in gist as follows: The postulates of STR are: (1) sameness of the laws of physics in all inertial reference frames (Principle of Relativity), and (2) sameness of the speed of light in free space in all reference frames. From these follow deductions like: (1) the Lorentz-Fitzgerald contraction of length of particles moving at the speed of light, (2) increase of mass of particle when in luminal and subluminal motion, and (3) slowing down of moving clocks in such circumstances.[190] If the speed of light need not be constant, or if there can be superluminal velocities, greater flexibility and consequent continuity of such values in Reality (matter and energy) should be recognized. STR and GTR need to cope with the requirement of the new principle of continuity of missing quantal values and values of speeds of particles. The relativistic deduction of the upper limit of velocities from the postulates (based on the criterial nature of the speed of light) should therefore face revision.

1.3.1.2 Principle of Relativity and Superluminal Velocities

Galileo first formulated the principle of relativity using Galileian Transformations. The laws of mechanics are formed for identical use in all co-ordinate

---

190 Rothman, *Discovering the Natural Laws*, 137. All the deductions from the postulates are based on the element, $\sqrt{(1 - V^2/c^2)}$. If c is extended indefinitely in the universe of all other island universes, the deductions will have different meanings.

systems moving uniformly in a straight line.[191] If the ether introduced in Galileian transformations[192] is motionless in one inertial system (K), then $c = 3 \times 10^8$ m/s irrespective of the direction. "[...] [I]n other inertial systems K' moving with velocity V relative to the ether (along the x and x' axes), the velocity of light is, as is obvious from the Galileian transformations, $c' = c - V$ along the x and x' axes and $c' = c + V$ in the opposite direction, etc."[193] This classical conclusion did not hold experimentally. Instead, the velocity of light was the same in all directions (but with respect to the very light signals). Einstein gave a solution of the confusion through his STR: that the Galileian transformations are only a meso-world approximation of what we call today as Lorentz transformations.[194] His equations predicted results experimentally true for the meso- and micro-worlds, implying thereby only constancy in the speed of light with respect to this island universe. Classical mechanics allowed observer-independence to objects because

---

191 Gribanov, *Albert Einstein's Philosophical Views and the Theory of Relativity*, 194.
192 In the words of Ginzburg, the essence of Galileian transformations is this: "If a given system is inertial, any other system moving uniformly in a straight line relative to it is also inertial. The generalization of this conclusion over all mechanical phenomena – the assertion that all mechanical phenomena occur absolutely identically in all inertial systems – is just what the classical, or Galileian, principle of relativity is all about. More precisely, the definition and application of the principle incorporates the quite definite prerelativistic assumption concerning the connection between the coordinates and time of events in different inertial systems. Thus, if one such system K' (coordinates x', y', z', and t') is moving relative to a given inertial system K (coordinates x, y, z, and time t) with a velocity V along the positive axes x, x' (the direction of which we assume to coincide), then, as assumed before special relativity, $x' = x - Vt$, $y' = y$, $z' = z$, and $t' = t$ (the Galileian transformations). The absolute nature of time – its independence of the motion of the reference system (whence the equality $t' = t$) – was, of course, assumed to hold in all reference systems in general." Ginzburg, "Supplement", 319.
193 Ginzburg, "Supplement", 320.
194 Einstein had independently developed a version of these transformations and published them soon after Lorentz did his. "The precise equations that link coordinates and time in the frames K' and K have the form $x' = (x - Vt) / \sqrt{(1 - V^2/c^2)}$, $y' = y$, $z' = z$, $t' = (t - (V/c^2) x) / \sqrt{(1 - V^2/c^2)}$ (the Lorentz transformations). If the relative velocity V of inertial systems is small compared to the speed of light c, the Lorentz transformations become the Galileian transformations; hence the degree of accuracy given by the parameter V^2/c^2." Ginzburg, "Supplement", 320–21.

it worked in the mesoscopic world. Now we find that even in the micro-world velocities equal to that of light are observer-independent.[195]

Lorentz sought to remove the incompatibility between Galileian transformations and the principle of relativity in electrodynamics and mechanics,

> [...] without rejecting the Galileian transformations by assuming that all bodies moving with respect to the ether contract. If a ruler whose length at rest relative to the ether is l_0 is of length $l_0 \sqrt{(1 - (V/c)^2)}$ when moving at velocity V, then we can explain why some experiments do not reveal the motion of bodies relative to the ether, and their results do not depend on the velocity of the Earth's motion with respect to the Sun. However, the contraction hypothesis is not adequate for all experiments; new facts kept coming to light which agreed with the relativity principle and required additional hypotheses to explain them. This was [...] an intolerable situation, and Lorentz had to show that for a body in *uniform rectilinear motion* (relative to the ether) the equations of electrodynamics allow for solutions which in a certain way correspond to the solutions for an identical body *at rest*. Correspondence is achieved by going over to new variables, x', y', z', and t', with the help of the Lorentz transformations, as well as the introduction of new (primed) electromagnetic field vectors. The field equations do not change as a result of these transformations, and they have the same form for the old (unprimed) and new (primed) quantities. This property is known as invariance, in the present case invariance of the electromagnetic field equations with respect to the Lorentz transformations.[196]

Invariance of the field equations in the light of the Lorenz transformations, along with the posited invariance of the speed of light,[197] helped in achieving among scientists quick concurrence with Einstein's STR.[198] Invariance of the speed of light within the backdrop of invariance of electromagnetic field equations and their combined application in the STR shows additionally only that observation and calculation of relative motion of objects by mesoscopic and microscopic apparatuses is now possible by involving the speed of light. Lorentz and Einstein thus found a solid argument to reframe the concept of reference frame by the

---

195 This is not because of Lorentz contraction or absence of simultaneity of events or absence of simultaneity of observation of events, but due to actual non-zero foreshortening of the particle in transit at near-photonic and perhaps even at superluminal velocities.
196 Ginzburg, "Supplement", 322.
197 Maxwell's experiment carried out by Michelson and Morley, showing the constancy of the speed of light, was initially in view of showing the reality or unreality of ether.
198 Einstein's 'Special Principle of Relativity' is the principle that "[...] the laws of nature are invariant (take the same form) in all inertial reference frames." *Routledge Encyclopaedia of Philosophy*, s.v. "Relativity Theory, Philosophical Significance of", by Michael Redhead.

velocity of light. This need not mean that greater reference velocities are experimentally impossible, and this fact has been experimentally proved later in the case of the so-called "nonlocality" event that we have discussed. Therefore, the relevance of the Principle of Relativity should be shown without its foundation on the fixed but epistemically intermediate value of c, by allowing space for superluminal velocities to work the same miracle. opening up the chance of there being STRs and GTRs for all possible realized and realizable worlds and all possible realistic criterial velocities.

This seems impossible within relativity theory, with its definitions based on the speed of light. Placing superluminal velocities in place of c in relativistic equations can do it, since *the Principle of Relativity does not have necessary connection with any constancy, source-independence and criterial nature of the speed of light.* The fact that the microscopic world we are dealing with is packed with experiments using electromagnetic radiation as the criterion should not preduspose non-acceptance of superluminal source-independent radiations. It might even be that if observed by superluminal velocities light quanta are not source-independent. Similar is the case also with relativistic mass and speed-related space and time measurements.

Even in other worlds with higher limit velocities, *when the velocity v approaches the limit velocity, the temporarily limiting character of it, used only as the measuring rod, will become omnipotent.* STR and GTR can be epistemologically sufficiently universal in the application of the laws of physics for all these worlds, only if the criterion-status of the speed of light is relativized and physical constants of velocity beyond that based on electromagnetism are postulated with respect to all possible island universes. The relevant theory in each world will be determined by the mass, angular momentum etc. of the respective Big Bang or blackhole explosion. Then it is easy to see that *concretist categories like spatiality, temporality, causality, mass, conservation etc., not merely physical constants, are the basis for the "universality" of the laws of physics*. Such universality applies only to the respective universe if their meanings and values are set for one universe.

If we read this conclusion in conjunction with the EPR experiment's "local" or causal interpretation of the phenomenon of superluminal velocities that allows dispersion beyond the maximum speed of light, and also with the principle of near-infinitesimality of wavicles of physical action derived in the sections on QM, *we obtain a sort of thoroughgoing continuity in the universe*. An energy value absent in this island universe may be present in another, and this adds up to the principle of continuity of values of speeds. That is, we may summate over all these values unto the cosmological ideal category of Reality-in-total, which I state conclusively after some more discussions on other more physical topics.

1.3.2 Space, Time, Mass and Causality from Newton to Einstein

Problems like wave collapse and measurement, wave-particle duality and non-visualizable probability, the EPR paradox (correlation / action-at-a-distance) vs. causality, velocity of light vs. superluminal velocities, infinitesimality of matter and continuity of wave-lengths, constants and other such values of wavicles physical action etc. concern the enterprise of physics and philosophy. The three key questions they pose to philosophy are: (1) Is it possible to address philosophical questions purely from within the mechanistic physics using traditional scientific categories? (2) Can we re-interpret the classical view of reality using the radical changes brought about beyond classical physics and quantum physics? (3) What metaphysical category/ies can effect such a re-interpretation? To discuss these questions we inquire generally into the classical, quantum and relativistic concepts of reality without neglecting possible effects of such concepts in realistic possible-worlds cosmology.

1.3.2.1 Classical and Reformed Views of Continuous Reality

Philosophical tradition till Locke has differentiated between primary qualities (absolute, objective, immutable and mathematical) and secondary qualities (relative, subjective, fluctuating and sensed). Primary qualities were attributed to substance as its intrinsic, essential attributes. The latter were effects of the action of primary qualities on the senses. That was the ideal of an objective entity based on the objectivity and invariance of primary qualities.[199] Soon came about the trend of reducing qualities into quantity, the history of which is well known. After we have relativised the Kantian phenomena-noumena distinction, we are in a position to substitute the questions of substance, primary qualities, secondary

---

199 Jiachang and Xinhe, "Relational Realism on Reform of the View of Physical Reality and Its Logical Manifestation", 360. The same page says further: "[…] the secondary qualities are binary (or plural) functions, neither reducible to the substantial entities, i.e., the primary qualities completely, nor ascribable to the state of human sense in and of themselves, namely (1) $y = f(x_1, x_2,...)$ where y are the secondary qualities, x_1 are entities and their essential attributes, x_2 are the states of human senses. Then, the y's are relative manifestations, or projections of x_1 or x_2, or relations between x_1 and x_2. But when the y's are the primary qualities, (2) $y = g(x_1)$ they have nothing to do, in their existence, with surrounding variables."

qualities reduction of the latter two into substance etc. by Reality and universals, by questions of theory (perception, ideas, universals, theories etc.) and of reality (that of which theories are), because it is impossible to address contemporary philosophical questions merely from mechanistically physical conceptions.

Lakatos analyses the core of physical research into theoretical and metaphysical components. The first constitutes the basic postulates and laws of a theory that comprises the physical foundations of research. But Lakatos does not go the way of continuity between processes and theory by differentiating between processes and their *qualia* or by showing some continuity in causal processes through the use of *qualia*. In classical physics Newton's laws of motion and the law of universal gravitation applied to mechanical phenomena. They are the basic conceptually sedimented framework of a world-picture or physical ontology of the times, serving as criteria of construction and evaluation of theories. In the absence of discreteness of quanta, the classical (Newtonian) concept of reality had absolute continuity. It is a case of mechanism that explains phenomena by *continuous* motion of particles with definite and objective properties in absolute space and time, according to exact causality mechanistically referable to all the past but in act referred merely to the immediate past.[200] The concept of continuity of matter and motion, Maxwell's field theory of matter and the view of space and time in STR may be used as reforms of the metaphysical component of classical categorial concepts. *The classical continuity of motion was based on the continuity of wave motion.* Today this is considered overthrown by discreteness of quanta of motion.

Yet, continuity in my opinion befits being treated as part of genuine scientific realism, provided we stick to the cosmology of many (or perhaps infinite) other gravitationally (or even anti-gravitationally) mutually connected Big Bang universes (however distant in extension-motion the physical influences be) in which there are missing values of quanta, velocity and universal constants. This sort of continuity is distinctly different from that in classical mechanics. It transpires in Reality-in-total with universals / *qualia*. Observation by means of electromagnetic signals is a problem in constrained realism that treats only of this world. Hence the absence of a principle of ontological continuity in STR and GTR needs bridging by limiting the present luminal foundations of velocity in STR to particular worlds. Now that all possible missing values of energy are assignable to Reality, we may consider all the depth-level continuity of values of

---

200 Jiachang and Xinhe, "Relational Realism on Reform of the View of Physical Reality and Its Logical Manifestation", 360–61.

any given finite extension-motion region by summation unto Reality. Everything that is, is in process within Reality, in which anything is in objectual-causal connection with the proper past of its contemporary world; and the *qualia* are in effect causal inheritances from the past. That allows postulation of thoroughgoing causal continuity in Reality, which I shall call the Ontological Principle of Excluded Vacuous Middle. We need to reinterpret in the present work the classical view of reality, using the radical changes brought about beyond Classical, QM and Relativity physics.

1.3.2.2 Beyond Scientific Causality and Categories

The classical concept of causality was deterministic,[201] based on actual individualized categorial observation of what were called spatial and temporal processes and mechanistic processual description of these observed causal objects and forces, as if there were to Reality only as much as what comes under the particularistic perspective. This is so somewhat also in QM, since particularistic measurement is a methodological limit in it.[202] So the classical concept of reality in QM is perspectively absolutistic (holding that the current perspective is exhaustive) of science, and not explicitly and thoroughly objectual-causal of all the inner and broader causal recesses of the whole Reality not yet captured in theory, because it was ridden with particularistic perspectivalism.[203]

The real-ness and mind-independence of the phenomenal-noumenal continuity of Reality and the relative mind-dependence of theory and observation of reality-in-particular demand abandonment of merely perspectival and deterministic causality. By the time of Newton the reductionist notion of exact space-time description of particles without use of the concept of forces[204] had substituted the animistic and medieval notions of cause in learned circles. The

---

201 Bohm, *Quantum Theory*, 150–51.
202 Due to the inherent particularism of measurement in Quantum Mechanics, "… the quantum theory resembles the Newtonian approach in its acceptance of the free particle as an elementary construct. On the other hand, it is unlike the Newtonian theory in its rejection of the deterministic concept of a mechanical particle." Sachs, "On the Elemen-tarity of Measurement in General Relativity: Toward a General Theory", 58.
203 Bohm, *Quantum Theory*, 152.
204 "The principle of economy of concepts would then suggest that, except as a convenient term lumping together the effects of many accelerations, the concept of force as the cause of acceleration ought to be discarded and replaced by the idea that particles simp-ly follow certain trajectories determined by the equations of motion.

perspectivally absolutistic aspect of determinism in Newton persists even today in a lot of astrophysics, physics, chemistry, biology etc. After the discussion of the problems of *wave-particle duality, the local-causal interpretation of the EPR paradox and the supposedly limiting character of the speed of light*, it is easy to see that perspectival absolutism is especially the case with the QM- and relativistic descriptions.

QM favours also approximate, statistical, instrumentalistic causality. The quantum mechanical variety of statistical causality is a phenomenalistic outgrowth from the epistemologically deterministic concept of observation, and so again it involves observation by source-independent and zero rest mass particles, without relativising also that form of measurement, which is itself perspectival absolutism of the present. *QM has a statistically perspectival absolutism* because of (1) its manner of making causes not only be observed using photons as probabilistic (2) but also its reducing the nature of Reality into something probabilistic *as if probabilistic causation transpired in processes*. Such then is interpreted as "[…] producing a qualitative tendency in a given direction."[205] The allegedly direct and complete determinations yielded by causal observation in Newton's laws of motion are inapplicable in QM, since momentum and position cannot be simultaneously observed within the perspectival level of observation in it.[206]

Here it is forgotten that this probabilistic phenomenalism to the detriment of objectual realism has arisen because of the comparability of velocities of the objects and the assumption of the criterial velocity of photons. The measurement, e.g. of momentum and position in Uncertainty, is by photonal limit velocity. The language of causal forces is thus avoided by Bohm in his QM language "[…] because the de Broglie relations yield a simple relation between wavelength and momentum. No such simple relations exist between the wave properties of matter and force."[207] Force and mass are described in particularistic measurements of epistemological space and time. Causality is taken phenomenalistically, deterministically and by perspectival absolutism. Hence, realistic scientific ontology must effect a fresh synthesis of scientific categories with Reality-in-total as the foundational cosmological category. Departure from the Newtonian and QM perspectival absolutism can produce the desired direction of effects.

    Thus, classical the-ory leads to a point of view that is prescriptive and not causal." Bohm, *Quantum Theory*, 151–52.
205 Bohm, *Quantum Theory*, 152.
206 Bohm, *Quantum Theory*, 152.
207 Bohm, *Quantum Theory*, 156.

For this reason it is difficult to accept the metaphysics behind where Bohm says that momentum and position do not *exist* with simultaneity and accuracy.[208] He does not recognize the possibility of a multiplicity of macro- and micro- and other perspectives and that the measured momentum and position are what they are in each instance due to the *measure of availability* of some real physical states (given in broad ontological commitment widely based on Reality) for the probabilistic mode of measurement. Caught in the midst of technical jargon and the mode of (anti-)metaphysics and analytic philosophy in vogue, he (and most others) was fixated in the camp against recognition of the infinitely and infinitesimally thorough causality prevalent in Reality. While resorting to a mistake in the anti-metaphysical attitude – by unconsciously substituting the ontological commitment in the expression 'exist' (of supplying an actual but definitionally non-determinate argument to the logical form) with 'be "observed" by mathematical instrumentalism' – and by being merely phenomenalistically causal about wavicles, Bohm contradicts the realistic purpose of his own causal hidden-variables version of QM.

We are in a position to advance the cause of realistic, infinite and infinitesimal causality and a mathematical and scientific practice commensurable with it in QM, STR, GTR and other fields of physical inquiry under the facility offered by our (1) synthesis of the Kantian phenomena and noumena, (2) refusal to accept existence of any actual wave-particle duality in wavicles under the cover of mathematical wave- and particle-notions, (3) making mathematical instrumentalism into something sophistic, by our expansion of values of velocities to superluminally causally local values, (4) relativizing of the alleged criterial character of the speed of light, and (5) acceptance of Reality as the Transcendent super-category by reason which the macro-, meso-, micro-, nano- and other ever-more minute layers of concretist-circumscriptive and instrumentalist interpretations are taken in terms of non-absolutist perspectival interpretations of aspects and parts of one and the same Reality. Thus we hold that classical, quantum and relativistic mechanics are *devoid of ontological commitment to all the depths of Reality*.

Chapter 1 has thus posited the epistemologically probabilistic nature of our ways of capturing Reality, through a study of the infinite and infinitesimal objectual causality in Reality and our episemic probabilistic access into it. This is a thoroughgoingly absolute causalism which I call ontological causalism about Reality that lends itself to epistemic-probabilistic knowing to human intervention – otherwise the rat would bump on the computer or box. Any depths and

---

208   Bohm, *Quantum Theory*, 152.

extents of extention-motion regions of Reality are in the final analysis causally determined in all their depths, *sub specie Realitatis*. This is not traditional ontological determinism because such ontological determinism holds particularistically that things / processes are determined from within local extension-motion extents and perspectival layers. Nor is it epistemological-particularistic determinism, since this version believes, in the manner of perspectival absolutism, in the capacity of immediate circumscriptive measurement to determine all sorts of causes.

To conclude, Reality-in-total is the cosmological-ontological super-category, one of the categories *par excellence* common to theories of infinite and infinitesimal causality in science and philosophy. This super-category synthesizes the merits of the cosmological contributions to thought by the scientific categories of Extension-Motion, Spacetime, Cause-Effect and Matter-Energy. Thus, sections 1.1, 1.2 and 1.3 complete our project of (1) reducing the quanta of action (the actuality of entities at given measured space-times of extension-motion regions) into causal actions, and (2) reducing the primary relationship of extensive connection into the terminology of attributes through their synthesis into a continuum that allows both entity-talk and attribute-talk. The concept of entity here is realistic, and the concept of attributes is not phenomenalistic but ontological and connotative. So, I argue that *the four scientific categories of Extension-Motion, Spacetime, Cause-Effect and Matter-Energy are concretist* from the point of view of Reality, and are inapt to making epistemological probabilism possible against the categorial background of the inductively totalised category of Reality-in-total. This allows us to give a final shape to the concept of Reality-in-total as an ultimate ontological category in the following section.

1.4 Ontological Synthesis of Categories of the Cosmos

1.4.1 Reality-in-total: The Ultimate Transcendent Category

The use of conceptual idealizations in scientific theory may be taken for granted. The ones considered so far to be ultimate in science are now subordinate to Reality-in-total. I hold that this is so in the most Transcendental and Transcendent manner realistically possible, by discussing the possibility of a synthesis of knowledge under this category through some philosophical conclusions from QM. The metaphysics of philosophical extension of QM cannot be deduced from or refuted by the minimal structure of QM. Its notion of 'empirical reality as the essential content' is only a first type of extension of QM. According to Dingguo, who ontologically summarizes D'Espagnat's philosophical extension of QM, 'empirical reality' has the following implications:

> (1) At any time, the center of mass of macroscopic bodies has the property of locality. (2) A property of a system must be defined operationally but counter-factually.[209] (3) The time evolution of the properties of a system in an ensemble is determined by the computation rules of a theory. (4) It is impossible to influence the past. (5) A theory of empirical reality is obtained by postulating (a) that replacing very large times by infinite times and/or very large particle numbers by infinite numbers is a valid abstraction; and (b) that on the other hand the possibility of measuring observables exceeding a certain degree of complexity is to be considered as nonexistent, even in matters of principle, even though this non-measurability does not follow from the theory and even the only way we have for making it compatible with QM [...] seems to be to ascribe it to some basic inaptitude of men.[210]

---

209 A counterfactual is a contrary-to-fact, subjunctive and conditional statement presupposing the possible falsity of the antecedent, e.g., 'If I were you, I would have...' Since their truth-value is not completely determined at the statement, they are hypothetical, and not truth-functional. *The Cambridge Dictionary of Philosophy*, second edition, s.v. "Counterfactuals". If a property of any system is defined counterfactually, the definition will show the inductive nature of the whole property. However, one commits here to the truth or falsity of a maximal set of statements or truths or entities called possible world which is somehow instrumental in the whole counterfactual talk. Possible worlds have in any case some originary and resultant connections to the real world via the origins of the universals they use – at least conceptually, and perhaps also in possible future realization from the time with reference to which it has been formulated.

210 Dingguo, "On the neutral Status of QM in the Dispute of Realism vs. Anti-realism", 310–11. For the original detailed discussion consulted by Dingguo, see D'Espagnat, *Reality and the Physicist*, 232–53 (Addendum). It goes without saying that D'Espagnat does not equate 'empirical reality' with 'independent reality'. He takes advantage of the axiom of empirical reality given in (5) in the body of this essay above, to define and delimit the concept of 'empirical reality': "[...] [T]his axiom limits the sophistication of the measurements that are taken into account, by referring to practical limitations in the abilities of the human species. The status of the concept of empirical reality therefore turns out to be a subtle and hybrid one [...]. [T]his notion clearly has much in common, if not with our naive idea about 'real things', at least with the view the most thoughtful minds have taken of what should be called real... In fact it allows us to use the verbs 'to have' (in assertions such as 'this system has this property') and 'to be' (in assertions such as 'this system is in such and such a domain of space') only in somewhat weakened senses, since in some cases some measurements that are not ruled out by any theorem, that we ruled out just 'by decree', would falsify what we then say." D'Espagnat, *Reality and the Physicist*, 250. In the end he

Lest such a concept of empirical reality end in scepticism and sophism, we need to transform it. Technically, by the ontological implications of empirical reality, we must allow extra-phenomenal actual existence not merely to macroscopic bodies, but also to actual energy particles within and from it.[211] This in conjunction with statement (5.a) (mathematical induction) and (5.b) (observability, extended for our sake from the micro- to the macro-cosmic), in the above quote, shows the relevance of the concept of Reality-in-total as the super-category under ontological, realistic, trans-perspectival-probabilist scientific ontology. It shows also that it is possible to re-define and synthesise the basic categories of specific events under the concept of Reality. Our epistemological probabilism of all that is physical-ontological, as a trans-perspectival theory, synthesizes perspectives and scientific categories within Reality. This yields a unified epistemological and ontological framework.[212] This is a philosophical antidote to the absolutism of empiricistic scientism.

Based on the elaborate Algebraic QM, Primas has constructed an ontological extension of QM. To quote Dingguo discussing and summarizing Primas,

---

defines the concept of 'empirical reality': "Empirical reality may then be defined as the set of all the subjects of the verb 'to be', taken in this weakened sense, whereas independent reality [...] would be the set of the subject of the verb 'to be', taken in the strong sense in which the realists of yore took it (supposing of course reliable sentences could be constructed with such subjects, account being taken of what we know now)." D'Espagnat, *Reality and the Physicist*, 250–51. D'Espagnat's note on this latter passage is that "[...] such words and expressions as 'to be', 'ontic states' and so on can be reconciled with other statements [...] only if [...] the verb 'to be' is given this weakened, human-centred sense." D'Espagnat, Reality and the Physicist, 271. To go along with the suggestions in this essay, the humanized (made dependent on humans) version of 'to be' is what causes perspectival absolutism, if such a concept of 'to be' is not treated as that of reality-in-particular. What obtains in QM is also a mere perspectival absolutism, if it does makes no room for continuous, infinite and infinitesimal causality in Reality. This latter variety of causality can be called 'objectual causality', which will be discussed in the following paragraphs. It violates the perspectival absolutism of the 'to be' of reality-in-particular.

211 "A scientific realist accepts the minimal ontological assumption that there is an actual world independent of human minds, concepts, beliefs and interests. Let us call it THE WORLD." Niiniluoto, "Queries about Internal Realism", 49. Categorially it is Reality.

212 "By extending QM in the algebraic framework, and endowing the observable algebra with a primitive status, it may be possible to speak of both classical reality and quantum reality in a united epistemological framework." Dingguo, "On the Neutral Status of QM in the Dispute of Realism vs. Anti-realism", 311.

(1) A strictly closed physical system without any concept of an observer is called an endosystem. (2) If the endosystem is divided into an observing and an observed part, we speak of an exophysical description. (3) The world of observers with their communication tools is called an exosystem. (4) The empirical interpretation of a physical theory refer[s] to our knowledge of the properties or modes of reactions of the system 'as we perceive them'. (5) The ontic interpretation of a physical theory refers to the nature of 'object itself'. It is irrelevant with respect to our knowing it or not, and it is independent of any perturbation by observing actions. In the view of Primas, the ontic interpretation of QM is a universally valid theory, an ontic interpretation allows us a consistent way of speaking as if we would refer to reality.[213]

The ontic interpretation is valid only if it is validated by epistemological probabilism of things ontological and objectual – a fact that most physicists do not get at, due to their concern for the pragmatic. By 'objectual' I mean 'with ontological commitment to the related processes in the world, which commitment is ontologically supported by summation of infinitely and infinitesimally detailed relations (universals) over Reality and reality-in-particular'. 'Objectivity' has always the overtone of epistemological adequacy of a concept or a point of view with respect to what is perceived as actual. So, 'objectual' means 'of actual objects (events) in their mind-independent otherness', based on an ontological commitment that sums over Reality through universals. As will be clear from the following and in Chapter 2, objectual realism depends not merely on the token entity or the particular, which is ensconced with the immediate universal, but in the final analysis on Reality-in-total and Reality-in-general which latter is the totality of all possible universals that we cannot dispense with in discourse. Hence, objectual realism is a realism of both Reality and reality-in-particular, the latter in a special way being made possible by their connexity via universals of universals of ... *ad libitum*, finally by the universal of all universals, i.e. Reality-in-general and also by the objectual fact of Reality. This is trans-pragmatic and trans-ontic.

Hence, objectual realism is a realism of Reality and reality-in-particular – the latter in a special way being made possible by the connexity of realities-in-particular via universals of universals of ... *ad libitum*, ending in as Reality-in-general (Chapter 2). It is epistemologically probabilistic over all Reality-in-total: no universal results absolutely from one entity or process but from a conglomeration of processes. The universals (including laws) of these conglomerations fade away into less relevance within the relevance of the immediate instantiated universals in question. When universals are objectually epistemologically probabilistic, any

---

213 Dingguo, "On the Neutral Status of QM in the Dispute of Realism vs. Anti-realism", 315.

theory is a perspective, and we can succeed making them ever better objectually probabilistic only over Reality. *Reality is thus the objectual, mind-independent, universal object that remains the highest category.* Every object is a transcendent to every other. But Reality is the highest Transcendent.

The concept of Reality-in-total as a category has also the capacity to be a conceptual instrument in ontological and scientific activity. Every transcendental is an abstraction / attribute over the many. We have already discussed Chisholm's manner of converting entifying language into attributive language beyond Frege, Russell and Whitehead. Purely entifying or purely attributive language does violence to Reality that is the result of maximization of ontological commitment. This maximization into Reality is not a blind infinitization of mere 'possibles' or 'possible worlds' but of worlds already realized in their own time. Since there are ever-deeper infinitesimal depths to reality-in-particular (circumscribable as sets of wavicles) and ever-higher superluminal velocities, even reality-in-particular is vertically inexhaustible by theory, where it remains just a member of one or more near-infinitesimal layers of the infinitesimal depths of Reality. These inexhaustibles totalise infinitely horizontally (i.e., at the level of the contemporary world of entities / events) unto Reality – with a clear totalization also over causal pasts of infinite depths at extension-motion regions.

The ontological *concept* of Reality works also as the most widely objectual Transcendent involved in the transcendental and Transcendental aspects of all theory. Hence, it is simultaneously also affected by the deepest possible Transcendental, To Be, which is most transcendentally co-implied in theory in the Transcendent Reality. The Transcendental Transcendent that Reality is, will therefore be the simultaneously Transcendentally affected and actually most Transcendent cosmological category of ontology. This category is not merely the most ontological concept, but deals with that to which all transcendentals and the Transcendental To Be have ontological commitment – i.e., the mutually continuous totality of all possible realized worlds, including the Divine, if It is-and-becomes. If the Divine is also creative, it is the highest and the most creative of everything in Reality. If creative and infinite, then the Divine is *infinitely* creative – not by converting himself to this world, nor by just shaping parallelly existing worlds, but by being positively originatively creative due to his infinitely intense structure in every finite extension-motion region which adds up to an infinity of infinity of … infinities. *If such is the justified concept of the* non-vacuously existing *Divine, It cannot be dispensed from genuine cosmological discourse, because it too accounts for the determinations / causal roots of the universe / multiverse.*

1.4.2 Mathematical Entities vs. Ultimacy in Denotability of Reality-in-total

D. Shapere (referred to by Lakatos and Zhengkun) speaks of 'idealization terms' or 'idealization concepts' in connection with fictional characters. According to him 'idealization' is a kind of fiction with sufficient scientific reasons. One reason and example for this is the idealization in the concept of 'point mass', which is a mathematical application in physics. He says,

> [...] the rationale for supposing their employment to be both useful and possible is clear. (1) There are certain problems to be solved – problems relating to the positions, velocities, masses, and forces of bodies. (2) Mathematical techniques exist for dealing with such problems if the masses are considered to be concentrated at geometrical points (namely, the geometrical techniques of Newton and, for later scientists, the methods of the calculus). (3) It is, as Newton showed, possible to treat spherically symmetrical bodies as if their masses were concentrated at their centers (and, incidentally, Newton conceived the elementary mass-particles to be spherically symmetrical). And, as for bodies not spherically symmetrical, they could be considered in the same way, provided the distances between their centers was large in comparison to their radii – a condition fulfilled, happily, by the earth-moon system. More generally, bodies could be considered as if their masses were concentrated at their centers of gravity."[214]

Here idealization is of particulars. As applied mathematics so idealizes objects for the purpose of denotation, it does not have a way of totalising those to which idealizations ontologically commit, except the method of mathematical induction. If idealization by the point instant (in macro-, meso-, micro- nano- and more infinitesimal layers of physics) is justified, then mathematical totalization over all possible objects in ontological commitment is also justified. This argument must be noted well as the point of transition from physical and mathematical particularism (applying mathematical induction for idealization) to Einaic Ontology (applying inductive totalization and generalization of results of ontological commitment, including universals).

We do not enumerate all the sub-level entities / processes, say, of a meso-object, in order to totalise unto the said meso-entity or process. By reason of the ontological commitment behind the idealization of the moon, geometrically as a point instant for calculations in celestial mechanics, the detailed processes within the moon are taken for granted and ignored, since the meso-level objectification of the moon does not need those finer details, and can still allow the details to be actual within the act of meso-level objectification. It is a general fact

---

214 Zhengkun, "Truth and Fiction in Scientific Theory", 272.

in concept-formation, definition, mathematical induction etc. The sub-possible levels are near-infinitesimally real, still not conceptually implied by the meso-level objectification, and yet we idealize unto the latter. This is also an infinitization. That is, even mathematical idealization for physical purpose does commit ontologically. The sub-possible infinitesimal levels are also idealizations of the same sort.

Similar is the case with higher totalizations *sub specie Realitatis*. Reality as idealized by mathematical induction over all possible objects (gross particles, micro-wavicles with point mass etc. *ad libitum*), with a sure trans-Quinean ontological commitment to "somethings" corporeal and ideal, maximized and infinitized, is also a way of categorial denotation. This is not mere particularistic induction but totalization by possible objects in ontological commitment. So, the mathematics of scientific categories should accept denotation by Reality too.

There is a dichotomy between the subject-object distinction in practice in Classical Physics and QM: between the actual existence of entities / wavicles and the mathematical, descriptive and instrumentalistic probabilism about their existence and properties. To bridge this difficulty in QM in line with the QM deliberations we have made, we need further extensions. Let the realities of QM probabilism be extended beyond the certainty of classical commonsense physics by *extending the criterion of truth as truth-probabilities (achieved in extension of semantic and ontological foundations), up to the farthest possible of categories, axioms and principles.*[215] Let this be the final systemic result of and solution for Bohr's complementarity-explanation, for no other solution is theoretically as best truth-probable as can be. The simple process of reduction of observational set-ups and phenomena to ever more minute micro-worlds is thus justified.

It is bizarre if it ends up in an irrational jump from objectual truth (where there are truth-probabilities that systemically imply ever broader ontological commitments bordering in each case upon Reality) to objective or subjective or objective-subjective truth (where ontological commitment is not explicit). The solution for this is re-radicalisation of the standpoints of macro-, micro- and other layers of actual worlds onto the totality of the ever-more ontologically probabilistic levels of observation, by the method of pushing the definitions of primitive notions and of axioms and principles[216] based on Einaic (Greek, Einai, "To Be"; so, Einaic means "at the level of the To Be of Reality") categories. The

---

215 This is justified by Gödel's Incompleteness Theorem. It is discussed in Chapter 2 of the present work, especially in sections 2.2.2.1 and 2.2.2.2.
216 See sections 2.2.2.1 and 2.2.2.2.

reason for this is that the cosmologically ontological category yielded from the physical sciences is Reality, which is the maximization or infinitization of what is objectual-causal by mathematical induction and ontological totalization. Hence, Reality is the most objectual of all we can speak of, and it is the most denotative and denotable of all. It is denotative as a term, and what it ontologically commits to is the highest denotable. This shows the ultimacy of Reality as the highest denotable, the highest Transcendentally Transcendent, category.

Is not such maximization and infinitization a mere abstraction? If Reality were to denote a mathematically inductively *generalizational* trans-finite entity, we would have philosophical doubts about the *existence* of Reality *as a universal*, for universals are not actual processes.[217] Our universals here and in the chapters that follow are not existing entities but entities' ways of being in process. They have a more-than-functional but less-than-existing-entities part in ontology. As ways of being of processes they are ontologically ways of processes, and not just mental abstractions without reference to processes. The mental counterparts of them, had by abstraction and idealization, are connotative idealizations. These again are with foundation in the ways of processes that are independent of the mental activity of perception and cognition. If we are justified in thinking of spatio-temporally measured extension-motion local totalities and of realized and realizable possible worlds, then it is also allowable to think of the totality of all possible worlds, which are connected to each other, by reason of the causal characteristic of anything that may even be superluminally connected.

Is it possible for anything infinite to exist as such? The infinite Reality is not in existence as a *completely* unified abstract or concrete substance. It is the totality of all connected and processual entities in existence. Connexity is not the same as absolute unity and non-difference of parts. Hence, Reality is a loosely mutually connected totality of all possible (i.e., rationally and scientifically thinkable) realized and realizable worlds and their causal roots in all possible processual

---

217 For example, the "being given" or "being there" of truths, mathematical entities, universals and *abstracta*. "The prevailing idea that the numbers are 'abstract objects' to which the numerical expressions of the language of arithmetic stand in an external relation of 'denotation', results from the hypostatization of possibilities, namely the possibilities of applying rules anew – as does the jargon of 'abstract objects' in general, which has been so generally and wholeheartedly accepted in modern analytic philosophy. This talk of abstract objects as a manner of speaking may be harmless and even practical in mathematics, where the technical details are what really count, but in philosophy, where this jargon is meant to have a literal 'ontological meaning', it creates nonsense." Stenlund, *Language and Philosophical Problems*, 144.

entities, even when these ultimate causal roots may imply the Divine. In 'all possible realized and realizable worlds', 'possible' merely means possible for thought in whatever manner involving abstract entities, and 'realized and realizable' means possible for causal-ontologically concretely informed thought to take in ontological commitment (seeing-that) by extension of objectual-causal reason and induction beyond the world that we know by perception.

Now remains the question of the denotability of the ontological referential commitment of terms that yield potential infinity. I argue in terms of our distinction between ontological and connotative universals. First of all, potentialities in all possible realized and realizable worlds are not merely conceptual but real, based in the causal ways of being of reality-in-particular and Reality and involving ontological universals. *In Reality and reality-in-particular the level of admission of existence is ontological commitment* – which need not be to any specific named entity in all its definitions but to some entity and to their associated formal entities: universals. *The level of admission of being in mathematical activity is based on formal independence from particulars.* Formal independence is not cosmologically ontological independence, but the former is based on the latter and is abstractly ontological. There is *no* absolute *theoretical distance* between them. That is, ontological commitment of the cosmologially ontological type is unnecessary for just admitting the being of the formal aspect of logical and mathematical truths and entities; yet cosmologically ontological commitment is implied behind logical and mathematical *abstracta* that are at work in cosmologically ontological processual existences; and exactly so vice versa.

Stenlund says something similar, but (1) without distinctions of the two levels in universals, without explicitly recognizing (2) that mathematical and logical entities and truths have ontological commitment (not an immediate formal or material reference) to reality in their conceptual formations and consequences, and (3) that mathematical concepts / constructs are also questionable from the point of view of the "reality" or ontological commitment behind them: from within the purview of the axiomatic framework of each such system, and not based on any direct innocent referentiality of their contents. Such realization will justify the place of cosmic ontology and ontology of logic and of mathematics in logical and mathematical activity. A similar situation exists in all theoretical endeavours and in the semantics of all sciences. Stenlund perhaps speaks of mathematical objects with ontological innocence, i.e., without considering ontological commitments prior to mathematical objects, thus:

> I would rather say that *there is no potentiality within the content of mathematical propositions* in the same way that there is no temporality. The distinction 'actual-potential' only

makes sense in connection with time, or development, or 'coming into existence', or – in general – in connection with *change*. And this is precisely what we do not have *within* the realm of the strictly mathematical content. In this conceptual sense, what mathematical propositions express is non-temporal, non-developing, unchanging, indestructible, and so on. In *this* sense one could say that mathematical entities are stable, or that they 'are outside time and space'. With what is perhaps too benevolent an interpretation, it could be said that this is what is correct in the platonist conception of mathematics. But then it must be understood, not as a metaphysical statement about some independent mathematical reality, but as a conceptual statement about the logical order of mathematical notions.[218]

I would say, the 'non-temporal' in mathematical entities is more 'not belonging to the temporal order as abstract entities' than really non-temporal in their foundations / oigins. Within the manner of the conceptual, mathematical and logical order it must be admitted that abstract concepts in mathematics do not innocently refer to physical extension or motion. This is exactly what points to the need and place of connotative universals in abstract concepts, whether ontological or logical or mathematical. The conceptual statement about notions of the logical or mathematical order, that they are independent of things actual-potential and independent of measure of motion (time) and extension (space), shows merely that something from the actual horizons reflects in mathematical notions at the level of the ontological commitment of the structures that mirror in them. Note also that ontological commitment is not correspondence to any specific thing, but to anything (generally) and something (specifically), i.e., at the level of the ontologically Transcendent, Transcendental, transcendent and transcendental *a priori* behind notions.

Now we are in a position to understand and approve the motive behind Stenlund's radical statement that suggests a solution to the question of abstract existence in a veiled fashion, and to (later) take it as demanding the distinction and ontological connection between ontological and connotative universals:

> Just as there is supposed to be a general notion of a set, of which ordinary, finite classes and infinite sets are two subspecies, so there is supposed to be a general notion of 'object', of which concrete, physical objects and abstract objects such as numbers are subspecies. But this is a conceptual mistake caused by mere linguistic similarities. It is caused by similarities in forms of expression imposed by the techniques of paraphrase of set theory and the predicate calculus. It is caused, for instance, by paraphrase into linguistic forms like 'the totality of all objects x such that …x…' and 'there exists an object x such that…x…', where the expression '…x…' is dealt with as expressing a function in

---

218   Stenlund, *Language and Philosophical Problems*, 146.

the mathematical sense. The set/element categories and function/argument categories are (erroneously) assigned the role of universal categories of logical grammar. Being an element of a set or an argument of a function (or a value of a variable) is treated as a paradigm for this general notion of an object.[219]

The reason why a set / argument / value of a function is treated as the general notion of an object is that particularism has no category between entity (process) and non-entity (non-process) to fix general notions / universals into. Names of things very well denote entities / processes, but names and universals themselves are not things, but are either occurrences within and through consciousness or are ways of processes, and are used instrumentally as objects. Universals latter should be based on processes – this is possible only when processes are bounded by ontological universals and ontological universals reflect themselves in their parallel connotative universals as ingredient in notions of all sorts.

Lack of realization of the distinction between and dependence of mathematical and logical notions on ontological universals (via connotative universals) is the blur at the root of what Stenlund argues in what follows: "[…] [O]nce something in mathematics is possible, then it is actual."[220] 'Actual' is perhaps too strong and has comparison with existing processes. A statement of the kind, "[t]he basic notion of possibility in mathematics must be 'what is actually within our capacity' […],"[221] takes away from the field mathematical the possibilities, with abstract ontological commitment, of occurrences in nature without our thought about them. Such a statement singles out a certain portion or region of processual extension-motion as fully circumscribable, and tastes absolutely particularistic. It wards off maximal ontological commitment to "something" ontologically total that involves ontological universals beyond the connotative universals theoretically directly involved in the mathematical notion at question. That is, somehow the place of Reality in mathematical objects is to be shown. Ontologically, the object of treatment mathematical particularism is in fact not any one fully circumscribed and localized activity, extension, motion, spatiality, temporality, development or change. Reality as such too has its mathematical aspect in quantities and mathematically manoeuvrable qualities.

If Stenlund were to say that, since he is now speaking of mathematics he would not vouch for any ontology, then he is being partial about the possible horizon for mathematics and limiting mathematics to mathematical activity

---

219 Stenlund, *Language and Philosophical Problems*, 145.
220 Stenlund, *Language and Philosophical Problems*, 147.
221 Stenlund, *Language and Philosophical Problems*, 147.

alone, he is dissociated from the axiomatic roots of mathematics in real ways of thinking. Real ways of thinking are based in real ways of being of processes. The axiomatic foundations happen to vouch for some sort of ontology, however particularized or generalized. What Einaic Ontology does is to show the basis of all discourse in things Einaic, of which the cosmologically most fundamental ontological category is Reality-in-total. Reality involves ontological universals, which reflect imperfectly in connotative universals, of which one type is mathematical (quantitative and qualitative) connotatives. If mathematical connotatives are ingredient in mathematical notions, then Reality, which is the cosmologically ontological categories of all connotatives, is implied also in mathematical notions – however directly or veiled. Hence, I opine that the statement, 'once something in mathematics is possible, then it is actual', is made at a level absolutely distanced from the axiomatically cosmic-ontological roots of any branch of mathematics.

Such comparison of mathematical possibles and natural actuals does not tell badly upon the notion of ontological commitment, via the impossibility of equalization of possibles and actuals in ontology, since their equalization is feasible only at the level of Transcendental and/or Transcendent maximal ontological commitments and their purely mathematical notions. This is just one little region of equalization, which does not require us to equalize all the two areas of possibles in mathematics and actuals in ontology. *In Einaic Ontology, when something is possible,* something *is ontologically committed to as actual but yet not defined to be this or that.* Thus, what is ontologically committed to need not be what the statement directly speaks of. Instead, what is ontologically committed to commits to something objectual-causally real with possibly realized or realizable ontologically actual internal constitution, or to something realizable but now in concepts with nomic-nominal and processual-verbal constitution. Insofar as Einaic Ontology eschews particularism and takes up the *sub specie Realitatis* point of view of its fundamental categories at the level of ontological commitment to Reality-in-total (and to Reality-in-general and To Be), therefore, the denotable (Reality) and denotatively connotative reality/ies (Reality-in-general and To Be) behind these categories are all infinitizations in the sense of idealizations. Here the issue is not of the validity of mathematical proofs but of *a priori*, most general, most total and ontological implications of any process within the "totality" of all processes, which, by the philosophy of *a priori* possibles, are defined only by partial circularity.

Thus, for example, the mutually relativised infinite values of possible subluminal and superluminal velocities ontologically commits to all sorts of wavicles of actual velocities, taking finite amounts of them at a time, though the total values happen to be infinite by the very nature of the principle of continuity. The principle of continuity has its basis on the fundamental principle of all rationality, namely, the Ontological Principle of Excluded Vacuous Middle, which has genuine application only in the science *sub specie Realitatis*. The Ontological Principle of Excluded Vacuous Middle has it that there should be some value of realities between any two different other values, because no values of processes are fully causally circumscribed in actuality. So, what is objectual-causally committed to by the two processes continues upon each other by objectual-causal mediation of other processes of values between them. These values are also not fully circumscribed. Therefore, there should be other processes of values in the sub-world of them, and so forth. This is no allegedly Parmenidean argument for non-actuality of change, nor for filling of spatial expanse. Instead, it is the rational basis for ontological commitment to processes of ever-different values in the widest possible setting of Reality.

Ontological commitment is no intuitionism (Brouwer), nor transcendental phenomenology (Husserl), where potential possibilities and infinities, as *abstracta*, have some purely formal *content*, 'content' being defined either in temporal development, or in the vacuum of possibilities given in consciousness's abstract referentiality, respectively. In Einaic Ontology all universals / *abstracta* behind ontological and mathematical notions are based on ontological universals – which are relationalities of actual processes based ontologically in all forms of similar processes *sub specie Realitatis*. Hence, all potential infinities, be they connotative or ontological, pertain to totalities of actuals by the maximal case of ontological commitment. *This is no merely mathematical infinitization, but an ontological infinitization* of those to which abstracta ontologically commit. Thus, by reason of their being *of* the totality of actuals given in ontologically universalized or totalised processes, all potential infinities in Einaic Ontology are actual in their totality, not in their particularity. But the conceptual realm of content of any abstract object in mathematics has its explanation in its Einaic Semantic reference and meaning based on Einaic Ontology. This reference of mathematical notions is not the same as what they ontologically commit to.

Reality has within it potential infinity in that it is never absolutely fulfilled in its extension-motion processes. If there is continuous creation, then the infinite new worlds and their processes will be added to its potential infinity. The denotability of Reality as the Transcendent potential infinity *par excellence* is guaranteed by the ultimacy of ontological commitment that backs it, and not by means of any functional necessity of the connotative universal/s behind it. The functional necessity behind the category of Reality-in-total, as the highest potential infinity, is that of the maximal case apriority of it, yielded by maximized Transcendent ontological commitment. Ontologically, necessity and necessary connection are most valid in ontological commitment. But Stenlund says,

> [...] there seems to me to be elements of Aristotelian realism left in current philosophical uses of the notion of potentiality. It involves an 'entification' of possibilities that leads to problems similar to the ones that occur in connection with the traditional notion of 'power', 'disposition', 'necessary connection'. I think that the use of this notion of 'potentiality' in mathematics, as in the concept of 'potential infinity', belongs to a way of thinking in times when there was no clear separation between pure and applied mathematics; when one could not distinguish the conceptual systems of pure mathematics from their applications in empirical situations; when mathematics was 'the science of quantity', as it is even in Kant's philosophy.[222]

It should now be kept in mind that mathematical and linguistic notions of infinite and finite sequences work at the level of idealizations. Idealizations in notions are connotatives, with content-level dependence on actuality. These are never to be likened to the Transcendentally Transcendent *a priori* that is Reality. Idealizations need not be merely at the level of finite terms of sets, but also of transfinite terms of sets. The connotative aspect of mathematical notions refers both to the actual, particular, process and to processes similar to it by maximization. Without simultaneously implicit reference, there exists no abstract term; and such implicit reference is also part of the nature of *abstracta*. That is to say, the particularity of the immediate content of any finite *abstractum* in any pure mathematical notion is not merely mathematically insulated within itself. It has ontological commitment via the Reality and reality-in-particular through which it is formed. All the more so is the case of applied mathematical notions.

Hence, there is no place in mathematics for absolute disjunction between the practices of pure and applied mathematics, because their *abstracta* are either identical or mutually connected. For this reason, *mathematical* abstracta *are ontologically on the side of ways of being of processes*. Einaic Ontology holds that even the naturalistic semantics of Quine and Davidson can be justified only by maximally ontological commitments unto Reality, reality-in-particular and the rest of Einaic ideals. This is so Einaic Ontologically and not particularistically. Hence, there isn't need for arguing that much classical stuff is present in ontological analysis of all mathematical notions blocking progress in mathematics.

Nevertheless, the entification of potentiality and the creation of vast terrains of verbal jigglery in contemporary modal semantics, modal logic and modal

---

222   Stenlund, *Language and Philosophical Problems*, 148.

mathematics, for the sake of substituting the lack of sufficiently realistically ontologically committed extensionality in modal logic, has been a denigration of the general ontological base of mathematics and logic. As regards this unwarranted entification, which at least should have been augmented by clear ontologies of causal possible worlds, Stenlund seems to be right. This state of affairs does not affect the maximization and infinitization of the highest categories in Einaic Ontology, since these maximizations are no conceptual entifications but only conceptual categorizations. Nor does it disparage the Einaic Ontological concept of universals and *abstracta* as ways of being of processes.

Though the conceptual activity of counting and differentiation of qualities and structures is impossible without ontological possibilization by connotative universals, it is a mistake to think that connotative universals implied in the mathematical notion of quantities / numbers (of all kinds), qualities and structures are what actually possibilize the ontological universals at the relational-processual realm of things. Similarly, the fact that mathematical activity is possible without recourse to the actual world does not imply that *abstracta* in mathematics are not actually made possible by actualities and their qualities in Reality. The processual aspect of Reality and reality-in-particular possibilizes connotative universals via ontological universals, and through that also mathematical universals and *abstracta*. Non-recognition of this fact has created controversies in the philosophy of mathematics, including absolute distinction between pure and applied mathematics, and between mathematical *abstracta* and ontological universals. It has consistently been forgotten in analytic-philosophical circles (1) that actuality (reality) is needed for pure mathematics at the level of possibilization of its entities, and (2) that the applicability of applied mathematics is thus possibilized at origin by reality, without direct reference to the results from applied mathematics.

The above discussion on mathematical idealizations shows that the theoretical basis for much essential differentiation between philosophical, physical and mathematical interests cannot be absolute. Stenlund makes an essential differentiation, and adds: "It is not just a difference in subject matter or in degree of rigour or generality, but a difference in kind."[223] 'Essential' does not mean here 'at the level of essences' if the ontological-connotative universal difference is not accepted. Then, should not the 'difference in kind' mean some difference in degree? Even such a difference brings into question the kind of the term 'kind', and we must accept the connotative-ontological differentiation as explaining 'kind'.

---

223   Stenlund, *Language and Philosophical Problems*, 133.

Recollect that these two kinds of universals are not terms or notions, but essences that are connotatively and ontologically implied, respectively, behind conscious concepts and terms and behind connotative universals. If there is no absolute difference in kind – which I believe is the case, at least since the contrary thesis is not provable *sub specie Realitatis* – they should have some commonality/ies. This we find in their common ontological commitment to Reality: (on the one hand) via ontological commitment to reality-in-particular and (on the other hand) via ontological commitment to connotatives and ontological universals and to their maximal cases, i.e. Reality-in-general and To Be, which will be studied in Chapters 2 and 3.

The *denotable apriority of Reality is not that of universals / relationalities* – which are ways of being – in processes, but of ontological commitment by processes to there being their totality. This totality is not a potential infinity, but an actual and objectual-causal infinity. It is actual-potential infinity, since it is ontologically the maximal commitment to what has been, and what will be, causally, in what is. The concept of the denotable Reality is a denotative concept of the totality of all actuals. *It is a possible one, only as it is possible to speak of such a processually existent denotable* by the denotative 'Reality'.

Reality can be conceived as a realized counterfactual-possible or a set of counterfactually once-upon-a-time possible but-now-realized worlds if counterfactuals are arguments that posit possibilities with reference to past states of affairs and have space for commonalities of causal strains of the past to act on the future of existing universes. If counterfactuals may be posited about the Divine's continuous creative advance, then most probably Reality is the totality of all the divinely realized possible worlds. The set of all non-realizable, realizable and realized possible worlds – possible with respect to counterfactuals formed with reference to the past of actualities – is thus Reality plus Reality-in-general (the set of all realizable ways of being of processes) plus imagined non-realizable realities.

In short, sets of entities of truths of logic, mathematics etc. need commitment to actualities – by ontological commitment to each and every thing real. Their ontological commitment is, first of all, to ontological universals, and only then to the actual entities / processes that these abstract and concrete idealizations might denote. **This must be the reason** *for the relatively higher formal independence that mathematics (especially pure mathematics) enjoys from physical facts and realities.* The theoretically more distant ontological commitment of mathematical notions is not reference in the ordinary semantic sense. It is ontological possibilization of tokens, particulars, totals, ontological universals and connotative universals, which have by Einaic Ontology their farthest and most ultimate ontological commitment to Reality-in-total, Reality-in-general and To Be. All of them are maximizations and infinitizations, but theoretically they do not take direct reference from abstract entities. The present chapter treats only of Reality as a pure infinitization, a cosmic-ontological category at that.

What has so far been said under the present sub-heading shows that Reality is nothing abstract. It is the most concrete of all by any form of ontological commitment. Its ways of being are concrete only to the extent that they are based in Reality. As is clear by now, we do not hypostatise[224] abstract nominalisations of relationalities. The interesting thing is that the processual aspect of even Reality involves the ontological universal, the Way of being that we call To Be. This latter issue will be discussed in Chapter 3, and the connotative and denotative aspects of concepts that refer to universal ways of being of entities / processes will be dealt with in Chapter 2.

---

224 Stenlund, in the grip of ontological particularism, is unable to recognize that there may be an *a priori* way of ontological commitment in which mathematics can be connected to reality. I repeat an important passage from Stenlund: "The prevailing idea that the numbers are 'abstract objects' to which the numerical expressions of the language of arithmetic stand in an external relation of 'denotation', results from the hypostatisation of possibilities, namely the possibilities of applying rules anew – as does the jargon of 'abstract objects' in general, which has been so generally and wholeheartedly accepted in modern analytic philosophy. This talk of abstract objects as a manner of speaking may be harmless and even practical in mathematics, where the technical details are what really count, but in philosophy, where this jargon is meant to have a literal 'ontological meaning', it creates nonsense." Stenlund, *Language and Philosophical Problems*, 146.

I need not further comment on this text, since our forgoing discussion has already shown the Einaic point of view regarding the denotability of the ontological commitment behind the notion of Reality-in-total: that Reality-in-total steers clear of such accusations. I admit the danger of hypostatisation of mathematical notions and their very counterfactual or otherwise possibility as formally "being there" without any existential recourse to actuality. This is clear in most modal ontologies and modal mathematics today. What I have tried to present here includes also the conclusion that what the mathematical philosopher Brouwer has done to mathematical and other notions involves naturalization of their intuitive explanations; and that what the mathematician-philosopher Husserl and analytic thinkers like David Lewis and others have achieved is a hypostatisation of mathematical and other abstract objects.

Conclusion

I have argued in favour not of a naïvely or epistemologically objective-subjective ontology out of the physical sciences, but for a scientific ontology of infinitely and infinitesimally objectual, highest Transcendent categorial entity, Reality. It is neither a classificational, nor a merely epistemologically *a priori* category. It is ontologically and cosmologically *a priori*: it belongs to the nature of things as a synthetic *a priori* category based on synthetic judgments.[225] It is not merely an ideal. Denotatively, it is the highest Transcendent. It does not mean in the epistemological (transcendentally transcendent or Transcendently Transcendental) fashion, but in a purely ontological-cosmological (Transcendentally Transcendent) manner. To that extent a category may be thought and spoken of in language, it is epistemologically expressible in a term. Reality is the cosmologically ontological ultimate – the final substance without which there is no discourse. It is synthetic not as something epistemically constituted, but as the Transcendent that constitutes the inner transcendent, transcendental and Transcendental nature of all that are thus predicable. By 'transcendent' I mean 'of the nature of any actual being / process beyond the immediately given'; by 'transcendental' 'belonging to any general attribute of beings and processes'; and by 'Transcendental' 'of the possible ultimate generality within Reality, i.e. the To Be of Reality'. It is the ontological category derived from cosmological discourse.

Reality-in-total, as the ultimate Transcendent category, is objectual-causal by reason of the maximal Transcendent ontological commitment, and when such commitment is had in consciousness denotatively it is merely idealized and represented by the term 'Reality'. It is conceived by means of conceptual idealization. Hence, the foregoing part of the present work on the objectual-causal Reality has to be followed up with its epistemological aspect in Chapter 2; and in Chapter 3 the purely ontological aspect with which it is conceptualized should be discussed. The present chapter has built the foundation for working these two aspects. Thus, the way to Chapter 2 has by now been cleared.

---

225   The context of *a priori* synthetic judgments is famous. It does not need elaboration. See Kant, *Immanuel Kant's Critique of Pure Reason*, 50f (A9ff and B13f).

Chapter 2. Ontological Categorial Transcendental of Epistemology

Introduction

As we discuss the transcendental and Transcendental aspects of system-building we need to construct an epistemological category that befits the new understanding of maximized substance as Reality achieved in Chapter 1. The present chapter begins by analysing the epistemological necessity of categories and essences for ontological talk, through a study of the tension between particulars and universals in contemporary epistemology *of* ontological inquiry. Thereafter we *inductively generalize* unto the epistemic category of all philosophic activity, i.e. Reality-in-general, as the final acausal and connotative universal of the objectual-causal background (Reality) of all thought. By 'acausal' I mean 'not committed to causal discourse'. In the process of our study we critically overcome the dearth of the general-to-particular approach in contemporary ontology caused by its near-total lack of self-enhancing comprehensive ontological systems of thought in 20th and 21st century philosophy. A critical study of the analytic tradition will facilitate the finality of the roots-in-the-skies approach that we lack.

Chapter 2 shows how, for the process of thought, the particular and the general are couched in connotative universals and Reality-in-general, the latter of which is the connotative generality of generalities that couches in human thought natural kind[226] particulars and ontological universals through

---

[226] The Neo-Platonist and Aristotelian logician Porphyry (c. 232–304) has systematized the Aristotelian categories / predicables of substance in his *Isagoge* ("Introduction"). The system is called the Tree of Porphyry. According to medieval philosophy's appropriation of Porphyry's division of predicables and non-predicables of the types of terms called in Latin as the *quinque voces* ("five words"), what we understand today as natural kinds may be thought to comprise the first two types of the terms, i.e. genus and species, which are predicables; but not the other three types of terms, i.e. difference, *proprium* and accident. As is common knowledge, a genus may be a species with respect to genuses broader than that genus; and a species may be a genus with reference to the sub-species of that species. A natural kind that is not the narrowest kind is a genus. A natural kind that is not the broadest is a species. A genus may be a species with respect to genuses broader than that; and a species may be a genus with reference to its sub-species. A *proprium* is never a species. It is co-extensive with a species proper to it. A difference is the difference of the species with its genus. It is essential only in such juxtaposition. So, it is neither a natural kind,

connotative generalities. Natural kind is a general term that comprises the notions of genus and species, which may be taken also as generalities and particulars – in short, generalities of various degrees. Reality-in-general, the most general case of a genus (generality), is, therefore, also a natural kind. Natural kinds keep within the access of human mind, and are within nature. They are not supra-natural concepts. They are necessary in discourse. But Reality-in-general is just the ideality of connotative universals. Hence, it is justified to conceive Reality-in-general as the ideal case of connotatives the notions of all natural kinds together. They are not the ideal case of tokens. The ideal of this latter is Reality-in-total. In Aristotle, the concept of substance is not that of a token. It is a general term (genus, natural kind) for all sorts of substances. Reality-in-general is epistemologically (as happening in consciousness) the proper predicable of Reality-in-total. Ontologically, To Be is the predicable proper of Reality-in-total. Hence, the highest token is Reality.

The truth-probabilistic aspect of knowledge and connotative universals is also taken care of by instilling the implications of Gödel's result in the very nature of connotative generalities and system-building categories. The probabilistic characteristic of connotative universals allows us to see why no system is conceptually fully self-consistent, and how to build systems of ever higher epistemological truth-probabilities by inductively generalizing unto the most general universals and represent them connotatively as the instantiators and potentials of the particular in the mind. Connotative universals are universals as reflected in conscious attempts. This means that ontological universals as ways of being of processes are not the same as connotatives and *the probability of connotatives in their representative character is not ontological but epistemological of Reality*. Thus, *the present chapter treats the epistemology of ontology*.

The attempt is also to record ontological universals, i.e. objectual-causal routes of processes in Reality, in laws of nature and connotative universals / essences, the latter being acausal entities and veiled reflections of objectual-causal ontological universals in thought. I qualify them for higher and higher truth-probabilities in systems by involvement of the connotative category of Reality-in-general (which represents the ideal of thought), along with that of the denotable category, Reality. An elaborate purely ontological interpretation of universals as ways of being

---

nor co-extensive with one. Accidents are not essential to actual things (tokens). *The Cambridge Dictionary of Philosophy*, second edition, s.v. "Accident", "Predicables", "*Proprium*" and "Tree of Porphyry". The Aristotelian categories / predicables were called *Praedicamenta*.

of processes will be made mainly in Chapter 3. This is presupposed in Chapter 2, since we have defined ontological universals previously.

2.1 Transcendental Categorial Orientation in Epistemic Actualities

2.1.1 Laws of Nature and the Epistemology of Essences

Questions of human knowledge have been answered variously in the history of western philosophy. One useful analytic way of answering throughout the two and a half millennia has been to posit things simply, commonsensically and objectively in the direction of thought-to-reality correspondence; and another way was to follow some form of idealistic coherentism. The history of long prevalence of various permutations of such indefinitely growing number of theories is well-known. One common set of threads that run through them is the particularism in the connections (1) between the conceptually connotative and ontological constitution of universals that belong to the ways of processes, and (2) between the meaning of particulars and the connotatives that we find to be both derivative and constitutive of concepts.

Today no one seriously asks the knaive sceptic question if knowledge is possible. It suffices to sum up the sceptic question of the possibility of knowledge and base it on "justified true belief"[227] based on law-like expectations, and by saying that the sceptic question is a pseudo-problem because, to constitute a genuine problem, (1) the question must give rise to at least two absolutely incompatible answers and (2) one should be consistently unable to determine which is correct

---

[227] All knowledge base ultimately on beliefs in certain cases justified by the law-like expectations supposed or concluded to surround them. Johnson alludes to "[…] the common-sense definition of knowledge as 'justified true belief' […]. For the ordinary person, knowledge is a type of belief, specifically a belief that fulfils certain conditions. The emphasis on belief reveals his practical interest in knowledge; he wants to be able to make distinctions among his beliefs that will allow him to arrive at successful decisions about how he should act. When he goes on to say that a belief, to be knowledge, must be true he usually has a simple model before his mind. His belief is about something; it constitutes a conviction that a certain state of affairs is the case. To be true the belief must correctly describe the state of affairs; it must mirror or reflect reality." Johnson, *The Problem of Knowledge*, 5. Clearly, there is a law-like background that facilitates the jump to name some belief knowledge.

of the two answers.²²⁸ These are not the case in scepticism. I omit the pseudo-question of knowledge. Instead, I stress the need of essences in ontological discourse and of the very rationality of essences, because the question of knowledge depends on claims of generalities or essential attributes that arouse in the person's belief law-like expectations. The questions of universals / essences has been hitherto been sand in the shoe in ontology and epistemology. We discuss the issue through the eyes of opponents of the concept of essences and pave the way to understanding the place of essences in laws of nature of all generalities.

Van Fraassen's "Essences and Laws of Nature" examines the concept of 'essences' through those of "counterfactual conditionals"²²⁹ and "laws of nature", and argues that "[…] the supposed objective modal distinctions drawn are but projected reifications of radically context-dependent features of our language",²³⁰ attempting thus to ferret out what he supposes to be the truth behind 'essences'. He begins by discussing counterfactual conditionals. Not merely universal generalizations, but in fact laws of nature and universalizations through laws of nature, yield some sort of warrants,²³¹ as in the following true singular counterfactual that expresses a law of nature: If this nickel coin were heated to n degree Celsius it would have melted, *because* nickel's melting point is n. He says that a physical necessity – not merely the object's being nickel (quality of / having the universal quality of nickel), but the necessity of its being nickel for it to melt at a certain melting point – works as a law warranting the truth of the counterfactual.

---

228 Johnson discusses this thesis at length. In the original: "For a genuine problem to exist there must be an issue or question in dispute, to which it is possible to give at least two mutually incompatible answers (*i.e.*, answers such that no more than one can be correct). In addition, it is necessary that those who address themselves to the question be unable to establish which of its possible answers is the correct one; otherwise the problem would be solved." Johnson, *The Problem of Knowledge*, 2.

229 When the if-clause of a conditional statement is in fact counter to fact, but is prefixed to a consequent, this argument uses certain additional premises as implicit background conditions and laws of nature. Such conditional propositions are called counterfactual conditionals. *Routledge Encyclopaedia of Philosophy*, second (1999) edition, s.v. "Counterfactual Conditionals".

230 Van Fraassen, "Essences and Laws of Nature", 189–90. Modal logics think in terms of necessity and possibility in possible worlds: "… [A] proposition is *necessary* if it holds at all possible worlds, *possible* if it holds at some." Chellas, *Modal Logic*, 3. Van Fraassen reduces possible worlds into contexts and objective modal distinctions into context-dependent features of language.

231 This he finally fails to explain the nature of, or explain away, since he has a non-objectual concept of causality and determinism.

If one gives up insisting on the natural law, then one uses attributes[232] taken to be essential and formed over law-like conclusions, as in: "This coin is essentially nickel, that is, it is nickel in every possible world (and hence, in every physically possible world) in which it exists [as capable of melting at a certain essential melting point]."[233] What it means is a law of nature: "This coin would melt in any circumstances in which it [exists and] is heated to n degrees, *because* this coin is necessarily made of nickel [if it exists at all], and all things made of nickel necessarily melt if heated to n degrees."[234] The concept of possible world is here an instrument to define possible and necessary truths.

Side by side with this discussion, a few comments in line with the thrust of Chapters 1 and 3 are in place. I call essences as nominal generalities (nameable qualities) which are in fact adjectival of processes capable of being re-formulated also as laws (i.e., nomic in quality) that pertain to entities / processes in their own right, independently of minds, but basically as strains or ways of being within Reality. Susan Haack argues: "… [T]he reality of kinds and laws is a necessary condition of successful inductions." For this she gives the following explanation:

> There *is* a connection between induction and natural kinds. My argument derives from Peirce, who held that the reality of 'generals' (meaning, roughly, 'kinds / laws') is a necessary condition for the possibility of the scientific method – for explanation, prediction, and induction. If there are natural kinds, the thought is, there are clusters of similarities holding together in a lawful manner; so the fact that observed things of a kind have had a certain feature is some reason to expect that other, unobserved things of the same kind will also have that feature – for this may be one of the properties in the knot tied by the laws of nature."[235]

For example, it can be a law that a certain species has a certain quality, stated propositionally. These, I shall show in this chapter, are not referents but ways of being resulting in general and epistemologically objectual-conscious abstractions, which strain nomically and ontologically from Reality into consciousness through conscious and less conscious activity, and do not merely occur in language from nowhere. They occur ontologically in Reality and are epistemologically active by veiled connotative reflection in thought. That is, nouns or adjectives can represent generalities. John Carroll calls them roughly as "nomic concepts"; and some epistemologists and logicians call them "natural kinds", which indicates

---

232 'Attribute' is a general name for properties, natural kind qualities, and relations.
233 Van Fraassen, "Essences and Laws of Nature", 190–91. My square brackets clarify the intent of the preceding phrase.
234 Van Fraassen, "Essences and Laws of Nature", 191. Square brackets in the original.
235 Haak, *Evidence and Inquiry*, 134.

a certain definitional characteristic that qualifies a class or species of tokens. A natural kind may be considered as any term that classifies groups of tokens. Further clarity on nomic concepts in Carroll is in place here.

> Because these concepts have almost universally been recognized as having a modal character and as inappropriate for use in a definition intended to tame the modality of lawhood, I give them a special name. I call them the *nomic concepts*. Be aware that the counterfactual conditional, lawhood, causation, etc., do not quite exhaust all the nomic concepts. Made-up notions explicitly defined in terms of the concepts just cited are also nomic. For example, [...] the nomic notion of *lawful sufficiency*: *P* is lawfully sufficient for *Q* if and only if *P* physically necessitates *Q*. There are also some ordinary nomic concepts that I have left off my list, ones that are very close cousins of the key nomic concepts; for example, *production* (a close cousin of causation) and *nonaccidentality* (a close cousin of lawhood).[236]

A natural kind is "[...] a group of objects which have some theoretically important property, or properties, in common." "Such common properties form the 'nominal essence' (Locke) or 'stereotype' (Putnam) associated with that kind." One takes the underlying set of properties (e.g. the deeper chemical structure of chemicals or the genetic structure of organic substances, including those currently unknown) to be the 'real essence'. Kripke and Putnam further argue that "[...] it is (metaphysically) necessary that something is a number of a given kind if and only if it has a certain real essence [...]". They conclude that "[...] there are certain *de re* necessities which can be discovered only empirically [....] Further, other philosophers have since argued that "[...] essentialism about natural kinds entails that the laws of nature which govern kinds are also *de re* necessary."[237]

Natural kinds may as well be stated in propositions. Propositional generalities (universals about facts and thus of states of affairs) tend to be general rules or laws of nature that are *finally*, concretely, but probabilistically (inductively in theory), relevant for theory *about* Reality. Universal generalizations and nomic generalities (laws) expressed in propositions are only circumstantially different in relevance with respect to measure of warrant of truth-probability: the one is modally contingently true, and the other modally by necessity. The necessity here may be expressed in terms of all possible worlds, too; it can be just true in the actual world too, in the sense of being realized.

---

236 Carroll, *Laws of Nature*, 7. The term 'nomic' derives in philosophy from the Greek *nomikos* (adj.), *nomos* (n.), "(of) a (discoverable scientific or logical) law (adj.)."
237 *Routledge Encyclopaedia of Philosophy*, first (1998) edition, s.v. "Natural Kinds" by Chris Daly.

> Laws have a *modal character* in that not every true proposition, not even every true universal generalization, is a law. For example, [...] I placed two nickels in my pocket. Because those pants will be destroyed in a fire tonight, those are the only coins that will ever be in that pocket. Then, there is the true universal generalization that all the coins in my pocket are nickels. Though perfectly true, this proposition is not a law. It fails to be a law because its truth is an "accident"; it is *accidentally true*. In contrast, consider Newton's first law of motion [....] Assuming for the moment that it really is a law, this Newtonian generalization is *not* accidentally true.[238]

The modal character of laws makes laws work in principle at necessity, not at possibility, i.e. not as accidentals. In short, laws are supposed to be the case in all possible worlds ("necessary"), and not in some of them ("possible"). The 'possible' here concerning natural laws presupposes the essential nomic nature of all the universals that go into the formation of a concept. Without concepts being nomically universal-laden, laws cannot be nomic; and laws are nomic concepts as they are named, i.e. nominals. Universals are nomically universally realized in all sorts of concepts. So, laws and terms of concepts have something nomic in common.

Laws and names require a criterion of relevance which, coincidentally, works as a criterion for possibilizing truth-probability about veridical perception, from which we might further know the ontology behind the epistemic nature of connotative (consciously working) universals that are ingredient in notions. For convenience and for the sake of distinguishing direct concepts from direct and indirect universals, I correlate them with perception as non-veridical (direct perception, sensually and contactually attained) and veridical (working towards the propositional level, resulting in the full sense of nomic concepts).

Renzong quotes L. Jonathan Cohen's condition for relevance (which I take also for truth-probability of reasons): "We say that a proposition R is relevant to the question Q if and only if R gives us a reason to accept or reject a proposition A as an answer to the question Q."[239] This condition for relevance is a generalization – like any others – which may also be stated as a law of nature within the purview of a given context. It should be noted that R is a propositional generalization, perhaps an inductive generalization, which used to be understood as a static essence. Nevertheless, we used to create nomic essences, which are often misinterpreted metaphysically. *If nomic essences are interpretable also as propositional ones, then why not use also nomic essences, instead of laws or inductions,*

---

238 Carroll, *Laws of Nature*, 1.
239 Renzong, "How to Know What Rises up Is the Moon?", 65.

by admitting that essences are generic theoretical niceties very much on par with laws as essences?

One can apply the condition for relevance (for truth-probability of reasons) also to the marginally more concretist cases of laws of nature implied by singular counterfactual statements (which present possibility through actuality / states of affairs, counterfactually). Application of the given condition for relevance is to be had only by stretching the very notion of essences, using their own depth-level condition/s for possibility (first level *a prioris*), condition/s for possibility of *a prioris*, and so on. The possible last of this train I call Reality-in-general – the generality of all possible generalities of generalities … – for which the ontological condition for possibility is Reality, which we have called the Transcendentally Transcendent category. It is Transcendentally Transcendent because it is characterized in theory by Reality-in-general which is the Transcendental of all transcendentals / generalities / essences. Counterfactual need not be actual. But if the Einaic categories are at the foundation of all processes and concepts, then notions and laws of counterfactuals too must be imbued by them – if not at the dimension of reality and actuality, then at least conceptually. Thus, everything counterfactual need be real and the reat are non-real in the sense of not being able to be realized in any manner, yet having real foundation in the formation of their formative notons.

There are connections between generalizations expressed in propositions and those in their sub-cases of nomic essences. Van Fraassen approaches the problem of essences by analysing the connection between singular counterfactuals (which present possible or necessary states of affairs while formulating the statement based on actualities at hand) and essences (which are propositional laws of nature stated as essentials / essences). Thus he questions the very need to talk of essences. He doubts

> […] whether the laws of nature of a world automatically contribute to the essence of things in it. […] Suppose that this coin is essentially nickel, that is, it is nickel in all possible worlds in which it exists. In addition, suppose that it is physically necessary that nickel melts at n degrees. It follows that if this coin were (to exist and) [to] have that temperature, it would melt. But does it follow that this coin has, not just physically necessarily, but essentially, the property that it melts if heated to that point?[240]

I submit that it is the ontological essences of processes (things) that contribute to our formation of the laws of nature; but laws of nature as formulated by mind are not the same as the ontological "there" of these laws. And the laws of nature

---

240   Van Fraassen, "Essences and Laws of Nature", 191.

as ontologically committing to there being ways of processes in a certain manner are already essences of processes. In the light of this argument, the question of the essentiality of laws of nature cannot be properly understood without coming to grip with the inductive-probabilistic status of laws of nature, connotative universals and their mutual connexity via universals of universals. (e.g., a generality about particulars / species / natural kinds which are all processual and their generalities imply involvement of broader universals; note that a particular is not a token process but a group of processes), and with ontological universals that are relationally active in processes and states of affairs. If a law of nature were directly derivable from a state of affairs, then it would have been with certainty. Then the essence derived thus would have to be absolute, metaphysical and non-relational beyond its own purview since it envisages no broader processual relations. Insofar as each derivation is further qualifiable with respect to other causal influences from beyond, the law of nature associated with one state of affairs and derived by implying causal universals of universals (which may be called qualities of qualities) and causal universals ingressing in processes in which other universals ingress is also inductive-probabilistic with regard to their connotative reflection (though not probabilistic in their ontological ingression in processes). That is, other similar or broader states of affairs, and then others etc. must be generalized to laws using universals in conceptual processes; and universals are in fact the causal ways of being of processes.

Though van Fraassen wants to do away with essences by using his non-determinationistic concepts of causality and laws of nature ('determinationism' in fact means non-causal or acausal in all its meanings and implications, and is distinguished from classical determinism and non-determinism), *his present intent of associating essences with laws of nature seems to me to be in place*. I would keep their alliance intact in order to justify nomic essences, which of course is contrary to his intention. If the connection between laws of nature and causality is also kept, then there should be some ways of connecting them. *These ways, at least for their partial reflection in thought, are to be called ontological connotative essences.* For this I relativise the two concepts from the objectual-continuity point of view of causality and determinationism. We discuss the continuity point of view of causality – derived from the QM- and cosmological demands of ontology and made epistemologically adequate – after discussing Carroll's concept of the same.

At the start of this chapter I quoted van Fraassen reducing the supposed objective modal entity called essences into projected reifications of radically context-dependent features of our language. The question he fails to answer is as to why they should be reduced into the linguistic context or the context of language, and not to the total context provided by Reality, the ground that demands the use of connotative essences of notions in discourse. *This latter connotative context of all contexts, which need not at all be reified for talking about essences, I call Reality-in-general*,[241] the connotative essence / universal of all essences, or conscious universal of all universals, active in laws of nature and in conceptual formulations of nominals of nomics / laws.

'Talking about' and 'reference' may be differentiated, the one as with a purely intentional commitment and the other as with a purely ontological commitment. To this effect, Rorty makes the distinction: "The clash is produced by the equivocity of 'refer'. The term can mean either (a) a factual relation which holds between an expression and some other portion of reality whether anybody knows it holds or not, or (b) a purely "intentional" relation which can hold between an expression and a nonexistent object. Call the one 'reference' and the other 'talking about.'"[242] I would re-interpret it: Reality-in-general is used to 'refer' in the mixed sense of both 'reference' and 'talking about', because objectual ontologically-committed referentiality yields 'talking about' in statements of even meagrely relevant truth-probability. Mere 'talking about' yields no objectual reference, which yields the criterion for error. I believe, therefore, that Rorty does not take into account truth-probabilities as the truths that need to be accounted for through the objectual referentiality of processes, which have connections due to ontological universals. Hence, the latter part of the disjunctive classification serves only for non-ontological, purely linguistic and radically subjectivist ratiocinations.

These are the universalised forms of laws of nature. If we can speak of laws of nature beyond the context of language and concepts, and base them causally on actual processes, we may speak also of ontological essences and base them on causes of any extension-motion extent. Its highest limit is Reality, and the

---

241 McKeon contends that "[e]xperience can be treated in many ways as a basic metaphysical concept." McKeon, "Experience and Metaphysics", 87. "Experience [...] includes the whole scope, from initial matter and beginning to ultimate accomplishment and ulterior ideal, of our thoughts, actions, and arts – all of our sciences, institutions, and values, together with the data and hypotheses, the communications and purposes, the cultures and expressions by which they operate. Experience is a continuum in which differentiation occurs as a result of actions …" McKeon, "Experience and Metaphysics", 88. In this context, Reality-in-general may be considered to be the epistemic ideal of experience, as it may be applied in all sorts of ontological exercise. It involves essences conceived as the very ideal epistemic formulation of laws of nature.

242 Rorty, *Philosophy and the Mirror of Nature*, 289.

conceptually grasped essential generality *par excellence* of the context and limit called Reality is the hghest connotative, Reality-in-general. That is, the scientific necessities of laws of nature, their formulation in counterfactuals etc. have finally resulted in extending the context of essences from language to Reality.

David Lewis has coined the term (sensible) 'counterlegal counterfactuals' for counterfactuals whose antecedents deny laws of nature. For example, "'If the world had been Newtonian …' or, 'If the melting point of nickel had been q degrees …' … or, 'If the strength of the gravitational field were to diminish drastically …'"[243] Formulation of the antecedents of a true counterfactual will imply the possibility of more correct concepts of instantiation and truth of "possibilities", if they are formulated to instantiate higher truth-probabilities. We do not bother here if counterfactual tuths are real *and* actual in the sense of being of the actual world. Our discussion dwells now at the realm of the initially formulated connotative generalities / universals that we use for counterfactual formulations. These universals have all been inherited from the actual world, and have some relevance to the laws of nature of the actual world. Universals are either just conceptual or also propositionally of ontological universals – all of which have inductively probabilistic relationship with the ideal of truth, Reality-in-general.

In short, the inductively probabilistic nature of the truths of laws of nature is a key to determining the connection between the physical necessity of the antecedent and the essentiality of the qualities implied in the nomic essences of the physical necessity which we use to name the physical necessity expressed in the respective formulations of laws of nature. Moreover, as seen in 1.4.1, objectual probabilism – which is minimally about any processes and maximally about Reality, and at the same time is not probabilism about the existence of the relevant causes whatever they are – is irreducibly based on iheritances of causality over ways of being of processes from many other past processes and finally from the Transcendent, Reality. Thus, not only laws of nature but even universals / essences are in fact inductive-probabilistic in their formative definition, based on Reality, and the former may be formulated in the shape of the latter, based on their objectual-causal appeal to Reality.

If we take laws of nature in their ontological rather than purely propositional sense, then causes and laws of nature will be seen intertwined and every law will be a statement of causal determination. The fact that every law of nature (except purely categorial ones in the sciences) is a law within a particularized context (which is set not merely in extension-motion but also by epistemic

---

243 See Van Fraassen, "Essences and Laws of Nature", 191.

considerations of apparatus-wise intellectual approach / extension / exhaustion) shows the supremacy of (objectual, infinite, infinitesimal) causality in Reality as a law over particularized contextual laws of nature that are prone to perspectival absolutism. But van Fraassen notes his antagonism to essences in the following manner:

> Assuming also that this object is necessarily or essentially nickel, I will reach conclusions about the properties it has with physical necessity. If essence is independent of physical necessity I won't know at all yet whether those properties are also essential. Hence the route of empirical inquiry is blocked. We might define the *sub-essence of* x *in a world* as the totality of properties implied, via the physical laws holding in that world, by its essence. Empirical access is then possible directly to the sub-essence only. The other alternative, that if some property is physically necessary in this world, then it belongs to the essence of everything in it, is therefore much more appealing. It removes that supposed limit to experimental knowledge about the essence. And indeed, I have found, upon reflection, the most marvellous possibilities for a physics based on essences if that alternative is embraced. As is well known, our world is not deterministic. This raises special problems for causal explanation for us.[244]

The absence of what I call causal-determinationism in van Fraassen's worldview is clear here. Van Fraassen proposes a sort of instrumentalistically vacuous physical necessity that is neither epistemologically deterministic nor ontologically causal-determinationistic. I presume that he is inspired by some moments in QM where causation used to be put to question. In Chapter 1 partially and in the whole of my *Causal Ubiquity in Quantum Physics* I have shown that QM need absolute causalism / causal-determinationism and that epistemological determinism involves absence of commitment to causation at some moments in nature.

By rejecting determinationistic causality behind an event, he means to overthrow the concept of essence as derivable from laws of nature. Exactly this sort of causality is what we prefer to transform under the concept of objectual-probabilistic realism, determinationism and ontology, in view of the rational facilities drawn up in Chapter 1 – i.e. into Einaic Ontlogical objectually determinationistic causality, by which Reality is the Transcendent object within which a particular, actual, causal event has lawlike bounds. These bounds are exhausted only if causalities from all the infinitesimal depths of all that have gone causally into the formation of an event are determined, or, at least, admitted in ontological commitment – the latter of which alone is humanly speaking possible. The fact is that complete conceptual representation or determination of all ontologically causal

---

244 Van Fraassen, "Essences and Laws of Nature", 192.

determinations transmitted by ontological universals via connotative universals is impossible. This fact does not prejudice objectual-probabilistic Einaic Ontology against just admitting a thoroughgoing activity of causality at all the infinite and infinitesimal causal dimensions (determinationism) in Reality, because (1) any other manner of explanation does away with the objectual nature of Reality and reality-in-particular and leaves us in mere Humeanism or Berkeleyanism, and (2) the objectual nature of Reality and reality-in-particular demands that there be infinite and infinitesimal causal processes in Reality not admitting epistemic exhaustion of them all, even the whole causal history of a wavicle.

The above implies that transcending the epistemically causal sort of determinism and perspectival absolutism is had best not merely by allowing commonsense existence of entities "out there", nor by making knowledge objective, but by admitting, as the foundation of both these, the objectual independence of Reality and reality-in-particular together as an infinitely and infinitesimally processual-causal, irreducible, ontologically theory-bounded, state of affairs. Ontological theory-boundedness is the cosmologically ontological universal-boundedness of the purely objectual Reality. This was the result of our discussions in Chapter 1. The epistemologically ontological aspect of it is the very aspect of connotative universal-boundedness under the aspect Reality-in-general, as seen from arguments throughout the present chapter. At the same time, it *does also prejudice* ontology against epistemological perspectival determinism of current level of epistemic exhaustion of information, which latter incurs phenomenalism. The continuous nature of the infinitesimal in Reality ill-disposes it to admit epistemic determinism which is perspectival absolutism and whose ontological concept of causality is purely acausal (i.e. purely conceptual, conceptually divorced from and non-committed to what occurs causally in things), epistemological (i.e. yielding absolutism of the given perspective) and does not encompass the ontology of actual and potential causal processes in things.

If one should speak of any law, then one should also speak of acausal universals, i.e. conceptual definitions of ontological properties in terms of general causal processes in processes, i.e. definitions of properties which (definitions) forget the exhaustive causal connections in their connotative appearance and formulation in mind and language and thus give a circumscriptively exhaustive impression. Causal processes of one and the same "spatio-temporally" or perspectivally circumscribed extension-motion realm are never fully determinable in all their extensions from within the respective extension-motion or perspectival extension of epistemic exhaustion, because the universals involved in the respective particulars and natural kinds are never fully confined within the given extension-motion and perspectival extension. Hence, any specific universal or law of nature is never fully determined or defined from within such extension. They have to be connected to ever-broader processes and universals.

This fact relativises the concepts of laws of nature and universals by the broadside of the spectrum of ever-broader universalities. Both are relativised; both have the same general field of abstraction; but the one is ontologically propositional (yielding a relation between two actualities) in effect and the other nominal / adjectival (yielding attributes universal to the many). By reason of the acausal reducibility of relations to attributes,[245] we may say that these are equivalent in intent and that both are relativisable by the broadside. This shows also the truth-probabilistic nature of true propositions, because ontological universals are always probabilistic in their connotative conscious instantiation and also relative to ever-broader ones. This allows space for truth-probabilities even in counterlegal and quasi-counterlegal counterfactual propositions,[246] which are based for their formation in ontological and connotative universals from within Reality. Ontological universals are not probabilistic; connotatives are.

If counterlegal and quasi-counterlegal counterfactuals are partial feignings of legal propositions, then most counterlegals constitute an unwarranted extension of more or less true logical or modal propositions in different circumstances. Therefore, the laws of nature and universals instantiated an instantiable in the antecedents of circumstantially true counterfactuals or factual propositons are themselves more of ontologically fully but connotatively probabilistically instantiated or instantiable, but ideally (acausally) formulated, universals (essentials) of actual events. That is, *ideally formulated connotative essentials of actual events are indispensable instruments to theorize.*

There is nothing just miraculously and vacuously produced between the causalities of the physical world, (1) since there is no absolute warrant (physical necessity) or absolute absence of warrant prescribed by perspectives – except the axiomatic or categorical necessity for purposes of science – and (2) since Reality is objectually causal in all its infinitesimal and infinite recesses. Reading these

---

245 Chisholm, *A Realistic Theory of Categories*, 45–49.
246 For comparative clarity of dependence on reality, of truth-probability concerning (1) counterfactuals, (2) counterlegals and (3) quasi-counterlegals, the following examples (true of but not "realized" in possible worlds), respectively, are of help: (1) 'If I had been the Minister of Education, I would have …', (2) 'If I had been an eagle, I would at least once have flown high enough to see the Everest', and (3) 'If the owl had acquired enough experience to click the mouse button of this computer now, it would have learned the essentials of computers'.

two facts together, we need a concept of causality that demands ideal (acausal) constructions in theory. Laws are themselves results of such perspectival ideal constructions. Van Fraassen explains physical necessity via a negative criterion for physical possibility: "If it is physically possible for x to exist under conditions Y, but x would under those conditions not be F, then being F is not part of the essence of x."[247] Though this is necessary, it need not be sufficient for absolute physical possibility. Here, therefore, one can be unaware of objectual ontological causality.

Van Fraassen's concept of conditions is just epistemically perspectival, not explained in terms of the real physical constitution of causal connections within physical events, and between the conceptual-linguistic expression and the physical event/s, under the perspective of causal issuing of universals from within the totality of all possible relevant causes upon reality-in-particular from within Reality. It does not have the ability to track any allegedly non-causal issuing of universals into the formation of any process. Nor is it capable of safeguarding physical-ontological influences within the infinitesimal (which is finite in any given spatio-temporal extension), and from a finite portion of the infinite recesses of the processes of Reality, by any other means. Thus, methodologically, van Fraassen fails at the very outset of his attempt to posit a purely non-deterministic (better, I name it a "non-determinationistic") and non-causal concept of causality, and thus to do away with theoretical uses of ontological universals and connotative essences.

Thus, essence will be defined in such a way that even being some F, which are not contrary to conditions Y but are necessarily implied by them, are included in x's *existence* over causally ontological commitment. Some F will be implied in x's existence under conditions Y. *The extra-remote conditions F, which play in consonance with the Y, are in fact pointers to the involvement of causal universals – ways of being of processes – different from those immediately involved. These point to the causally real internal constitution of the processes actualised in the past (universals of the causal past), and will be causal in some way in the future (universals of potential future).* Their admission as fact is what makes the definition necessary and sufficient, by summarily and necessarily implying causally deeper and boader universals at play in the very universal / attribute at issue, because no universal has its own existence without ever deeper and broader ones, which alone dispose them from the past of processes. This points to the possibility of formulating a spectral theory of necessity and possibility with higher and lower values of both in any truth or universal.

---

247 Van Fraassen, "Essences and Laws of Nature", 193.

The F are nothing but forms of existence of x abiding by *conditions internal to Y*, by reason of universal causally formed from the infinitesimal recesses of reality-in-particular and Reality, which express themselves internally but hiddenly in Y. Entities x are in fact reality-in-particular. They are such by reason of conditions internal to Reality, some of which express themselves hiddenly in conditions Y. Thus, relevant causality that are from the finite but ever-widening extension-motion recesses of Reality – recesses immediate to reality-in-particular (here, x) – should be theoretically included in the definition of the inner possibilities of x's existence under conditions Y, if the definition is to be adequate. This is the Einaic Ontological way to explain causal conditions and essences, where each universal or truth has its own spectral value of necessity and possibility.

This means that the sub-essences of an essence / universal are not obtainable merely by mathematically probabilizing physical necessities or laws given by a certain empirical perspective. They are entirely dependent on the wider number of processes, and finally on Reality and the connotative and ontological universals of all universals that pertain to Reality. This conclusion about nomically formulated essences and sub-essences shows their function beyond the immediately given in any current perspective and in reality-in-particular. *They tend to generalize not merely based on the perspectival hinterland and vicinity, but also based beyond the empirically physical immediacy of reality-in-particular. This yields genuine idealization, always based on the spectral values available for necessity and possibility in truth-probability.* No more is possibility alone present in any proposition; nor necessity alone, since the possibility and necessity obtained is merely from a perspectivally fettered extension-motion neighbourhood. Things perspectival within the extension-motion neighbourhood of reality-in-particular do not exhaust what are nomically brought together in essences.

Carroll does away with the "metaphysically significant" definition of laws using his concepts of nomic and non-nomic commitments in epistemic notions. Presupposing for our purpose that laws and causes are connected via ontological universals – which are for us the ways of being of processes –, noting that modals can come only in spectrally valuated mixtures of possibility and necessity (along with contingency and other modals), and keeping in mind also the resultant Einaic revamping of all that is empiricistic, we read Carroll introducing the epistemic notions related to laws of nature and the necessity of causality for any limited or wide perception and cognition:

> Returning to the question of what concepts can be used in the definition of 'law of nature', let us consider another concept: perception [....] [M]ost philosophers [...] have drawn the same conclusions about some other notions like action, reference, and even

such a basic metaphysical notion as persistence (identity over time). The modal character of these concepts is commonly recognized because, as many so-called causal theorists have convincingly argued, there are some easily specified and extremely plausible connections between these assorted concepts and causation. For example, with regard to perception, it is clear that nothing perceives anything else unless there is a causal connection between the perceiver and the entity perceived. Regarding persistence, no single material entity exists at two distinct times if there is no causation linking an entity that exists at one of those two times with an entity that exists at the other time.[248]

This way of connecting causes and laws of nature through perception (and also action, reference, etc.), themselves combined with the trans-concretist function of nomic concepts, works against van Fraassen's and Carroll's concepts of causation and perception of causes. Even in perception and in all that result from epistemic / conceptual activity, there are strains of causation that ingress into every one of their aspects, say, the F (that, as I have proposed, are *necessarily* implied in conditions Y), which an x can be when x is under Y. *The Y are not a multiplicity of atomic conditions. Each Y is a result of acts of conglomerations of actual conditions accruing from within the respective reality-in-particular, which have causal connections also from without.*[249]

---

248 Carroll, Laws of Nature, 7.
249 Unaided by further analysis, one might remark how bizarre the ontology behind Carroll's concretist idea of "being a table" is in the following passage, if its ever-broadly ontological universals are not found to be active in the connotatively conscious concept of it: "What is not often recognized, nor its importance always appreciated, is the range of concepts with nomic commitments. It may be that this is often missed because the connections between many of our ordinary concepts and the nomic concepts are not always as apparent as is the connection, say, between perception and causation. Consider, for example, the mundane and ordinary concept of being a table. At first glance, one might think that nothing is a table unless it *supports* other things. But, this proposed connection is obviously incorrect; there are tables that were built, and then destroyed, before they ever supported a single thing." Carroll, Laws of Nature, 8.
    He continues in a different vein, as if he wanted to get every strain of causal involvement from beyond reality-in-particular immediately specified for one to determine what a nomic concept is, and what is not. Soon he veers clear of the concretism of the ontology his concretist epistemology presupposes: "A better, though probably not perfect, suggestion is that nothing is a table unless it is *capable* of supporting other things. It is not crucial, nor is it even very important, that we give a precise and interesting statement of the ties between tablehood and any of the nomic concepts. I suspect that the relationship between being a table and the nomic concepts is much like the relationship between causation and lawhood: Any very

Now we analyse the actual absence of anything of causality in Carroll's comments on these epistemic but allegedly causal notions, where he uses his distinction between nomic and non-nomic commitments:

> Following tradition, I do not count perception, persistence, or any of these other concepts as a nomic concept. Introducing some new terminology, we might say that though they are not nomic concepts they do have *nomic commitments*. It is the distinction between the concepts with nomic commitments and those without, not our earlier nomic / nonnomic distinction, that is metaphysically significant. Only a definition of lawhood that uses just terms free of nomic commitment could explain away the otherworldly character of laws.[250]

To explain away any otherworldly character of laws, he wants to do away with any nomic commitments of epistemic notions that involve causal notions. The best way to defuse the other-worldly instrumentalistic possible-world element in laws (e.g., claims in favour of trans-mental but non-existent "being given" of maximal sets of nomic concepts and truths *as natural kinds*) need not be divesting epistemic causal concepts of nomic commitment, if we understand 'nomic commitment' as necessary in cosmological and ontological talk of the phenomena-noumena continuity that Reality is. Any non-nomic commitment in laws should however be causal, as Carroll already agrees. Else, it would incorrectly have to be non-causal (not acausal, which "for the time being does not speak of causation").

If causal, it cannot be merely epistemologically or scientifically deterministic or mathematically instrumentalistic, because determinism and instrumentalism involve perspectival absolutism of the present systems and extent of knowledge. Perspectival absolutism blocks the progress of human knowledge, too. According to the extension-motion continuity-concept of causality, causes are always referable by processes, not by entities. Causes are always external to the effect, external to the extension-motion position of the entities considered as causes.

> interesting connection is difficult to specify. Still, as is the case with these two nomic concepts, it is easy to state a fairly uninteresting and weak connection. It is absolutely clear that nothing is a table unless it exhibits at least one dispositional property. Since no dispositional properties are exemplified unless there is also at least one law of nature, nothing could be a table unless there is at least one law." Carroll, *Laws of Nature*, 8–9. Although by being concretist, he does realize the most general, trans-concretist, implication of nomic commitment in the aforementioned passage. But he would further defuse nomic commitment from laws and nomic concepts, in order for him to divest laws and nomic concepts of 'metaphysical' overtones.

250 Carroll, *Laws of Nature*, 7.

Any cause within an entity is active towards other parts of the entity or towards further ones, not upon the causal part itself. So, a cause is external to those parts that are its effect. Such externality can go to the limits of Reality. *Thus, we must stress the causal effects, via ontological universals, from the past of deeper portions of Reality as active in actual reality-in-particular.* Therefore, natural kinds, which are reality-in-particular, are properly referable by nomic concepts only if ontological universals (included conceptually-connotatively in nomic concepts) proper to Reality are recognised within them. These are derived from causal processes of natural kinds and their properties from without the entities considered for causality and laws of nature.

Analysis of the properties of a natural kind allows formation of nomic concepts, and thus also of counterfactuals. *Insofar as nomic concepts are not equated with members of natural kinds, we are sure to evade Platonic essentialist metaphysics.* Natural kinds are particulars (not tokens), i.e. possibles with a certain measure of necessity involved. Mere possibility, as non-existent, is the limiting notion of all possibles. According to Armstrong, "A merely possible state of affairs does not exist, subsist or have any sort of being. It is no addition to our ontology. It is 'what is not'. It would not even be right to say that we can *refer* to it, at any rate if reference is taken to be a relation."[251] Mere possibility involves absolutely no necessity. The ontological commitment of a natural kind is not a mere possibility. It works in theory under the guise of the name of an actual token process which is active within the framework of all potentials for instantiation in tokens. That is, a natural kind that is the set of many token processes, is by its very fact imbued in ontological universals. If the process of instantiation is also naturalized, we do not have to reify the abstract universal / nomic as natural kind, but can refer it to processes / events in its ontological commitment, via propositional reference, and neither by Platonic correspondence, nor merely by coherence of the theory that has given rise to the respective laws of nature.

In short, *if we have a nomic concept (of a natural kind or essence) that refers only propositionally (i.e. through laws of nature), we can automatically have nomic commitment (i.e. commitment to laws of nature) through propositional reference, not through concept-to-object correspondence.* That is, propositional reference ("sense") is always ontological commitment to processes, because the propositional routes of formation of nomic concepts are always natural processes. This is not the variety of reference which is by word-meanings. Therefore, the potentials

---

251 Armstrong, *A Combinatorial Theory of Possibility*, cited in Lycan, "Armstrong's New Combinatorialist Theory of Modality", 5.

for instantiation as natural kinds, i.e. ontological possibles (essences active in matters of fact reflected by nomic concepts), are possibilities with a certain measure of nomic necessity. We do not reduce laws (by causality) into analytic individual essences / possibles / Platonic essences, but discover the universals, as word-senses, from laws of nature by sentence-senses that ontologically commit to processes.

Absolute necessity in nomic concepts is propositionally expressed as in the sole absolute identity: 'To Be is To Be', which deals with the most abstract but most instantiable case of *nomos*, "rule", in Reality. *The absolute ontological necessity of To Be in Reality differs from mere possibility. Mere possibility without necessity, reflected in nomic concepts via misplaced connotative universals, is roughly a* **source of error**. This is via transposition of nomic universal concepts to improper connotative reflections of processes, and thus through formation of improper propositional reference. So, any proposition that employs 'connotative reflections of ontological potentials for propositional instantiation', applying a certain mix of possibility and necessity, does not assign a Platonic, metaphysical, dualistic, correspondence for the nomic concepts of the proposition. Counterlegal conditionals employ at least one clause wherein at least one necessary possible (potential / universal) of the reality from which the counterlegal takes origin conceptually is not mixed with necessity.

Nomic concepts ontologically commit to ontological universals that are ideal causal (and what other is there, which does not pertain to the causal?) connectives of Reality and reality-in-particular. If divorced from reality-in-particular and Reality, they are pure, abstract objects. If not, they are theoretical niceties that are referent to processes in reality-in-particular, as also deriving from Reality. But they are not theoretical niceties corresponding to objects but to extension-motion strata of processes. The strata are not real unless their unision in a token process is supported by the unison of the corresponding ontological universals. Universals are possible, where necessity is spectrally appropriately incurrent. *Therefore, mutual confluence / collusion of modalities (i.e. of possibility and necessity) is the epistemological and ontological condition for the possibility of ontology that does not reify essences.* Pure potentials (ontological universals expressed in connotative universals ingredient in nomic concepts) are pure and abstract, insofar as they are isolable from processes. *The confluence of possibility and necessity is actuality derived*, since (1) from no token bereft of possibility is a natural kind derived and (2) it is not derived if it were to be derived without *always and in each instance* relating it to and from *many other* tokens that always involve extension of the processes of the tokens by both possibility and necessity.

That is, strains from and to Reality are always implied in tokens via natural kinds. So, nomic concepts are propositionally connective strains of possibility in actualities / natural kinds.

What are connecting strains? This may be explained in a manner that connects the ontological (entities made out of processes) with the linguistic and conscious (nomic concepts of processes), by naturalistically noting that the way of naturalizing epistemic activity is without success if they are not connected via Reality (i.e. by implication also via the infinite and infinitesimal causality in Reality). That is, Reality (with nomic ontological generalities within) is the Transcendent reason, which ontologically allows (i.e. it is the explanation of) the natural process of propositional formation of nomic concepts with nomic commitment. Confluence of possibility and necessity allows essence to be defined in such a way that even being some F, which are not contrary to conditions Y, but are necessarily implied by them, are included in x's existence *sub specie Realitatis*. This is so, because propositions about any process imply and involve possibilities and necessities from beyond and unto themselves, without which connotatives in discourse do not arise. Hence, nomic essence-terms and nomic laws, which are essential for theory, have their base in Reality, which is causally continuous and Transcendentally Transcendent.

No amount of reductionistic, eliminativistic, instrumentalistic, pragmatic, scientistic explanations of nomic entities works with the deeper issues (e.g. reasons for the physically and biologically developed intellectual vision in humans) involved in causality, except when the *concept* of (not our having reached all) causality is exhaustive of causal strains from the relevant causal recesses of Reality. The notion of the exhaustive and objectual causality of the infinitesimal and infinite in Reality determines all shapes of mutual influence of events over others. The ontological strains (nomics as ways of processes stated in laws) of causality that issue from events to events are *somehow* reflected in theory. These strains are ways of being of processes, i.e. ontological universals.

"A *theory* is a set of hypotheses positing […] entities and properties."[252] The concept of entities is in fact as a linguistically and logically reified form of events, and the concept of properties is as such a reification of causal strains from within and from without events. Ontologically, what is absolutely present and existent is not entities nor events, but Reality in and through token processes. This is because the To Be of Reality is the highest of all ways of being of processes. *Hence, theory about a few things is not merely of a few entities and events, yet, it is of them*

---

252   *The Cambridge Dictionary of Philosophy*, second edition, s.v. "Philosophy of Science".

within the context of nomic properties or laws, which are ultimately derived from events transcendent to reality-in-particular. The truth of anything, as the ultimate warranted assertability about it, is ultimate only when traced through continuous infinitesimal causality from within all relevant portions of Reality. What connect the truths of reality-in-particular with those of Reality are ontological strains / essences active as nomics / universals.

2.1.2 Transcendental Categorial Orientation in Particularism

The epistemology of the categorial orientation in the question of laws of nature may be understood only at an attempt to analyse the nature of universals within Reality. The tension is now between the epistemological aspect of the ontology of Reality and of reality-in-particular. The former includes the latter, which is augmented and brought to fruition by the former. Therefore, we examine one such epistemology of reality-in-particular and show how it can be brought to fruition by augmenting it by the epistemology of essences and by arguing that scientific ontology's naturalism is authentic only if causally characterized essences are brought into play.

Epistemological concretism of linguistic and scientific analyses exists in many analytic thinkers. Strawson gives it voice in his individualist descriptive ontology. According to him, there are three kinds of universals (general things):

> (1) Examples of the first class are such partitive nouns as 'gold', 'snow', 'water', 'jam', 'music'. These I shall call *material-names* and what they name, *materials*. (2) Examples of the second are certain articulative nouns such as '(a) man', '(an) apple', '(a) cat'. These I shall call *substance-names*, and what they apply to, *substances*. (3) Examples of the third are such abstract nouns as 'redness', (or 'red'), 'roundness', 'anger', 'wisdom'. These I shall call *quality-* or *property-names*, and what they name, *qualities* or *properties*.[253]

It is interesting to note that for Strawson all these universals have ontological commitment to tokens and/or kinds, which, by reason of what he calls "placing" or "feature-placing", are universal-bounded by instantiation. Note also that even near-tokens like a man [kind (2) above] are instantiations of natural kinds: 'a man' is materially near to 'the man Wittgenstein' in number, though one is indefinite and the other is definite. Discussion of the instantiation of kinds may be initiated by stating that in the case of kinds or generals the formula "the … of …" suffices.[254] This yields individual instances (some materials and substances),

---

253 Strawson, "Particular and General", 216. "Anger", a state, is in fact classified with the third, without detriment to the discussion. See footnote 7 of the same article.
254 Strawson, "Particular and General", 217.

whose notion is that of "[…] a logical compound of the notions of a feature and of placing."[255] In effect, this intertwines the concept of a particular with the concept of a general feature. The gist of Strawson's individualistic ontological concepts of particulars and generals is expressed thus:

> *It is a necessary condition for a thing's being a general thing that it can be referred to by a singular substantival expression, a unique reference for which is determined solely by the meaning of the words making up that expression; and it is a necessary condition of a thing's being a particular thing that it cannot be referred to by a singular substantival expression, a unique reference for which is determined solely by the meaning of the words making up that expression. This specification of mutually exclusive necessary conditions could be made to yield definitions by stipulating that the conditions were not only necessary, but also sufficient.*[256]

The necessary condition for generality is stipulated here in connection with instantiation in many particulars; and of course the particular is made inextricably linked to plural expressions beyond a singular substantival expression. The two conditions here are really commendable in what they would have accomplished. *The major defect I find in such an effort is the lack of continuity from the particular to the general*, and especially from one general to another and to a more universal family of generals.

If a singular being cannot be referred to merely by one singular substantival expression or concept, the reason for it is its processual connection with other entities in its proper past (other processes are in its present are not directly, but only through their pasts, connected to it causally). For this reason it must be ontologically agreed that the the existence and explication of particulars are dependent on those of the ever more inclusive whole. But see how Strawson's particularism or individualism goes counter to the real import of his own criteria in what follows. Strawson's thesis on objective, individualistic, ontology runs thus:

> We think of the world as containing particular things some of which are independent of ourselves; we think of the world's history as made up of particular episodes in which we may or may not have a part; and we think of these particular things and events as

---

255 Strawson, "Particular and General", 223. Simply, 'placing' is feature-placing in a context. See also Strawson, *Individuals*, 202–203. Aristotle speaks of 'connecting' as the epistemic manner, which may be interpreted as something similar to 'placing' and 'feature-placing'. Barnes, *The Complete Works of Aristotle*, Vol. 2, Metaphysics, Book I (A), 980 b 25ff., and 981 a 1ff. It is true that a specific context is a must for placing and connecting. Nevertheless, any particularism in it, forgetful of the possibility of the total context within its most general universal/s, disempowers discourse.
256 Strawson, "Particular and General", 228.

included in the topics of our common discourse, as things about which we can talk to each other. These are remarks about the way we think of the world, about our conceptual scheme. A more recognizably philosophical, though not clearer, way of expressing them would be to say that our ontology comprises objective particulars. It may comprise much else besides.[257]

See how the categorially objectual nature of the particular within the framework of Reality is being mistaken by taking the specific definition of it as finally objectual. If particulars and tokens must be objectual, the discourse of them must be bounded over universals to further reality-in-particular and to Reality. This must be the final implication of his criteria if pushed to all their implicatons, which he seems to forget.

In Strawson there is no serious talk of the train of ever-wider generals and totals. This I find is the major defect in his ontological accomplishments, which bars him from acceding to Reality-in-total, Reality-in-general and the To Be of Reality. His basic approach of limiting the horizon of tokens to particulars and generals is clear: "Given a true general-thing name, like 'gold' or 'wisdom', the question of the criteria of identity of its instances cannot be answered until the kind of instance is specified, by such a phrase as 'a *piece* of gold' or 'a wise *action*.'"[258] A 'kind of instance' is not a particular, but a contextual particularizer of a general, by "placing", and nothing more – understandable in the metaphysics-hostile context in which Strawson has brought in universals / generals. He shows how the contextual placing / feature-placing of a particular implies the use of generals:

> To recall, first, some vague, figurative and unsatisfactory terms I have already used: the schema suggests that the notion of a particular individual always includes, directly or indirectly, that of placing, whereas the notion of a general thing does not. Now placing is characteristically effected by the use of expressions the *reference* of which is in part determined by the context of their use and not by their *meaning*, if any, alone.[259]

The concept of a particular is such that it is infused with generals. But the concept of a general thing (natural kind) is not further placed in a broader kind. This is an arrest of the very movement of ontological commitment beyond terms and in contradiction to his admission of the theory-laden nature of tokens and particulars / natural kinds. Strawson's propaedeutic remark about the notion of

---

257 Strawson, *Individuals*, 15.
258 Strawson, "Particular and General", 227.
259 Strawson, "Particular and General", 228.

an individual or particular elucidates the theory-laden nature, the tangle with generals / essentials, of particulars:

> (1) The idea of an individual is the idea of an individual instance *of* something general. There is no such thing as a pure particular. (This truth is too old to need the support of elaboration.) (2) The idea of an individual instance of φ is the idea of something which we are able in principle (a) to distinguish from other instances of φ; and (b) to recognize as the same instance at different times (where this notion is applicable).[260]

This statement and the context of my remark that Strawson arrests his own openness to universals can open the door for inquiring into what exactly these ever-more generals are and how their limit mays be generalized. The categorially objectual nature of the particular and the token, which is possible only within the framework of Reality, is so by reason of the token's status as instance *of* something particular, which, again, is a similar instance of something more total by reason of the generals instantiated therein, and so on *ad infinitum / libitum*, and thus, as infinitely universal-bounded, and yet actually instantiated in tokens of a broader or limited kind of universal. This is in Strawson forgotten or relegated or downgraded, because Strawson takes the specific, singularly universal-bounded definition of the particular as finally "objective" – and his purpose of constructing an analytic ontology of individuals is served.

The objectual nature of the particular is generated solely within the framework of all instances of feature-placing (and nomic / acausal commitment) upon it within the backdrop of Reality. This may look far-fetched, but most important for success of theory. Such maximizing and infinitising "placing" is feasible only if it is made in more and more general universals and finally in the dimension of its purely Transcendental finality, i.e. To Be, which sort of wider placing may be done in an ontologically connotative fashion only by Reality-in-general, which must be the final dimension of infinity of infinity of … instances of feature-placing on tokens through ontological universals, and realized by means of connotative universals ingredient in universal concepts in mind and language.

Further, this latter is done in fact only epistemically and by using the acausal (conceptually idealized) generals (connotative essences) in lieu of the causal strains (resulting in ontological universals) of Reality. This is possible only by involving ever-greater breadths of essences, which have their shaping in the ever higher limits of Reality. Through this it is also clear that processes and their groups are ontologically more important than ontological universals and their groups. Unfortunately, Strawson does not treat of the connection between

---

260 Strawson, "Particular and General", 218.

generals, generals of generals, etc. Without such a treatment one must think that for Strawson one single feature-placing general or one set of parallel generals exhausts the universals within the concept of a thing in its processes, and yields the token and the particular as objectual – his notion of objectivity being vague. I propose in this chapter to resolve this issue by Reality-in-general (the ontologically connotative generality of generalities) as the most *general* epistemological category of ontology – this conceptual universal of conceptual universals being the purely connotative conscious counterpart of To Be.

Hartry Field puts succinctly the plight of thought about particulars and universals, when the former are left without the positively universal but non-reifying version of the facility of mathematical Platonism. I convert Field's Platonism into Platonism of the connotatively highest *dimension* of Reality-in-general through involvement of ever-broader ontological generalities. Field says:

> It may be thought that the difference in the explanatory role of mathematical entities and physical entities is enough to motivate a restriction of inference to the best explanation to the latter. The position would be (1) that we should literally believe in the existence of electrons and their properties as postulated in our physical theories, since there are good explanations in which they are assumed causally relevant, and there is no obvious prospect of eliminating them from explanations; but (2) that we shouldn't literally believe in mathematical entities, since there are no good explanations *in which they are assumed to be causally relevant*, despite the fact that there is no way of giving explanations that avoids postulating them in an *acausal* role. The problem is that if one takes this line, then the properties of electrons that one literally believes in can't include any properties that require mathematical entities for their expression; so if mathematics is not eliminable or close to eliminable, there are going to be very serious limitations on stating explanations in terms of electrons without going beyond what one believes. Perhaps one can maintain a belief in electrons and a belief in those of their properties which are describable without mathematics, on *something like* inference to the best explanation grounds, without a belief in the explanations one gives; but it seems to me a very delicate position to maintain. Consequently I am inclined to think that unless a very substantial amount of explanation involving electrons can be given in a mathematical entity free fashion, the prospects for maintaining realism about electrons without maintaining probabilism are dim.[261]

Probabilism cannot be causal (Chapter 1). The remaining alternative is epistemological probabilism of universals about ontological processes. That is, acausal mathematical Platonism at the epistemological level about ontological universals should go hand in hand with causal realism about ontological universals and entities / processes.

---

261   Field, *Realism, Mathematics and Modality*, 19–20.

Even Field – with all his realizations about the inevitability of Platonism – does not realize the need to extend it into the purview of both remote ontological and connotative universals / abstract objects together, in order to explain the objectual-causal and ontological reality of the electron and its connotative reflection through conscious activity. Not connecting an ontological general to other relevant but remote ever-more general universals while particularistically explaining even the notion of a particular / natural kind / species (say, electron) or a token (say, a specific electron) by immediate causality from causal elements schemed out from the immediately more general ones – this is a massive betrayal of the very admission of the inextricability of the general from the particular in the very particularistic explanation. This is evident in all sorts of analytical ontologies. Lack of distinction between ontological universals (ways of being of processes) and veiled connotative reflections of them in consciousness and language are a debilitating state of affairs in analytical and phenomenological ontologies.

Though we may avoid connecting the notion of a particular to that of an acausal Platonic (general / abstract) entity, and thus attempt to safeguard the givenness of the particular with a simple ontological commitment to its objective nature, *what we tend to forget is the objectual fact that ever more universal generals are at play in simpler general/s, which, immediately and conspicuously, but insufficiently, instantiate the natural kind / particular in question in Reality.* As a result, the objectual nature of the entity / process, which belongs to it properly from within Reality, and not merely within reality-in-particular, remains incomplete. Freeing the particular from the acausal purity of a mathematical entity is to be attempted by the objectual Platonism of placing the particular in the ever more universal essences of processes beyond the particular. These greater essences are the epistemic counterparts of the ever-wider contexts in which the particular is placed, although these ontological essences are farther away from immediate contexts of states of affairs. It is immediately useful in the sciences, if one is after formulation of more holistic and inclusive ways of doing science. These ensure theoretical adequation of any given process by the influences of the causal continuity that Reality maintains with the particular. The ontologically inspired connotative composition of concepts and the analogy of essences that follows in the explication of the thing within its ever-wider contexts ensure the truth-probabilistic nature of human knowledge.

Strawson – though from within his dear particularism that speaks mostly of (1) materials, (2) substances and (3) properties and qualities – shows the essential connection between particulars and generals: "[…] [A]t least in the case of some materials and some substances, we can regard the notion of an individual instance as partially explained in terms of the logical composition (of the two notions of a feature and of placing). When we turn to properties and qualities, we may make use of a different kind of explanation which is also, in a sense, the completion of the first kind."[262] Problems due to the acausal way that we take to think generals (general properties, states, qualities etc.) may be avoided by taking them (1) as of almost the same level of logical complexity as individual instances (tokens), because conscious events are token processes and the connotative qualities / universals in compex nexus of concepts do somehow tend to represent the complex processes outside, and (2) as ascription to individual instances within a given context of placing / feature-placing.[263] This is what most particularist ontologists do.

There remains the deeper question of the connection between the more general essences, laws and totals that are widely different from but partially similar to those of them immediately involved in a particular. Another question is as to how we shall integrate the particular with the general. Strawson suggests a way out, which might help in connecting the particular with the general, with the general of generals etc. beyond the approach he has himself adopted here:

> The notion of placing a feature is taken as basic, as consisting of the logically simplest elements with which we are to operate … [N]either of these elements involves the notion of an individual instance, nor therefore the notion of certain types of general things, such as properties and species; and it is shown that the idea of operating solely with these simplest elements can be made intelligible for certain cases. (Features in fact of course belong to the class of general things; but so long as we remain at the feature-placing level, they cannot be assigned to it; for there is nothing to contrast the general with.) From this basis we proceed by composition and analogy.[264]

It should be noted that composition and analogy are applicable not merely for any (allegedly) pure particulars, but for those particulars that are by nature intertwined with generals. In short, Strawson admits essences in the individual

---

262 Strawson, "Particular and General", 223.
263 For a particularist application of universals on par with actuals the following serves well: "It is natural, rather, to regard those general things which are properly called qualities, conditions, etc., as belonging at least to the same level of logical complexity as the idea of individual instances of the kinds we have so far been concerned with: to regard them, that is, as feature-*like* things, the incidence of which, however, is primarily indicated, not by placing, but by their *ascription* to individual instances of material or substantial features the incidence of which *is* primarily indicated by placing." Strawson, "Particular and General", 223.
264 Strawson, "Particular and General", 224.

ontological sense which, unfortunately for him and fortunately for us, admits also a relatively more top-to-bottom composition, i.e. composition from and analogy to the generality of generalities, to the less specific of generalities, and then to the objectual particular as such. The last of the series, namely the particular (not the token), is perceivable as objectual merely by reason of the causal strains of Reality, through the acausal involvement ultimately of Reality-in-general, in its notion (and the Transcendental involvement, ultimately, of To Be) – but all these in ontological commitment. It is thus to be concluded that *'an objectual particular' means a particular within the totally objectual umbrella of Reality, its ontological universals (including To Be) and the connotative generalities that pertain to it (including Reality-in-general).*

Strawson's admission of the inclusiveness of strains of relatedness *to and from* other entities (events), in the very notion of the individual, does not admit of the indefinite strains *from* all of them from the causal past of the particular along with its immediate ontological universals. The strains of relatedness from the particular (event which is not yet recognized by him as something propositionally referent) to others (again, events) do extend their causal relatedness to indefinite amounts of others in the related past of the particular, but need not be to an infinite expanse of them. He seems not to have immersed (to be read in terms of 'placing' / 'feature-placing') the concept of the particular in the most total and the most general, in order to imbibe all relevant universals from within the widest possible relevant context of the infinitesimal, infinite and continuous concept of the causal Reality. Naturalizing epistemology should therefore take place by thoroughly *imbuing particulars and generals through processual causality*, which allows a *top-to-bottom, past-to-future constitution of the particular*, which used to be mistaken in the tradition of ontology as reifyingly essentialistic metaphysics.

If deeply analytical of causal strains from the totality of Reality, every experience of a concrete entity / event / process may be *found to* possess inherent strains from particulars and totals – may be found through the involvement of the connotative generality of all generalities, i.e. Reality-in-general, in concepts / notions. The purity of the Platonic mathematical / ideal object is from the absolute and idealized distance from actuality *assigned by the naming, characterizing and measuring character of consciousness*. Actuality is never fully subsumed under "pure particulars" or "pure generals", for there are no such independently pure universals except To Be, and all the rest have some measure of both ideality and concreteness inbuilt in them. The ideality inbuilt in To Be is maximal. Reality-in-general, as the connotative generality of all generalities, is the best conceptually connotative reflection of the To Be of Reality. This is the highest basis for connection between epistemology and ontology.

All general connotative reflections (connotative universals) of general objectual qualities (represented in processes by ontological universals), together, imply a world of possibilities different from reality-in-particular, i.e. Reality-in-general, as active in ontological reasoning. *Reality-in-general and the various connotative universals build up conceptual conglomerations of possible worlds which have their direct ontological reflections in the respective nexuses of ontological universals, constituting ontologically relevant **possible worlds**.* Individually, essences are instantiators of substances (Strawson's type 2 general entities).

Ontological universals are the qualitative causal strains as such of processes. These qualitative ontological universals together are not Reality-in-general or connotative universals, since the latter are specifically acausal and Reality-in-general is infinitesimally and infinitely acausal. Ontological universals and connotative universals always connect beyond individuals. Individuals do not exhaust the ontological depths of possible perception of reality-in-particular (in the sense of the Strawsonian particular) through Reality-in-general. Reality-in-general, as the generalising confluence of generals in consciousness, is a world in itself, properly instantiating its own particular, i.e. Reality, in thought. In other words, acausal (connotative) generals / essences can be further generalized acausally by mathematical induction, not as the repository of all essences but as the inductive generality of all connotative generalities. This is Reality-in-general. (The same may be done also about ontological universals and culminated at To Be, which see in Chapter 3.)

It seems strange how most twentieth century thinkers went against system-building and confined thought in consciousness, ideas, language, mind, cognition, humans, society, religion, science etc. and rarely thought of integrating concept with nature, form with matter. And if they did so, they rarely went beyond scientism and naturalism to ontologize against the backdrop of the essential connexity between the particular and the general within a schema that permits all possible widening and deepening of ontological universals / ways of being of processes from Reality under its To Be. I propose that the categories of Reality-in-total and Reality-in-general are two of the most general of such a system-building schema, which *avoid all forms of reifying metaphysics* by way of its more tenable epistemological probabilism in ontology. The said aspect of probabilism will be shown using Kurt Gödel's work in Section 2.2.2.

In the proposed categorial scheme, Reality-in-general properly *instantiates* to thought Reality-in-total under To Be. If (1) connotative essences are general instantiators of particulars to mind and possess generality only with respect to certain particulars and their ontological universals, and if (2) there can be gradation of such generalities, then we can imagine also of positing the generality that instantiates Reality to thought, i.e. Reality-in-general, as the carrier of the genuine ingredient of particulars to thought. There is a difference between Reality-in-general and To Be: the latter instantiates both, i.e. Reality ontologically to processes (i.e. assuring the ultimate influence of final causal roots on particulars) and Reality-in-general epistemologically to mind and language; and the former instantiates Reality in its To Be epistemologically to mind and language. Instantiation over To Be is so, because of *the supra-categorial nature of To Be*, which has its reason for instantiation in its nature as ontologically of Reality and epistemologically via Reality-in-general in consciousness and language.

Therefore, mere particularist description as in analytic thinkers does not suffice for genuine epistemology and ontology. To repeat what was said many times, the theory-ladenness of particulars in mind has been shown to issue not merely from respective generals, but from the generality of generalities of ... *ad infinitum*, i.e. Reality-in-general, which is the complete epistemological instantiator of Reality to mind via ontological universals. Reality-in-general is the connotative generality proper to Reality, which latter is the particular that uniquely instantiates Reality-in-general to mind. They differ in their particularity and generality, i.e., Reality is the particular proper to Reality-in-general, the latter being merely general at the epistemic-epistemological level. There will be no sound philosophy that accepts merely the Strawsonian, Quinean, particularistic, analytic and near-sightedly minor-essence-bounded[265] entity as the only entity. If Reality is a substantial inductive totalization, then it should work as the most suitable ontological categorial particular in place of the concretist 'particular'. The ontological category proper to the epistemology of the concretist 'general' shall be Reality-in-general, i.e. the general that instantiates the totally processual substance, Reality, to mind.

The above shows that even analytic and individualistic ontologies rest on the Transcendent and Transcendental categorial dimensions. Strawson's concept of categories is that of the possibility of a working scheme, not categorial dimensions that subject themselves to broadening. "What mostly concerns philosophers is the conceptual scheme, or general framework of thinking about reality and the

---

265 This sort of particularism, especially the Quinean variety, will be analysed and criticized extensively under different sections in Chapter 3.

world, and even scientists have to start their inquiry within a framework."[266] On the other hand, Popper holds that a framework is "[…] conscious of the fallibility of all our methods, although it tries to replace all our theories by better ones. This is, admittedly, a difficult task, but by no means an impossible one."[267] For any scheme to be made possible, there is a combined maximal categorial dimension behind them – that of Reality and Reality-in-general. If this is not imbibed in particulars via the seemingly distant essences that pave the way for accruing the causal roots on the future, then there are no more particulars like atoms, electrons, quarks, fields etc.

Evem if we dwell solely and continuously on the possibilities involved in the great ways of connection or mutuality of actual entities, there will be enormous possibilities for thought. This enormity misleads us to believe that all that there are to particulars are just what transpire (causally in a physical-ontological manner, from them ontologically through universals of all kinds, and acausally in a conceptually connotative manner) within what is spatio-temporally specific to some actual instantiations of those particulars. This is nothing but perspectival absolutism of the present. One often forgets that causal strains from beyond and outwards in the domain of Reality may be imbibed in the theory of these particulars only through Reality-in-general the universal of all universals. Reality-in-general is the acausal theoretical generality of all the causal strains of Reality. Causal strains are theorized only via acausal, epistemic, qualitative and nomic elements called essences, universals or laws, defined simultaneously both analytically and synthetically.

Instead of working such syntheses, we find analytic philosophers busying themselves with conditions of truth in subject-predicate presentations. Observe the way in which the question of subject-predicate distinction and that of sortals[268] in analytic ontology are dealt with in isolation from the final cosmologico-ontological and ontologico-epistemological ideals, Reality and Reality-in-general. Sortal is consistently kept in its status as a particularistic acausal, never in any connection with the whole depths of the causal-natural in 'Reality' and the depths of the epistemological acausal in 'Reality-in-general'. *This is due mainly to*

---

266 Strawson, "The Theory of Property and the Theory of Reality in Quantum Mechanics", consulted by Renzong, "How to Know What Rises up Is the Moon?", 64.
267 Popper, *The Myth of the Framework*, 60.
268 Durrant, *Sortals and the Subject-Predicate Distinction*, 1. John Locke in his *Essay* (III iii 15) used the term first, to mean sorts of material and sorts of object. Charlton, *The Analytic Ambition*, 163.

the two and a half millennia old mis-identification of connotative universals with ontological universals.

Michael Durrant argues that "[…] whilst the categories of 'name' and 'predicable' are indeed *exclusive* they are not *exhaustive* and that logical theory has failed to take due cognisance of this."[269] He emphasizes that "[…] *both* the category of the name (particular [proper] and general) *and* the categories of the predicable and predicate *presuppose* the category of the sortal."[270] Here sortals are not merely connotative universals, but ontological universal strains that work in them. Just as names and predicables / predicates presuppose sortals, so also speech, discourse, reference, description etc. presuppose identifying and specifying by reference to sortals. Hence, all forms of subject-predicate distinction in practice presuppose sortals. A sortal is any symbol with a principle to distinguish and count particulars, in its own right relying on no antecedent principle or method of distinguishing or counting.[271] Sortals, particulars, universals etc. presuppose higher totalities and generalities (as we have seen) and finally also the categories of Reality and Reality-in-general.

Hence, all thought must be ideally revamped in line with the categories of Reality and Reality-in-general. What is important here is not merely theoretical application of the more general scheme of categories to all possible causal and acausal entities (which has been the traditional range of particularistic ontological thought), but also application of the very dimension of more and more general universals within the purview of Reality. Reality and Reality-in-general are not classificational categories. They are the very highest dimensions of cosmological and epistemological realities applicable as categories in ontology. These make possible the overthrow of particularism and the subsuming of the particular in probabilist-universalist discourse. With this background, the following analytic and other concretist conclusions look theoretically simplistic: "It is the particular which is the starting-point and end of human cognition, but not the universal. […] The ontological reason is that the world is composed of numerous particular individuals, and what we have to deal with are all particular individuals."[272] How simple would it then be to philosophize! This is the aftermath of non-recognition of the highest cosmological and epistemological categorial *dimensions* of ontology. Such non-recognition of the very theoretical depths of the particular can render thought impotent by going round the particular.

---

269 Durrant, *Sortals and the Subject-Predicate Distinction*, 275.
270 Durrant, *Sortals and the Subject-Predicate Distinction*, 275.
271 Durrant, *Sortals and the Subject-Predicate Distinction*, 277.
272 Renzong, "How to Know What Rises up Is the Moon?", 64.

We saw so far Reality-in-general as the connotatively transcendental-universalist dimension of discourse (being the highest connotative ingredient of concepts) and as the highest epistemological category of Einaic Ontology. This can be effected only after making it epistemologically probabilistic and ontologically realistic. What follows now attempts this by probabilisation of all connotative universals as ingredients of concepts.

2.2 Transcendental Categorial Dimension of Probabilistic Essences

2.2.1 Categorial Confluence of Causality, Laws and Essences

The *epistemological tension between the particular and the universal cannot be resolved epistemologically*: acausal universals cannot be explained acausally for integration with the causal reality-in-particular and Reality. The 20th century analytic tradition has attempted mostly linguistic-semantic-universalistic[273] resolutions of the problem, for fear of incurring metaphysics. Strawson and others have extended it up to the ontology of reality-in-particular. These remain merely non-ontological. So, linguistic-, semantic- and particularistic-universalistic essentialisms must be transformed into an ontological one that makes all the former possible in a more probabilistic key.

David Armstrong works out a full-blown semantic universalism / essentialism based on a causal, physicalist,[274] analytic particularism. In logic, 'particular' means anything that may be denoted by 'some'. Particularism bases itself on the bound variable 'some', but forgets its logical cognates 'everything' and 'nothing'. The fact that these latter two are beyond the ontological reach of particularism is reason enough to look for an ontology and epistemology of that (Reality) which includes also the latter by way of the To Be of all that may be known. Armstrong thinks that it suffices for ontology that 'particular' means anything that may be denoted by 'some'. This is too general and does not give space for universals of

---

[273] The universalism here is linguistic-semantic, because Armstrong's universals are based on the immediacies of 'talk', not on the broadest possible of them. The linguistic-semantic kind of meaning may be satisfied with immediate universals.

[274] Though it is not possible to air here a rebuttal of pure physicalism, I outline in this chapter certain epistemological niceties for a physical and transcendentally physical (not trans-physical) ontology, by extending the particular-to-universal question into a question of particular-to-universal-to-universal … of all universals. This does not change the physical state of affairs, but differentiates the meaning assigned to 'physical' from the level of the perspective of the total and the general.

any greater and greater width being ingredients of particulars. Lately he has done it against the backdrop of his epistemology of truthmakers: "The idea of a truthmaker for a particular truth [...] is just some existent, some portion of reality, in virtue of which that truth is true."[275] The major demerit of this sort of particularistic (and happily, non-reifying) essentialism is that the whole causally actual past is not well accounted for within the present and the possible in truthmakers. The roots-in-the-skies method has not been perceived as a valid solution in line with the tradition of system-building. In my opinion, the categories of Reality-in-total and Reality-in-general are the cosmological and epistemological keys to its resolution without incurring essence-reifying metaphysics.

Though without insisting on Reality-in-total, Reality-in-general, and the physical-ontolgoical category of Extension-Motion (instead of Spacetime), Sfendoni-Mentzou discovers the absence of transcendence of the domain of the actual in Armstrong's concretist idea of laws of nature:

> [...] [A] consistent attempt at a realist account of laws of nature grounded in universals cannot be given, if trapped in the narrow domain of actually existent law in spacetime; what is further needed is an appeal to the infinitely rich reality of that aspect of law, which is in potency of determination, so that the relation between law and its instantiations will be accounted as a relation of *quidditas* to *haecceitas*, or potentiality to actuality.[276]

She puts in gist Armstrong on the confluence of laws of nature and universals:

> The central idea propounded by Armstrong is this: Laws of nature are relations between universals. The idea that laws of nature link properties with properties and things with things, N(F,G), holds, in virtue of a *de re* necessity linking the relations between universals and the uniformity it produces. Furthermore, relations between universals are treated by Armstrong as universals themselves, so that he can 'assimilate the relation

---

275   Here he holds also that the relation "[...] is a cross-categorial one, one term being an entity or entities in the world, the other being a truth." Another such possible cross-categorial relation, according to him, is that of difference. The theory of truthmakers accepts a sort of realism (I would say, highly particularist realism) for particular truths. "There is something that exists in reality, independent of the proposition in question, which makes the truth true. The 'making' here is, of course, not the causal sense of 'making'. The best formulation of what this making is seems to be given by the phrase 'in virtue of'. It is in virtue of that independent reality that the proposition is true. What makes the proposition a truth is how it stands to this reality." Armstrong, *Truth and Truthmakers*, 5. I would call even Armstrong's ontology as analytic-semantic, because he too is busy with the immediate universals at play in particulars.
276   Sfendoni-Mentzou, "The Reality of Thirdness", 90.

between law and positive instantiation of the law to that of a universal to its instance.' His realism, therefore, which is grounded in a realism about universals, makes him, as he claims, an opponent of the regularity theory on the one hand, and a defender of the idea of necessity involved in laws of nature on the other.[277]

In regard to such an understanding, Armstrong works his way to causal necessity based on universals and laws of nature, against the Humean idea of causality:

> [...] that 'the law is exhausted by the fact that the observed Fs are Gs, and the unobserved Fs are Gs'. By contrast, he holds, 'the law involves an extra thing' the presence of which 'serves first to explain why all the observed Fs are Gs, and second, to entail that any unobserved Fs there are will be Gs'. The basis of such necessity can 'be found in *what it is to be an F and what it is to be a G*', namely, in F-ness and G-ness. If it is a law that all swans are white, then the necessity is grounded in swanhood and whiteness. This is what Armstrong means in claiming that 'the necessitation involved is a law of nature in a relation between universals'.[278]

This work on laws of nature and causal necessity too smells particularism: if from the law that all swans are white the specific concept of necessity as grounded in swanhood and whiteness is concluded, it is apt merely to the immediate context. But the general concept of necessity of a law cannot be so specific of just the thing-hood and quality-hood of two things or types of things. The citations here from Armstrong show how particular-bounded his universals are.

We have thus discussed Armstrong's distinguished manner of treating off what he takes to be particularism from his own viewpoint of analytic ontology, and I note my reservations about his partiality to particularism. He avoids the objectual nature of infinite and infinitesimal objectual-causal relations in Reality as is available while generalising ontological universals of all universals (which is To Be because To Be is the deepest universal in its own right) ever broader, and of generalising unto the connotative universal of all universals (i.e. Reality-in-general, which latter alone can properly instantiate Reality to consciousness and language). His criticism of Stoutian particularism is nevertheless particularistic. He deals with the extreme particularism of G. F. Stout:

> [...] [I]f a curtain and a carpet are [...] "both red," then not only do we have two numerically different things – the curtain and the carpet – but we also have two numerically different *rednesses*. This has nothing to do with the fact that the two objects may be two different shades of red. Let the two objects resemble exactly in their shade. According

---

277 Armstrong, *What Is a Law of Nature?* cited in Sfendoni-Mentzou, "The Reality of Thirdness", 76.
278 Sfendoni-Mentzou, "The Reality of Thirdness", 76–77, citing also Armstrong, *What Is a Law of Nature?*

to Stout, there are still two numerically different rednesses. Orthodox Realism would say that the two objects were of the *identical* shade of red: the objects are different but they have the same property [....] Stout applies the doctrine to relations as much as to properties, but [...] it will be convenient, yet will not affect the argument, if we restrict ourselves to properties.[279]

Stout thinks of particulars as the stuff of the world. The physicalism implied in it has to be supplied with further qualifications – which in turn can temper down the perspectival absolutism of the concept of 'things physical'. For Stout these particulars are "thin" particulars in comparison with "thick" or concrete particulars "constituted out of abstract particulars". Strawson and Armstrong and many analysts mean by 'particulars' not the bare tokens / instances of any ontological-processual universal, but tokens inextricably embedded in universals. Stout is almost an exception. Even his doctrine applies to relations and properties. Strawson calls 'abstract / thick particulars' as 'particularized qualities'.[280] Armstrong gives two celebrated reasons why particularism is insufficient:

---

279 Armstrong, *Nominalism and Realism*, 77–78. The realism of universals is clarified thus: "[...] Realism about universals is the doctrine that there really is such a thing as a generic identity, identity of nature, which cannot be analysed away. Any theory which denies this may be challenged to explain the symmetry of resemblance." Armstrong, *Nominalism and Realism*, 49. The principle of identity of indiscernibles is discussed here: "Keith Campbell has pointed out that the Resemblance theorist can attempt to meet the charge of lack of economy in principles by giving an analysis of the identity of particulars in terms of resemblance. He can assert, intending it as a logical analysis of such identity, that a is identical with b if and only if a resembles b exactly. Two different particulars, he can say, necessarily fail of resemblance at some point. At the top of the resemblance-scale, the resembling things are necessarily one. The symmetry of identity is then derived from the symmetry of resemblance." Armstrong, *Nominalism and Realism*, 49. "However, this attempt by the Resemblance Nominalist to even the score depends upon his being able to maintain that it is logically impossible for two things, in Realist language, to have exactly the same properties and relations. This is a version of the principle of the Identity of Indiscernibles [....] I simply assert that it is impossible to give an account of the identity of particulars in terms of resemblance. If this is correct, then those who fail to base the symmetry of resemblance upon the symmetry of identity sin against economy principle." Armstrong, *Nominalism and Realism*, 49–50.
280 Armstrong, *Nominalism and Realism*, 78. He mentions Strawson's classification of particulars in his *Individuals*: "In general, we perhaps have most use for the ideas of particular events so framed, less use for the ideas of particular conditions or states, least use for the ideas of particulars which are simply cases of qualities or properties. But we do say such things as 'His anger cooled rapidly' [....] Some philosophers, no

First, it seems clear that the very same particular cannot instantiate a property more than once. To say that *a* is F *and* that *a* is F is simply to say that *a* is F. Given the Identity view of properties, this is immediately explicable. For a Particularist, however, an ordinary concrete particular is a collection of Stoutian particulars. Why should not this collection contain two Stoutian particulars which resemble exactly? But this will be equivalent to saying that the concrete particular has the same property twice over. The Particularist can only meet this difficulty by introducing an ad hoc principle forbidding exactly resembling Stoutian particulars to be parts of the same concrete particular.[281]

Only partially do these apply to Strawson, because he does couch particulars in universals – in my opinion, insufficiently, i.e. without recourse to the world of universals of universals of … *ad libitum*. The second argument, he holds, is more convincing:

> The second argument depends upon the premiss … that Particularism about properties and relations must be supplemented by an Immanent Realism. Only so can we explain how Stoutian particulars can be classed and sorted. Let us suppose, then, that a particular *yellowness*, say *the yellowness of this lemon*, has the universal property, *being a certain shade of yellow*. The *redness of this tomato*, however, does not have this universal property. The question is: is there any reason present in the nature of the first Stoutian particular, but lacking in the nature of the second Stoutian particular, why the first particular has this universal property? If there is no reason, then it seems that the Stoutian particulars, in abstraction from their universal properties, are mere bare particulars. This would make different Stoutian properties of a thing, in abstraction from their universal properties, indistinguishable from each other, which seems absurd. But if the nature of the first Stoutian particular is not bare and so "fits" *being a certain shade of yellow*, while the nature of the second particular does not, must not the nature of the two particulars already be something universal? Could not the nature of *the yellowness of this lemon* be duplicated in another lemon? But then *yellowness of this lemon* is not, after all, a particular but a universal and no further universal is required.[282]

The Stoutian particularist view of properties and relations cannot be defended because it stops at the immediate universal, the rationale for which should have been applied also to posit further universals more general than the first, *ad libitum*. Note that Armstrong and Strawson too do not hold any further universals about such universals – Armstrong's may be more general than the Stoutian ones – that they have as additional instantiations and potentialities in the very causal antecedents of the formation of the universal in question. For Einaic Ontology

---

doubt, made too much of the category or particularized qualities." See Strawson, *Individuals*, 168–69.
281 Armstrong, *Nominalism and Realism*, 86.
282 Armstrong, *Nominalism and Realism*, 86–87.

these universals are ontological, with conscious instantiation in thought. Without inclusion of both these in ontology we are not able to talk of things as such and of thought about real processes. In short, the only other possibility for inclusion of those more general universals in the process of the given entity/ies is to posit properties and relations as universals, universals of universals etc. as distantly (but surely) comprehensive of all the other possible and actual universals. This trend has the final conceptual limit, Reality-in-general. To Be instantiates both Reality-in-total and Reality-in-general in an ontological sense and is already the highest universal that instantiates Reality-in-general in a purely epistemological sense.

It is *a special theorem* of Einaic Ontology that Reality-in-general does not instantiate To Be adequately; and that the connotative instantiation of To Be in consciousness does not correspond to the instantiation of an ontological universal of universals. This is because Reality does not have equality with another reality within it for Reality-in-general to be the instantiation of the highest ontological universal in Reality. That means, what is in universals, over and above perception of instantiation, is not merely necessitation by laws of nature, but also necessitation of the objectual, trans-particularist, dimension of Reality and Reality-in-general, which transpires from the realm of To Be in the case of Reality.

Armstrong's truthmaker-theoretical way of treating off particularism also does not suffice for necessitating these, because truthmakers are particulars, which are yielded not merely in any direct reference, but in truth-level implications equivalent to particularistic ontological commitment. Reality and Reality-in-general are necessitated by the infinitesimal and infinite aspects of objectual causality. *In exclusion of Reality and Reality-in-general, there is no fully justifiable ontological universal ingredient in partial causality (which is causality circumscribed by immediate universals) or partially trans-immediate universals (universals of universals, without Reality-in-general). In exclusion of Reality and Reality-in-general, theory would be either technology, or near-sighted science.* We can have both these in causal immediacy or intermediacy and still admit infinitesimal and infinite objectual causality beyond them, which are to be treated in Einaic Cosmology and Einaic Ontology in relation to Einaic Epistemology.

Analytic ontology posits ontological entities not from the point of view of Reality but from that of the immediately given meaning system, which is studied satisfactorily in ordinary semantics, grammar and science pragmatically, not giving heed to the farthest recesses of causal roots. Insofar as analytic ontology posits ontological entities, processes and particulars linguistically, semantically and scientifically, the objectual nature of Reality, the theoretically most feasible realist aspect of ontology, is at stake. To integrate the objectual-realist aspect of particulars and Reality into things conceptual, we should find the final essentialist aspect of specific laws of nature and Reality. This aspect is Reality-in-general.

Sfendoni-Mentzou stresses something similar in her ontological (not epistemological) concept of 'potency of determination' and explains it as the relation of *quidditas* (Aquinas, "whatness") to *haecceitas* (Scotus, "thisness"). I join her with an argument slightly different from but akin to hers – hers being against anti-realists who insist on the concept of 'particulars' to be the self-consistent way of describing reality and holding only a barren similarity between laws of nature. Particulars instantiate universals, but these are not explained in terms of the totalised objectual causality of Reality. So, as we discuss Armstrong's concept of causality, I have been studying Sfendoni-Mentzou, who has a critical and reasonable idea of the same.

If analytic thinkers (Strawson and Armstrong) hold sufficiency of immediate essentials or laws to describe particulars, the question of potency for instantiation by essences is incomplete, because *future instantiation is not infused in their concept of essences*. Armstrong's stand on universals of universals is extendable to Reality-in-general, the epistemological aspect of future instantiations. Armstrong comments on universals of universals, without distinction between ontological and connotative universals:

> [...] [U]niversals can themselves fall under universals, that is to say, that properties and relations themselves can have properties and be related. A universal which falls under a universal is *ipso facto* a particular as well as a universal. It is, however, a *higher-order* particular. But the doctrine of Particularism [...] is the doctrine that the properties and relations of first-order, or ordinary, particulars are themselves *first-order* particulars.[283]

Armstrong goes beyond such particularism, but stops short of the broadest limits of his exercise of extension. *This, again, is particularism, because Armstrong does not permit us to enter the world of the dimension of universals of universals of ... ontological and connotative universals*. Realists of analytic persuasion do tend to go beyond uniformity between such simple two-way particulars – which are simultaneously universals and particulars – and also beyond uniformity between laws of nature that are also universals (since they are universal truths instantiated in processes), and link them by really Stoutian particularist necessity at a slightly higher order. Two-way connotative (epistemological) particulars have a limit, Reality-in-general. This latter itself is a two-way particular / universal, since it is a limited instantiation of the deepest, broadest and highest of all ontological

---

283 Armstrong, *Nominalism and Realism*, 79.

universals, i.e. To Be, without direct clarity as to its cosmologically ontological (Reality) and epistemological (Reality-in-general) bases. Reality-in-general is the highest possible case of connotative two-way particulars. Hence it has its own part, superior to the mere 'universals of universals', in ontology.

That is, *the crux of the problem of realism and antirealism* hinges even today on the problem of the reality of what the tradition has called the ontological *fundamentum universalitatis*,[284] which is the To Be of infinitesimal and infinite objectually causal Reality. The *fundamentum* of infinitesimal and infinite objectual-causal nature formulated in an acausal manner is Reality-in-general. The foundation of universality in thought is epistemologically this generality of all universals, universals of universals etc., because one should not stop by any criterion to fix up with a universal of immediate universals in thought. One is thus empowered to think Reality and reality-in-particular in terms of its highest ontological universal, To Be. The universals that refer to distant but more and more objectually inevitable causal necessities in Reality should move via the Transcendental To Be through its showing in consciousness, i.e. Reality-in-general.

The epistemological problem here is related to that of ontological identity. See how McMullin mixes ontological and connotative universals in the last statement here below:

> A distinction must be drawn between the abstract property (e.g. whiteness) and the individual instantiation of the property in a thing (indicated by the predicate 'is-white'). Φ-in-a is the latter; it is not identical with φ-in-b. Both of these are individual and concrete. They are the real grounds in a and b which permit me to predicate φ of both objects. These are not things but aspects of different things, and they are not identical. We can speak of *identity* only at the abstract level of meaning.[285]

The analytic's immediate abstract level does not suffice, and one cannot set the level of abstract meaning (of which connotative universals derived conceptually from ontological universals are ingredient) upon the immediate or mediate ontological universals of universals in processes. *Meaning rests on perception of universals of any breadth.* This points to Reality-in-general as the final epistemological possibilizer of identity and meaning. The *dimension* of inexhaustible meaning/s akin to the infinite, infinitesimal and causal Reality, and active in the ontological universals and their meanings in mediate and immediate ones, is had only through Reality-in-general. The question of the ontological possibilizer

---

284   Sfendoni-Mentzou, "The Reality of Thirdness", 75.
285   McMullin, "The Problem of Universals", cited in Armstrong, *Nominalism and Realism*, 78.

hinges around ways of connecting this problem with the objectual nature of reality-in-particular and Reality via Reality-in-general. This is forgotten by both McMullin, Strawson and Armstrong alike. I hold that this connection that determines identity can be made only by way of the objectual (not merely objective), infinite and infinitesimal causality in Reality. Identity in discourse is somehow a reflection of processual identity (the actual structure and various processes of which we do not treat in this book).

This allows us to call identity-determining universals as ways of ontological causal relations in Reality. All other sorts of epistemic (acausal) relations must be explained in terms of thoroughgoing, direct or indirect, ontological-causal relations, because *otherwise there is no real connection but only consciously qualitative connection (consciously named into acausal nominal forms) between what are physical-ontologically connected. If so, identity of particulars cannot be determined ontologically and epistemologically.* Purely acausal relations are not ontological, do not directly concern the world, and are not within the *fundamentum universalitatis* of the processes under discussion. The variety of causality Hume discussed is an acausal connection in concepts and without reference to processes and their universals. *Humean causality blocks any identity due to its dispossession of connections – be it ontologically identifiable ones or conscious determinations of it.*

Observe how Armstrong speaks of David Lewis's counterfactual theory of causality, which is based on Humean supervenience,

> [...] the doctrine that, in this world at least, all causes and laws supervene, and supervene without ontological addition, upon particular matters of fact. These particular matters of fact do not, for Lewis, include any singular causal relations or primitive nomic ties between selected properties. Given this, regularities in the world would seem to be the only truthmakers for causal and nomic truths that this world affords. In a wide or loose sense, Lewis is a regularity theorist.[286]

The Humean acausal theory has ostracised the causal. Lewis's regularity theory does not allow for there being necessarily actual causal processes with physical-ontologically connecting links between all the really possible future worlds and their probabilistically best realizable laws. Simple regularity of one or another sort provides for identity of laws. This is not necessarily based on Universal Causality in Hume and Lewis.

Armstrong's singularist theory is opposed to the supervenience theory: "[...] the truthmaker for the sequence being causal is the holding of a dyadic relation,

---

286 Armstrong, *Truth and Truthmakers*, 125–26.

the relation of singular causation, between the token events themselves. This relation is, using terminology introduced by Lewis, *intrinsic to its pairs*."[287] Beyond the supervenience- *and singularist backgrounds of identity / endurance*, we have constructed the finally cosmological and epistemological dimensions of an objectual, continuous view of causality based on the categories of Reality (the token *par excellence*) and Reality-in-general (connotative universal *par excellence*). According to Armstrong, "[…] we think that there is some very close connection between causes and *laws of nature*", provided we take it ontologically, i.e. with respect to "the truthmakers of such statements"; and if laws of nature are considered ontologically rather than propositionally, causes and laws will "seem to be intertwined", and every law becomes a statement of causal determination or "probabilifying" of something further.[288] Such connection is itself causally ontological, many instances of which, in farther extension-motion expanses, are another layer of causal connections of the past or future, and so on, extending the identity of a process and its connotative determinates ever wider unto Reality and Reality-in-general.

The facts: that laws of nature (even scientific-categorial ones like causality) are laws within a context (set not merely in extension-motion, but by epistemic considerations of apparatus-wise intellectual approach or extension), and that acausal expressions of laws may be interpreted in terms of the causal connections that have ensued them over ontologically physical connections mediated through qualitative universals from past processes – these show the universality of the Law of Causality and a two-way (strictly particular-to-general and general-to-particular) implication within the matrix of simple or scientific-categorial laws of nature like causality and the Universal Law of Causality. Then there should be *a causal way to contextualize all possible acausal laws of nature: nomic concepts*

---

287 Armstrong, *Truth and Truthmakers*, 126. Armstrong discusses also whether the intrinsic relation holds necessarily or contingently. "The orthodox analytic view until recently is that the relation is contingent. But if we admit properties into our ontology, and if furthermore *powers* are an essential component of properties, then we seem to get a necessary connection *in re* holding between token causes and token effects. Again, even if we deny that properties are powers, but do allow them to be universals, causation may come out as necessary. […] [T]he attribution of properties to particulars in states of affairs may well be a necessary one. This would hold for polyadic states of affairs as well as monadic ones and also, it seems, for higher-order states of affairs, such as the particular sort of connection between universals that I hold to constitute a law of nature." Armstrong, *Truth and Truthmakers*, 126.
288 Armstrong, *Truth and Truthmakers*, 126–27.

and nomic propositions. Everything acausal has a necessarily causal interpretation, given the objectual nature of infinite, infinitesimal causality in Reality. This is determinationism, not near-sighted and perspectivally absolutistic determinism. Nevertheless, the whole process of casualization takes place via the presupposed ideal generality of ontological and connotative universals behind all ontological and connotative universals, not merely through immediate universals. Reality-in-general is therefore the universal that is already presupposed by every individual causal process that ends in partially nomic connotative acausal universals, which derive in consciousness from partially ontologically committing universals.

Acausal "processes" are the conceptually definitional aspect of occurrence of universals to consciousness, and not processes; and they have to be explained in terms of the causal body- and brain processes that generalize unto acausal universals as their connotative ingredients. Connotatives are nothing but the conscious generalities that instantiate at the relational-processual aspect of the constant tendency of certain generally more similar types of neuronal communications to repeat themselves in somewhat partial identity and slightly partial difference of extension-motion regions, and acquire intensity. In short, the repeating themselves and intensity of similar neuronal vibrations are in fact the causal processual part of concept-formation in consciousness; and the relational-instantiative part, which is the *tendency* of similar neuronal vibrations to repeat in form, is the connotative part of consciousness. The tendency is not a thing / process; the tending is a process. *The tendency* here is the mental meaning of what is processual in conscious processes, sensing, perceiving, imagining, conceptualizing, thinking that/of sensing, … thinking, etc. The instantiation is of the relationally ontological universals of objectual processes that are objects of formation of concepts. This is the crux of the epistemology of objectual-causal ontological processes in consciousness.

By reason of brain- and cognitive sciences, it is clear that what we understand as a concept is a processual aggregate of continuous neuronal activities in a certain way. *This 'certain way' is so important in the connection between consciousness and the particular natural processes it is about at a given stretch of extension-motion, that names and predicates in propositions somehow represent to itself and to other consciousnesses the processes and qualities that play in Reality in variously adumbrated ways.* Names of things and qualities in formulation are incomplete specifications of the actual ingression of processes and ways of processes occurrent in particulars, and given in consciousness. 'Processes' and 'ways of processes' point beyond the actual names and qualities, from beyond which the processes in question have their causal inheritance. This pointing beyond is in ontological commitment, and it is *connected not merely to the particular/s* in question, but to the particular/s and their contexts together, identifying or placing a universal or quality or feature in context. (Recall here Strawson's concept of 'feature-placing statements', which ends in particularism.[289]) The context is not merely the causally inherited processes, but also their ways, i.e. ontological universals. Names and predicates are a mixture of connotative reflections of ontological universals in consciousness.

---

289 Strawson begins his inquiry about universals by first discussing the familiar sort of universals, called "feature-universals" or "feature-concepts", and the statements that he calls "feature-placing" statements. He suggests the following examples: (1) 'Now it is raining', 'Snow is falling', 'There is coal here', 'There is gold here', 'There is water here'. He argues: "The universal terms introduced into these propositions do not function as characterizing universals. *Snow, water, coal* and gold, for example, are general kinds of stuff, not properties or characteristics of particulars; though *being made of snow* or *being made of gold* are characteristics of particulars. Nor are the universal terms introduced into these propositions sortal universals. No one of them of itself provides a principle for distinguishing, enumerating and reidentifying particulars of a sort. But each can be very easily modified so as to yield several such principles: we can distinguish, count and reidentify *veins* or *grains*, *lumps*, or *dumps* of coal, and *flakes, falls, drifts* or *expanses* of snow. Such phrases as 'lump of coal' or 'fall of snow' introduce sortal universals; but 'coal' and 'snow' *simpliciter* do not. These sentences, then, neither contain any part which introduces a particular, nor any expression used in such a way that its use presupposes the use of expressions to introduce particulars. Of course, when these sentences are used, the combination of the circumstances of their use with the tense of the verb and the demonstrative adverbs, if any, which they contain, yields a statement of the incidence of the universal feature they introduce. For this much at least is essential to any language in which singular empirical statements could be made at all: viz. the introduction of general concepts and the indication of their incidence. But it is an important fact that this can be done by means of statements which neither bring particulars into our discourse nor presuppose other areas of discourse in which particulars are brought in. languages imagined on the model of such languages as these are sometimes called 'property-location' languages. But this is an unfortunate name: the universal terms which figure in my examples are not properties; indeed the idea of a property belongs to a level of logical complexity [....] This is why I have chosen to use the less philosophically committed word 'feature', and to speak of 'feature-placing' sentences." Strawson, *Individuals*, 202–203. This ends in particularism, since universals are not given their natural facility to connote beyond themselves. Hence, the context remains just the empirical features, and not the systemic connections.

Ways are ontological essences, because of Reality-in-general, which is one of the ways but in consciousness (i.e., ontologically less perfectly) and transpires as the generality of generalities occurring connotatively and connecting Reality via reality-in-particular with consciousness. *By connotative I mean 'of the conscious manner of noting together'. It refers to the conscious activity of involving ontological universals in referring by ontological commitment.* For consciousness, universals are connotatives of ways of processes. Reality-in-general is a conscious connotative, a "connotative way of processes". Another such "connotative way", similar but subsidiary to Reality-in-general, is causality – a nomic essence, an ontological law subsumed under Reality, of relations of multiple particulars. Multiples in mutual causal relation form an event, a state of affairs, with the involvement of respective universals. Likewise, all universals, ontological or connotative, involve multiple particulars – a fact that clarifies the nature of universals as never belonging to a unique thing, except in the case of To Be that belongs properly to Reality.

In what was discussed about Strawson's concept of feature-placing and the implicit particularism, he seems to favour the view that there is nothing extraneous to the multiple antecedent entities / processes – except the immediate semantically satisfying universals and second-level universals of universals – occurring in the name of cause and effect in the semantic perception and formulation of causes of events. He fixes ways of being of beings merely in the immediate sense of the term 'multiple', not by attempting to include all possible ways in the discourse of multiples of multiples, their multiples etc. such a position is blind to Reality and Reality-in-general. What makes the analytic ontologist limit discourse to the immediate matter of fact?

The epistemological insufficiency of merely physically immediate causality to ontologize is the epistemological argument for the ontological necessity of essences from beyond the immediate. To use – against his will – one of Armstrong's own physical-causal arguments to prove this point, I take recourse to his study of perception from the point of view of a contra-ontological Argument from Causation that has troubled epistemologists much. The Argument has confounded perceptions non-veridical (pertaining to what is causally prior to truth in perception) and veridical (pertaining to truth in perception): It

> [...] has confounded two quite distinct things: (*a*) perceiving an X or perceiving that an X is Y; and (*b*) the causal conditions which bring about this perception. It cannot be denied that perception occurs when, and only when, a certain very complex process begins in the object perceived and ends in the brain. But what warrants have we for identifying this with *perceiving*? May it not simply be the necessary, or even the necessary and sufficient, *precondition* of perception: that which must occur if perception is to occur, but

which is not to be identified with perception itself? The Argument has done nothing to show that there is any identity here. And the fact that we know very well what perceiving is long before we know anything about these complex processes, suggests very strongly indeed the two are *not* to be identified, however closely they may be related. [...] The beating of the light-waves on my eyes and brain makes me see, but seeing is not identical with the beating of the light-waves.[290]

He now advances two reasons to show that it is justified to wonder why the Causal Argument has any persuasive power.[291]

(1) First, a mere stimulation of our optic nerve or cortex by instruments causes perception that is indistinguishable from veridical perception, although no physical state of affairs corresponds to the perception. This implies that,

> [...] we have no direct perception of physical states of affairs, but only of the last links in the chain from physical object to perceiver, that is, sense-impressions. [...] When our cortex is stimulated in such a fashion that we have a non-veridical perception, we should not say that there is *any* immediate object of perception involved. We should only say that this cerebral stimulation causes us to think falsely. [...] We will simply have discovered ways of inducing false beliefs or inclinations to believe that we are perceiving.[292]

Thus, solving such simple problems of illusion will clear also the problems connected with the Argument from Causation.

(2) Second, suppose that one holds one end of an iron chain which is shaken up and down, and says, '*The chain* is moving'. His immediate and justifiably tactual (touch) experience was of the movement of the link of the chain immediately within his hand; and the physical process from the link to the brain may again be taken to be a sub-chain of which only the last link is within justifiable tactual perception, and so on. Still, there is a certain sense in which the whole iron chain exerts a causal effect on the brain. So, Armstrong holds that we should reject the Argument from Causation,[293] meaning argument from physically immediate non-ontological causation that does not connect to objects perceived.

Armstrong admits that these show, by the particularistic ontology of causation, that there is also an epistemological connection beyond the immediate link in the chain of respective causal process/es in the act of perception. Moreover, there is no way of fixing the immediate link, since causation even at that level can be taken to tend ever more infinitesimal. This fact points to the place of time-gap between cause and its perception. However small, there is always *some* time-gap

---

290   Armstrong, *Perception and the Physical World*, 142–43.
291   Armstrong, *Perception and the Physical World*, 143–44.
292   Armstrong, *Perception and the Physical World*, 143.
293   Armstrong, *Perception and the Physical World*, 144.

between the lighting of the lamp and the non-veridical perception of it; and it is possible to assert at the sight of the sun that "[…] eight minutes ago there was a very bright hot object at a great distance from me in more or less the direction that my eyes are now pointing."[294] But our impression is always of something in the present.[295] In the case of a dying star that sends signals, our perception is of the death of the star in the past, but immediate perception does not perceptually account for the time-gap.[296] "We tend to make it a logical necessity that what we now immediately perceive must exist now."[297] The difficulties here may be solved by saying: "[…] [T]he trickiest problems are raised, not by the presence of a time-gap in perception, but by its *absence* in immediate perception."[298]

This line of thought, nevertheless, demonstrates the fact that perception is of the object/s or event/s or process/es proper only if physical (causal) communicability, and not immediacy of measure of extension-motion, is the criterion. Additionally, there is no one causally influential past that is infinitely distant in time from any process. Hence, all possible time-gaps may clearly be held to be finite. All ontological universals, being applicable to tokens of sets of respective qualities and, in the final analysis, being based in causal pasts for qualitative affinity, are ways of being of processes within the ambit of possible mutual communicability of processes. Hence, whatever the spatio-temporal distance between an effect and a cause, between the immediate physical link and the links that are deep in the spatio-temporal distances in the past of a process, they all have importance in the system of causal pasts. This is studied solely through ontological universals, universals of universals etc., which are explanatory of all that the specific process is, and not through some particularistically immediate universals of the causal vicinity of a process. This is part of the systemic approach of Einaic Ontology that bases itself on Reality, Reality-in-general and To Be; and *it works not only from the primacy given in analytic and process ontologies to the particular and the actual entity but also in primacy to the total and the general.*

Ontological universals reflect in a veiled fashion in the connotative universals that are ingredient to concepts. Consciousness is a holographic instrument capable of ever more universally realistic conscious reflection of the ontological ways of being (universals) of prcesses that are epistemologically distant to the extension-motion immediacies physical-causally available to consciousness. The

---

294 Armstrong, *Perception and the Physical World*, 146.
295 Armstrong, *Perception and the Physical World*, 147.
296 Armstrong, *Perception and the Physical World*, 148.
297 Armstrong, *Perception and the Physical World*, 151.
298 Armstrong, *Perception and the Physical World*, 152.

holography involved here is not merely of perception or spiritual perception but of ontological universals. All sorts of systems approach and the more general systemic approach point to the roots-in-the-skies approach insofar as "epistemologically distant universals" are considered as rational reflections of ontological ways of being (universals) of the processes at hand, which have ontological commitment deep into the farther recesses ("skies") of causal influences upon those processes. This demonstrates the epistemological insufficiency of physically immediate causality in ontology. I call this argument *the epistemological argument for the ontological necessity of essences* based in causalities from beyond the immediate. Such is admissible by pushing the argument of Armstrong in the context of his study of the Causal Argument and the Time-Gap Argument.

Ordinary science and technology are busy reaping the benefits of the roots-on-the-earth approach, following which has vitiated much of philosophy in the Newtonian and post-Kantian era. This method has its own merits, but these are to be implied in the roots-in-the-skies method by the way we build up scientific or technological systems. System at the confluence of Reality-in-total and Reality-in-general already includes all there are, provided they are subsumed under To Be, which should guarantee all the goods of individuality too. It should include the particular in such a way that instantiation guarantees the individuality of the particular within the respective totals. This is assured in Einaic Ontology by rendering the categories of Reality-in-total and Reality-in-general non-totalising at the individual or partially collective levels.

Any conceptual system totalises. Why not build it in such a way that it is open at every instant for renormalization of the very definitions of these categories? Our concept of Reality does not totalise by perspectival absolutism, but instantiates by the uniqueness of the individual, which is guaranteed by Reality-in-general by its natural and continuous re-connection of ideally fixed universals (that pertain to particulars) to ever different and ever wider universals (which, again, assuage their own and the pertinent universals' particularity by means of other ever wider universals), and finally guarantee this process by Reality-in-general itself. This non-totalising process is realized in Reality. Hence, Reality-in-total is not a category that totalises in the traditional sense. The probabilistic status of conceptually *idealized* universals, universals of universals, and finally of Reality-in-general will further enhance the universal-bounded totalising property of Reality-in-total. Further, we need to turn to To Be in Chapter 3, where the three concepts will be brought to greater systematic unison by giving to To Be a meaning slightly different – by reason of Reality and Reality-in-general – from what philosophers have hitherto assigned to it.

By reason of our ontological systemic approach, ontological universals are not merely connotative resemblances or mere abstract indicators or names of processes in consciousness. They are also ontological potentialities for instantiation in the actual present and future of processes – yielding part of their rationale from the past. These are ontologically reasonable and inevitable because ontology as such and the ontology of knowledge are impossible without positing potentialities as ontological ways that are neither actual entities, nor unreal vacua. This is why there exists today a vast literature in this region named modal logic and modal ontology, but mostly without recourse to the conceptual origin and nature of the worlds in possible worlds.

Universals are real potential ways of being perceived at the level of connotative generalization over large portions of extension-motion – which, as ways, are primarily in things and only secondarily in consciousness. In short, the ways of being of connotative universals in consciousness are *subsequent and subordinate to* the ways of beings that are ontological universals. The objectual-causal nature of Reality comes forth justified only through these. This fact does not seem to meet with justification, or even the well-deserved notice, in Armstrong. Hence, we need to introduce the concept of potentiality into such theories of essences.

Potentials are potentials with respect to Reality and Reality-in-general. Both these must be incorporated through the concept of instantiation and without detriment to the involvement of potentiality, without which there is no universal at all, nor ontological talk. According to Peirce, there is something that constitutes potentiality, the category of Thirdness, in reality: "The *will-be's*, the *actually is's* and the *have beens* are not the sum total of the reals. They only cover actuality. There are besides *would-be's* and *can-be's* that are real."[299] If instantiation may be emphasized, then potentiality – which is the uninstantiated sum of universals – should equally well be stressed. That alone fulfils the reality of connotative universals as veiled rational reflections of ontological universals / ways of being of processes and as niceties posed for theory by extension-motion requirements.

---

299 Peirce, *Collected Papers of Charles Sanders Peirce*, 8:216. Here reality is not restricted to the realm of actual, individual, instantiated, beings; much more is real, not completely determined: this "more" does not present itself "in actually objectified form" (Peirce, *Collected Papers of Charles Sanders Peirce*, 6.365), as concrete spatio-temporal existence. This mode is potential being, a "capacity for realization," "some inherent tendency to actuality, which if not thwarted, leads to final completeness of being" Peirce, *Collected Papers of Charles Sanders Peirce*, 6.365. See also Sfendoni-Mentzou, "The Reality of Thirdness", 87.

The reality-based nature of ontological universals (possbles) justifies the conception of possibles as not at all belonging to but only related to the realm of actual entities. This is also the justification for possible-worlds talk, if it hinges on reality-based possibles. This fact does not end up in justifying any absolute bifurcation, from Reality, of possibles (1) as the ontological essences / universals at a realm of reality, which are neither actuals nor non-entity, but ways of processes; and (2) as the connotative and acausal universals of consciousness (which, more happily, have their justification in Reality-in-general) and active in theory. The first realm of reality, i.e. of ontological universals, is of the ways of being of beings, which are the *sine qua non* of existence of processes; and the second realm is of the way of being of beings in consciousness, which is the *sine qua non* of ontological talk given in ontological commitment. In short, both the groups are ways of being, but the first set is of first order ways and the second is of the totality of second, third, and higher orders of ways of being of beings in consciousness. To both the groups belong both actual instantiation and potentiality.

With the introduction of *ontological universals with a sound potential aspect*, it is possible to speak acausally also of the wider ontologically causal realms (spoken of in terms of laws of nature) of processes, *representable by potentiality for future instantiation*. Now generalize such potentiality over the past and the future together, extend such generalization to the connotative realm of universals ingredient in concepts – and thus there is the generalization of their connotative universals over Reality. This is called Reality-in-general. However, instantiation in processes is by ontological universals that belong primarily to Reality and secondarily to parts of Reality.

Sfendoni-Mentzou criticises Armstrong's concept of instantiation and laws of nature that are divorced from the concept of potentiality: "What will be criticized is his overemphasis on the principle of instantiation on the one hand, and his rejection of the idea of potentiality on the other. By contrast, the burden of my analysis will be on potentiality as a *sine qua non* for a realist theory of laws of nature."[300] Potentiality is ontological, it means always being in the process of ingression in the sense of following beaten tracks of evolution and transcending or at times transgressing beaten tracks by causal instantiation from processes via their universals farther than the ones directly at hand. Potentiality as such is not connotative. We may have ontologically conscious reflections of such potential ontological universals in consciousness, which is also an ontological form of instantiation in consciousness. Otherwise the processes of connotative

---

300 Sfendoni-Mentzou, "The Reality of Thirdness", 76.

instantiation, that gives rise to theory, would have been ontologically purely acausal vacuous actualities. Hence, propositional laws of nature and concepts in the mind are ontologically more and/or less sufficient valuations of ontological ways of being of processes. I have not circumscribed here the nature of their connection, but just argued that they are connected and *somehow* produce theoretical reflections of Reality in consciousness and language.

This much has been said about ontological potentials and their relation to connotative potentials. We cover also their purely epistemological aspect here. The epistemology of Armstrong's causal theory does not seem to hold the promise of thoroughly favouring objectual-causal realism via the category of Reality-in-general that generalizes conceptually over instantiation and potentiality. Sfendoni-Mentzou attempts a close-heel study of the scholastic revival of what she calls anti-realistic essentialism in Armstrong's account of universals as laws of nature, and exposes the disguised nominalism implied in his thought. "The central idea propounded by Armstrong's alleged realism is that laws of nature are relations between universals, which link properties with properties and things with things."[301] Nominalism holds that there are no objective universals, except "resemblances"[302] holding between particulars that are not themselves physically identical. Genuine realistic ontology does not hold any physical, actual-particular existence of universals. Universals appear not merely in the immediate vicinity of classification of a few given things without regard to others in their extension-motion past and future. *This is a philosophical problem that peripherally unites and then inhibits nominalists and many realists from taking universals seriously.*

Causally acceptable classification is possible only if there is real partial persistence over extension-motion. The aspect of process that allows perception of persistence over extension-motion persuades at once both nominalism and realism.

---

301 Sfendoni-Mentzou, "The Reality of Thirdness", 75.
302 Wittgenstein's 'family resemblances' of language games may be termed nominalistic. For example, Wittgenstein, *Philosophical Investigations* 67: "I can think of no better expression to characterize these similarities than 'family resemblances'; for the various resemblances between members of a family: build, features, colour of eyes, gait, temperament, etc. etc. overlap and criss-cross in the same way.—And I shall say: 'games' form a family." See also the numbered remarks that follow, especially 77. Family resemblances, as Wittgenstein holds, are not potentialities with a trans-object, ontological, objectual-causal existence in the intellect, with foundation in the ways (universals) and the Way (the universal, To Be) of Reality-in-total. Note also that Bambrough argues that Wittgenstein steer a middle course between nominalism and realism in his theory of family resemblances. Bambrough, "Universals and Family Resemblances", 274ff.

The nominalist stresses the otherwise merely acausal and epistemological-classificational aspect, stressing the *non-actuality* (which unconsciously stresses instantiation and potentiality) *of universals* and, on this way, classifying them as non-entity (due to the unconscious opposition of non-actuality with unreality and impossibility). Realist metaphysicians stress the aspect of causal persistence, some realists even concluding unconsciously to the existence of universals.

Scientific ontology must reduce universals as instantiating the potential ways of being of processes, and reduce their concepts into a mixture of the ontologically validated reflections of ontological universals, i.e. connotatives, with conscious elements in brain activity. To circumvent the problem of merely acausal, nominalistic and extension-motion-persistence metaphysical interpretations using the tension between universals and particulars as a reason against ontological and connotative universals, the explanation of both ontological and connotative universals must be made objectually causal.

Armstrong makes a causal explanation of laws in a way that allows Sfendoni-Mentzou to critically overcome the same in an instantiating-potential manner. She makes an in-depth study of Armstrong's meagrely causal interpretation and critically transcends it in favour of meeting the Aristotelian and Peircean demands on universals and particulars, which, for Einaic Ontology, involve the two varieties of universals. In Armstrong's concept of causality within the context of his theory of truthmakers (Sfendoni-Mentzou does not use the theory of truthmakers to study Armstrong's concept of causality) there is close connection between causes and laws of nature in the ontological sense. The nature of the relation between the token events in causation may be seen as necessary (not by being merely regularistic, but singularistic[303] and intrinsic). Insofar as a cause bestows a positive causal power, say, in the case of a first ball coming into contact with a second,

> [...] each member of the totality of the relevant properties of the first ball (at the beginning) is different from the property of being in contact with the second ball. If, in addition, we think of causality as 'the cement of the universe', the central relation that lawfully binds things together in the world, we will not be very impressed by these general states

---

303 Causal regularism denotes the Humean (and Lewis') variety of non-processual ontological concepts of causality that holds that only a mere regularity may be perceived between the token events. Singularism about causation, as in Armstrong, may be put thus: "[...] the truthmaker for the sequence being causal is the holding of a dyadic relation, the relation of singular causation, between the token events themselves. This relation is, using terminology introduced by Lewis, *intrinsic to its pairs*." Armstrong, *Truth and Truthmakers*, 126.

of affairs as terms of the causal relation. We will want the cause to be something *positive* – a particular's having some positive property – and equally we will want the effect to be something – again a particular's having some positive property. This suggests that the ontological terms of the causal relation should be restricted to states of affairs, events in a somewhat restricted sense.[304]

A simple "general" relation not involving processes from beyond from the past at any stage cannot any more be a causal relation. It merely involves restricted states of affairs at any given extension-motion region without mention of further causal aspects in the wider matters of fact. Causality is for such general-relation accounts a relation between entity-like processes. Discourse about entity-like processes is also language-bounded. These relations are entity-like (named) for the language world, but not for the purely objectual nature of processual Reality, in which reality-in-particular is couched as processually connected unto other particulars. Reality, as the finally objectual token, the final total of all that are infinitely and infinitesimally causal, has to somehow be allowed to justify the causal relation between the terms that represent states of affairs involved in causal relation. Hence, concepts of entity-like terms require causal past-ward extension for ontologically more justifiable explanation, and these can be had only by involving universals beyond the given immediacy.

Universals involved in specified causally active parts in processes are not particular things nor are they unreal. This fact points to the 'real'-but-not-'actual' and instantable-potential nature of ontological universals in contrast to particulars and to the basic feature of the metaphysical mode of essences or the common nature (the indeterminacy implied by potentiality) of reference of essences as *ens reale*:

> The first thing to remember is that common nature *per se* is not a *res* but a *realitas*, which is neither particular nor [sic] universal; it exists objectively only with its individualizing conditions, and is conceived by the intellect as universal. It is the mind itself which endows it with its character of universality, for of itself it is indifferent, a mere *natura*: "Horseness is just horseness, neither of itself one nor many, neither universal nor particular." Accordingly, it does not point to any particular individual case and this is exactly what constitutes the indeterminacy of the *ens reale*: "it can be real without being determined to exist in any one thing". We are thus led to the fundamental character of essence or universality *per se*. It can be defined as a mere possibility of existence, just as when it has become existent it is the possibility of a universal.[305]

---

304 Armstrong, *Truth and Truthmakers*, 144.
305 Sfendoni-Mentzou, "The Reality of Thirdness", 86. She refers here to Scotus, Avicenna and Moore.

As against this view of Sfendoni-Mentzou there is Armstrong's moderate realism, which he calls *a posteriori* realism. Instantiation as the nature of the universal *per se* is stressed here. What exactly Armstrong means by 'existence' of the instantiated law, anywhere in the context of his realism, is difficult to answer, since there is no space left in his theory for the real-ness of the ontological universal in causal pasts and futures, which are apart from particular instantiations. A careful consideration of Armstrong's principle of instantiation as divorced from potentiality shows that he leans towards nominalism,[306] ending up in mere semantic and acausal satisfactions in ontology, which is bereft of a tenable objectual theory of Reality.

A distinction between 'objectual' (at the level of Reality) and 'objective' (at the level of knowledge of the particular within its immediate setting) allows us to accept the objectivity / reality of the (ontological) universal as instantiating potentiality, as understood by Sfendoni-Mentzou. She does not provide sufficiently for the final epistemological aspect, Reality-in-general. In itself, a common nature has no instantiated actuality, it is a universal in the intellect, but it has a foundation in Reality and reality-in-particular, since the intellect "veridically" (not merely directly perceptually) objectualizes from the realm of Reality through connotative essences and the most general Reality-in-general.

Universals *can* be seen epistemologically as intensional entities, meaning (1) the connotation or intension or meaning of a linguistic expression, i.e. "the (individual) concept conveyed by a singular term, the property expressed by the predicate, and the proposition expressed by the sentence" and (2) as distinguished from denotation and extension which consist of [ontological commitment to] things signified by the expression.[307] I argue in this chapter that there is not merely a particularistic, but a universalistic, ontology behind the epistemology of universals. That is, the actuality of the common nature of universals is not merely epistemological or particularistic, but steeped in the bosom of actuals: i.e. reality-in-particular and Reality, always via the ontological universals of processes that reflect themselves in *the intensionals qualitative of actual particulars and Reality*, namely, connotative universals, universals of universals etc. and finally Reality-in-general.

---

306 Sfendoni-Mentzou, "The Reality of Thirdness", 77–78. What she understands as nominalism is what I have called the merely 'semantic' study of the relation between universals and particulars. Only what is objectual can be objective. The contrary is not necessary.
307 *The Cambridge Dictionary of Philosophy*, second edition, s.v. "Intension" and "Intensional Logic".

Reality-in-general and connotative universals are the results of mental objectualization. They do not "veridically" objectivate universals, but do so objectually in the case of entities through universals of processes. The intellect is unable to satisfy itself by objectivation at the level of reality-in-particular, since it can objectualize at the level of Reality and reality-in-particular alike. Without objectualization by the categories of Reality and reality-in-particular via universals and Reality-in-general, there is no discourse at all.

Objectualization is not merely the work of the intellect alone. Reality and reality-in-particular are in process with the intellect currently, they have been so in the past of both the respective reality-in-particular and in the causal roots of the intellect, in which the intellect existed as reality-in-particular, and will be active from within the treasury of the past proper in the future forms of existence of *parts* of the same intellect. This whole process is holographic of ntological universals – a fact that is least treated in the ontology of epistemic and other conscious activities. In short, *Reality and reality-in-particular are active through ways of being of processes, in the intellect, by means of effects of reflection of ontological universals (ways) in connotative shapes in the very processes that were minutely conceptual in the causal past of the physical parts of the intellect.* The intellect is the present processual culmination in the train of the on-going process of feeling and thinking in terms of Reality-in-general.

Hence, the intellect is capable of attaining truth-probabilities in the most genuine sense only if Reality-in-general is involved. This aspect of essences, as particularizations of the highest connotative (Reality-in-general) of the highest potentiality (To Be), is not taken into account in the confluence of realisms of laws of nature, essences and causality in Armstrong's or any other analytic ontology.[308] The purely objectual aspect of Reality (To Be) is epistemological and epistemic (i.e., in perceptions, feelings and thoughts by mentality – in its conscious, pre-conscious and unconscious aspects). The purely objectual aspect of Reality is seen as potential by perception, feeling and thought at the level of the highest connotative ingredient – Reality-in-general – of all thought, and at the level of ontologically fully realized Reality. Such is the Reality that is seen as individuated via other ontological essences in reality-in-particular. Only at the realm of Reality-in-general can thought individuate To Be in concepts and propositions.

---

308 In Armstrong there is at least the admission that universals may be extended up to universals of universals. This is a welcome idea, not to be encountered in most other analytic thinkers. Hence, it is safe to conclude that positing the objectual aspect of Reality-in-total and the consequent satiation of universals in Reality-in-general are almost totally absent in analytic ontologies.

That is, Reality-in-general is the second, epistemological, category *par excellence*, epistemically instantiative and potential of Reality in consciousness. This should be understood against the background of the fact that To Be is ontologically instantiative and potential of Reality and epistemologically in Reality-in-general.

The connection between ontological universals and To Be on the one hand and connotative universals and Reality-in-general on the other is worth studying. This is feasible only in an elaborate Einaic Epistemology, which is beyond the reach of the present work of introductory type. Before going into the connection between Reality, Reality-in-general and To Be, let us do the spade-work for getting at the probabilistic nature of connotative universals, so that ontological commitment to ontological universals will give greater acceptability to the fact of universals. This facilitates a systemic approach to the three highest categories.

2.2.2 System-Building vs. Absolutism and Probabilism in Essences

Ontological universals are not entities self-subsistent to any extent like actuals. They are neither actual entities nor unreals, but real potentialities instantiable and dependent on actual entities (that point to the past that has bestowed actual causal effects on actual entities), including the widest actual entity Reality. The dependence of theory on actuals and Reality makes universals ontological niceties. An additional feature of universals, not very much given sufficient thought to in the history of philosophy, is the fact that they are potentials always subsistent on other ontological universals including To Be. We now discuss the mutually dependent and relativising manner of ontological universals and the consequent probabilistic implications of connotative universals.

While analysing the infinitesimally microscopic and the infinitely macroscopic in existents, we have discussed why probabilism in QM (like the Uncertainty / Indeterminacy Principle, Planck's constant, etc.) basically remains perspectival absolutism and why it incurs the same mistake when it goes for a non-causal interpretation by absolutising on an alleged break-down of *all forms* of causality, laws of nature etc. at the micro-, nano- and other levels. In substitution of this extreme state of affairs in QM, (1) I have proposed in Chapter 1 objectual-causal realism from the microscopic and macroscopic levels of Reality, and (2) I have defended in 2.2.1 confluence of objectual, infinite and infinitesimal causality, laws of nature and universals in the region of the infinitely and infinitesimally objectual-causal Reality and its connotative ideal, Reality-in-general. Together these allow instantiation and possibilization of every specific process within the systematic categorial framework of the two categories of Reality and Reality-in-general.

Two questions about the nature of categories are to be dealt with before finally accepting Reality-in-total and Reality-in-general as two flexible maximal categories of the best possible of systems: (1) What is the extent of tenability of system-building on the foundation of categories? (2) If universals are not absolutes, but instead ontological notions probabilistically built on ever broader universals and ever more truth-probabilistic propositions, how are systems possible? The objectual nature of Reality and the possibility of connecting Gödelian deductive systems with formalized physical theories can be an answer for question (1). The mutually relativising nature of universals may be derived from the concept of pushing primitive notions and axioms, yielded by Gödel's Incompleteness Theorem, and its influence on system-building can show that an answer to question (2) is possible, if the place of universals and categories in systems is made compatible with Gödel's result, combined with the epistemological implycations of the Indeterminacy Principle, against the background of the objectually causal Reality and the broadest connotative universal Reality-in-general.

The following is thus a prelude to acceptance of the truth-probabilistic capacities of humans in the field of knowledge in general and scientific activity and philosophical system-building in particular. The tradition of metaphysics has badly mistaken epistemological probabilism as a mere inability of human mind "to have all knowledge". Our ontology goes a long way to demonstrate that probabilism is the combined name for (1) human intellect's inability to have no absolutely justified knowledge about anything and (2) the inability of Reality and reality-in-particular to give itself to any connotatively characterized conscious absorption of ontological universals and To Be as such by absolutely direct replication in connotatives. Such a state of affairs has to be incorporated into the nature of all philosophical activity which we have by applying the Einaic Ontological implications of Gödel's theorem. Determinisms of all sorts have resisted (1) above, i.e. epistemological probabilism. Our determinationism facilitates both (1) and (2). We take care against epistemological determinism and merely epistemological definitions of probabilism, which resist determinationism and the Ontological Principle of Excluded Vacuous Middle discussed in Chapter 1.

2.2.2.1 Gödel's Theorem and System-Building

Any mathematical or logical system begins by defining primitive notions[309] and fixing axioms based merely on these notions. It is somewhat the same in

---

309 For example, David Hilbert's programme of axiomatization of geometry (1899) employed 'point', 'line', 'lies on', and 'between' as primitive notions. Their meanings

purpose-built ontological systems. They tend to reify the denotations of their denotables, absolutize their denotatives; and consider the axiomatic propositions as self-evident and the procedural results (conclusions) as solidly evident by two-valued logic because even multi-valued logics must formulate their primitive notions and axioms using two-valued logic.

One of the axioms of Euclid's work, *The Elements*, was the parallels postulate.[310] Later doubts about any self-evidence of the postulate (doubts due to the infinite asymptotic regions of the uni-dimensional space involved at the meeting of two parallel lines) used to be discussed by attempts to derive it from the other presupposedly self-evident axioms. Non-Euclidean geometries[311] found alternative expressions for the parallels postulate without deducing the parallels axiom from the other axioms, and unconsciously made diversity in system-building achievable by showing that the feasibility of relativization of at least one axiom of geometry will produce new systems according to the shape of the parallels axiom. This showed the way to proof of the impossibility of proof of at least some axioms from within the available clarity of primitive notions that are theoretically intuitively set prior to the axioms of the system. Thus, it became clear that the business proper of pure mathematicians and logicians is to derive systems and theorems from already postulated genuine-looking and intuitively rational assumptions.[312]

Consequently, a set of fundamental categorial notions can intuitively give rise to a set of highly rational-looking axioms, and we have each time a new mathematical system. The primitive notions of such re-deepened systems will border with qualitative notions, and there arise disciplines that need not even be the science of quantity that mathematics is wont to be. Mathematics today is thus a structural science more abstract than the mathematics that used to be taken as the science of quantity, because it should now abstract from the qualitatively primary and also from the qualitative notions and axioms of its results. Mathematics is now more formal because the validity of theorems in it now lay purely in the structure of propositions as based on the qualitatively primary notions and axioms, and not directly on objects and processes in this world.[313] The

---

      are defined by the axioms they enter into. All other commonsense meanings of the same are to be shunned for the system to continue. Nagel and Newman, *Gödel's Proof*, 12–13.
310 Efimov, *Higher Geometry*, 16ff.
311 Efimov, *Higher Geometry*, 85ff.
312 Nagel and Newman, *Gödel's Proof*, 9–11.
313 Nagel and Newman, *Gödel's Proof*, 11–12.

ontological aspect of such theorems can still make ontological commitments, but this is the subject proper not of mathematics but of metamathematics and ontology of mathematics. For this reason, non-ontologist mathematicians claim a sort of formal purity, not available elsewhere, for mathematics, even while seeking frameworks for re-deepening primitive notions and axioms and creating new mathematics.

Gödel's proof showed the same possibility regarding arithmetic, using Russell and Whitehead's work, *Principia Mathematica*, as a system that makes endeavours to render arithmetic a foolproof system based solely on primitive notions and basic logical axioms:

> A system is *complete* when every one of its formulas, or its negation, is provable. Therefore, to prove the incompleteness of the system it suffices to exhibit – or prove the existence of – a well-formed formula of the system, such that neither it nor its negation is derivable in it. Finally, a system is *decidable* when there is an algorithmic (mechanical) procedure through which we can determine, in a finite number of steps, whether or not each well-formed formula is provable in it, i.e., whether or not each formula is one of its theorems.[314]

Further, Gödel's "Über formal unentscheidbare Sätze der *Principia Mathematica* und verwandter Systeme I" showed that if there exists a "fundamental pair {A,T}" of the set of true statements for the language of arithmetic, and if there exists "a nonnumerable arithmetic set", then "[…] there does not exist a complete and consistent deductive system for the fundamental pair {A,T}, of the language of arithmetic."[315] For formal logic this means that "[…] for a large class of formal systems there must inevitably exist an undecidable statement, i.e., a statement such that neither it nor its negation can be derived from the axioms of the given system."[316] In ordinary language it means that in any formal system not all true propositions, whether axioms or deductions therefrom, can be proved or disproved from within the system, and these can be proved from other formal systems that re-deepen the conception of some of the primitive concepts of the system in which the few propositions are not provable.

Something intuitive (deduced rationally from any system/s that may allow the conclusion from the outside) should play at the roots of the primary notions and axioms, the propositions (axioms) being undecidable (by reason of the Theorem)

---

314 Rodríguez-Consuegra, "Realism, Metamathematics and the Unpublished Essays", 18.
315 Uspensky, *Gödel's Incompleteness Theorem*, 30–31 (Paraphrase by Uspensky).
316 Uspensky, *Gödel's Incompleteness Theorem*, 84.

but true by assumption (by any system/s from without). Since all systems base themselves on logical first principles – all of which may be brought to veridical probability of axiomatic truth – the result of the Theorem may be extended to formalized physical and ontological theories too.[317] Any formalized physical or ontological theory has some propositions that should be demonstrated by systems outside it, which use some notions alien to the system under consideration. Hence, "[…] to complete the meaning of a formalized physical theory, it is necessary to make some philosophical extension of it."[318] This brings in elements that are intuitive with respect to what are reasonable within the system at hand, but rational with respect to what are reasonable in system/s from without. In short, the relatively more intuitive aspect of the foundations of systems cannot be overstressed. Note here that intuition of notions more foundational to existing basic notions is not irrational by common admission.

If there are different primitive notions and axioms in mathematical, logical, physical or ontological theory, different abstract connotative universals are at work at their foundations. The more the abstractness of the system, the greater is the urgency of the question "[…] whether a given set of postulates serving as

---

317 Gödel argues clearly that, just as physical propositions have content, so also mathematical propositions have content. This puts mathematics and physics on par. Gödel, *Unpublished Philosophical Essays*, 216. In the editor's notes to the text of Gödel's unpublished work, "Is Mathematics Syntax of Language?, VI", Rodriguez-Consuegra elucidates this point further: "We are told about "immanent existence" as being attributed by the positivists to the "correct mathematics." Moreover, Gödel adds the following strong assertion: "[…] these mathematical objects and facts [e.g. infinite sets or properties of properties] cannot be eliminated (as, e.g., infinite points in geometry can), since there always remain primitive mathematical terms and axioms about them, either in the scientific language or the metalanguage, where 'axiom' here means a proposition assumed on the ground of its intuitive evidence or because of its success in the applications." Therefore, Gödel concludes that the positivistic point of view regarding mathematical objects is inconsistent when the positivists speak about their "pseudoexistence"; the difference with physical objects "primarily lies in their different *intuitive character*", but their roles in the formalism is similar, so they could be regarded as "irreducible hypotheses of science, exactly as the assumption of a field and of the laws governing it."" Gödel, *Unpublished Philosophical Essays*, 220. Gödel puts the shortest form of the argument for content in mathematical propositions thus: "If mathematical intuition is accepted as a source of knowledge, the existence of a content of mathematics evidently is admitted. If it is rejected, mathematics becomes open to disproof and for this reason has content." Gödel, *Unpublished Philosophical Essays*, 182.
318 Dingguo, "On the Neutral Status of QM", 310–11.

foundation of a system is internally consistent, so that no mutually contradictory theorems can be deduced from the postulates."[319] Moreover, since the primitive definitions and consequently also the axioms can always mean differently in different systems, deductive systems are no more merely formal-syntactical operations proceeding from primitive notions, axioms etc. In that case, the notions and conclusions that originate in any system are deductively true only after ratifying the constitution of the initial conditions. Prior to that, the generalized notions and propositions have senses that are possibly also inductive (i.e. depending on external conditions, as in the laws of nature).[320]

Such lack of absoluteness in inductiveness and deductiveness gives free play for the insufficiency of deductiveness and inductiveness in themselves, and thus also for truth-probabilities in the axioms and results of any system, all of which prove to be inductive and deductive in varying degrees of admixture of inductiveness and deductiveness. Therefore, the probabilities of truth and error in axiomatic principles and as a result of use of primitive notions (except in the case of the meagre level of error and the higher level of truth in expressions and statements of the broadest possible of categories and related axiomatic principles of all thought) are indices of the combined effect of heightened or diminished tenability accrued from the separate spectra of values of inductiveness and deductiveness. Thus, mutual contradictoriness is possible in systems, because probabilities are inbuilt at their very sources. This fact tells upon the nature of propositions, notions and truth-probabilities worked out by their use in the admixtures of inductiveness and deductiveness in propositional conclusions.

---

319 Nagel and Newman, *Gödel's Proof*, 14.
320 The possibilities of confluence of induction and deduction in systems were foreseen by Gödel: "It might be objected that the analogy between mathematical intuition and the supposed additional sense breaks down insofar as the general laws holding for the supposed second reality could be disproved by further observations. However, the same would happen for mathematics if an inconsistency arose. For a disproval of observed laws also is nothing else but an inconsistency between different methods of ascertaining the same thing since empirical induction and the application of laws of nature also are such methods. The 'inexhaustibility' of mathematics makes the similarity between reason and the senses [...] still closer, because it shows that there exists a practically unlimited number of independent perceptions also of this 'sense'. In reply to another possible objection it should be noted that, exactly as mathematics, the second reality, although implying nothing about the facts in the first reality, nevertheless might help us to knowing the latter, e.g., if it contained schematic pictures of all situations possible in the first reality." Gödel, *Unpublished Philosophical Essays*, 208–9.

Gödel holds also that the primary notions (which he mentions under the name 'meaning of the terms') and axioms have some inherent connection with the real world. By reason of the inherent connection, mathematics as an abstract and formal science finds application to *what he conceives as "objective" reality*; but, as I propose, *formal science finds application to the objectual-causal world as is presupposed by ontological commitments of primitive notions and axioms*. This connection is only perceived and described. According to Gödel, "What is wrong however is, that the meaning of the terms (i.e. the concepts they denote) is asserted to be something man-made and consisting merely in semantical conventions. The truth I believe is that these concepts form an objective reality of their own, which we cannot create or change, but only perceive and describe."[321]

What he calls objective points to the particularistic world, beyond which (broader / more total than which) is Reality, which I have characterized objectual. If the "objective" are treated not merely as particular tokens or objects, we will have a world without ontological universals. That is, Gödel does not couch the objective in ontological universals or Reality or its To Be, without which the final possible edges of systems after systems, in principle, are not comprehended by way of discovering the real *fundamentum* for couching the farther limits of possible systems. In short, Gödel's Incompleteness Theorem opened the way for pounding (1) the alleged absolute formality of mathematical and logical systems by showing loopholes in the formal nature of deductive systems, (2) the alleged absoluteness of any definition of the senses of concepts and (3) as he may not have realized, also the self-consistency of definitions of ontological notions.

Here we should resist the common temptation to trace the roots of such non-absoluteness of systemic formality and definitions of senses of concepts crudely to consciousness, language, social praxis etc. – as in particularisms of analytic thought, phenomenology and hermeneutics – which are not sufficiently capable of thoroughness by possibly exhausting the limits of thought ever better. Instead, for maximum possible thoroughness we should trace those roots to the very Reality. This is ontologizing indirectly, using Gödel's result. Hence, our way of procedure here will be such epistemologizing of connotative universals as probabilistic at the level of Reality-in-general, ontologizing of ontological universals as non-probabilistic (due to their ideality) at the level of Reality, and thus truth-probabilizing our knowledge of Reality.

Insofar as any ontology based on the Einaic categories is about Reality it is also a system, because it rests on possible limits of totalization of background

---

321 Gödel, *Unpublished Philosophical Essays*, 144.

realities of any process at study through the epistemological universal proper to it, i.e. Reality-in-general. Are these two categories equivalent to the primitive notions of deductive systems? They are similar, but they are the very ontological and epistemological generalizations of all possible notions and can work as primitive notions for systems. These too cannot resist furtherance of conceptual basis, as is mandatory for primitive notions. *The merit of these categories* is that they have imbibed into their very nature the probabilistic character of effects of pushing of possible ontological categories, and so, need only definitional and explanatory augmentation due to their cosmological and epistemological dimensional finality. System-building upon these foundations are most possible, because these foundations offer the highest possible cosmological and ontological totalization and epistemological generalization of all, and generate reality-in-particular by instantiation and potentiality, based on probabilistic ontological and connotative universals, as is required by application of Gödel's result on this system of categories. It is necessary, because discourse can be justified only on the best possible of foundations, and not on an absolute lack of foundations, according to Gödel's result.

2.2.2.2 Gödel's Result vs. Probabilistic Universals and Categories

What Gödel mentions as the connection of the formal to the actual, through primitive notional and axiomatic foundations of deductive systems, implies nothing but what I term as *the intertwined nature of induction and deduction*, especially at the roots of eminently inductive or deductive systems. I believe that Gödel is not sure of the farthest ontological implications of his work. He limits the implications of his results merely to the "incompletability or inexhaustibility"[322] of mathematics – thus exhibiting no awareness of ontological commitments to ontological and connotative universals, tokens and totals behind the primitive notions – and insists on the truth of consequents in contrary-to-human-epistemological-nature statements that involve the pragmatic assumption that undecidable proposition/s at the roots of systems are somehow to be accepted as absolute truth, with some sort of contact with existing and seemingly necessary practices:

> E.g. *some* implications of the form: If such and such axioms are assumed then such and such a theorem holds, must necessarily be true in an absolute sense. Similarly any theorem of finitistic number theory such as 2 + 2 = 4 is, no doubt, of this kind. Of course the task of axiomatising mathematics proper differs from the usual conception

---

322   Gödel, *Unpublished Philosophical Essays*, 129.

of axiomatics in so far, as the axioms are not arbitrary, but must be correct mathematical propositions and moreover evident without proof, there is no escaping the necessity of assuming some axioms or rules of inference as evident without proof because the proofs must have some starting point. However there are widely divergent views as to the extension of mathematics proper, as I defined it. The intuitionists and finitists e.g. reject some of its axioms and concepts, which others acknowledge, such as the law of excluded middle or the general concept of set.[323]

The stumped recourse to pragmatism evident in his finitistic theory, I believe, is inspired by direct semantic reference to reality-in-particular, without recourse to the Einaic categories presupposed by ontological commitment behind all truths and statements that involve particulars and tokens. I hold that here is evident a defect of Gödel's ontological horizon. Within the context of Gödel's Incompleteness Theorem, the finitistic attitude of the *Completeness Theorem has, for its horizon, a given system*. The proofs here need not be the exclusive candidates for determination of absolute truths. To see the truth or otherwise of the undecidable proposition/s of system-1, it has to be made part of a system-2, by use of the primitive notions and axioms of which it must be argued further to demonstrate the said proposition. This too is not ultimately decided, because system-2 is itself based on other undecidable proposition/s of at least a system-3 ... and so on.

As Gödel's Theorem shows, any truth is a truth when decided in the context of the primitive notions and fundamental propositions of a system. The lack of absolute truth-probability in any system, thus, works towards the possibility of every truth being a counterfactual conditional ('If it were the case that ... then would it be that ...') of a certain kind. The condition is incapable of an absolutely necessary satisfaction of the Einaic categorial connections of the primitive notions and axioms. The reason is that the Einaic categorial connections are real and processually existent with respect to Reality and only real and non-existent with respect to the irrelevant universals. Thus, every conditional offers a different truth-probability from an angle different from that of another, because the connection between the Einaic categorial dimensions and the primitive notions is widely insufficient to attain the highest truth-probability. Therefore, the more the axioms are pushed backwards by pushing the universal-bounded meanings of the primitive notions implied in the axioms, through further systematic meta-justification of meanings through broader universals and study of the relevant processual realizations, the more is there truth-probability for the relevant affirmative or negative of any proposition in view. An additional reason for this is

---

323 Gödel, *Unpublished Philosophical Essays*, 130.

the Einaic Ontological necessity of connecting primitive notions of systems with the dimensional categories of Einaic Ontology.

The insurmountable problem here is that the meta-system in which all these (quasi-) and other counterfactuals are members will require other conditions for truth of propositions, based on adequacy of connection between the Einaic categories and the very primitive notions of the given system. These are provided by assumptions proper to a system-2, which is made counterfactually conditioned upon assumptions of a system-3, and so on. Thus, we can speak of the absolute truth all the more worse of a counterfactual law statement, under the implicit particularistic conditionality of respective antecedents, merely because the ontological universals (ontologically) and the connotative universals (epistemologically) implied in the primitive notions always connect the processes that the primitive notions ontologically commit to, unto other broader ontological processes, and finally unto the Einaic categories – of which Reality-in-general is the mother of all possible worlds.

This line of arguments comes up with the question of the need of different metamathematics and metalogics, which is a question within logic and ontology. As regards first order mathematics, the absence of proof of validity of all and every one of the propositions of a given system is catastrophic in view of absolute consistency. Gödel's Completeness Theorem for logic came in 1930, stating that sentential logic – the logic with no quantifiers – is complete;[324] and his Incompleteness Theorem for arithmetic came in 1931. In the words of Rodríguez-Consuegra, Gödel has shown, as sequel to the second theorem, also that "[…] it is impossible to find a proof for the consistency of such systems, because a sentence asserting that consistency would be one of those undecidable sentences."[325] This resulted in abandonment of ideal formalism for mathematics. To read the two theorems together, absolute formal consistency can only be a presupposed consistency, not a proved one. That is, the consistency at the level of proved consistency is that of ontological commitment to the objectual implications / presuppositions of the primitive notions that appear in the axioms.

In 1904 David Hilbert had proposed what is called "Hilbert's Programme" in order to save mathematics from the crisis of contradictions in set theory. Hilbert's Programme had two parts: (1) formalization of mathematics, which was very much supposed to have been done by Russell and Whitehead's *Principia Mathematica*, and (2) consistency-proof of a formal theory. The latter means

---

324 Rodríguez-Consuegra, "Realism, Metamathematics and the Unpublished Essays", 19.
325 Rodríguez-Consuegra, "Realism, Metamathematics and the Unpublished Essays", 19.

"[...] the consistency of a formal theory, i.e. probabilizing of a proof-figure ending in a contradiction $0 = 1$."[326] To facilitate the same, Hilbert assumed the "finite standpoint": "We can finitely operate on a concrete figure given before us, and infer a general statement as a Gedanken experiment."[327] Bernays (Hilbert's co-worker in Göttingen between 1917 and 1933) invented "[...] the axiom of dependent choice as the part of the axiom of choice necessary to analysis."[328] This is useful within questions concerning completeness of mathematical theory.

In 1930 Gödel's doctoral thesis had proved the Completeness Theorem that states "[...] a theory is consistent if and only if it has a Model."[329] A model is within the system, and obeys the axiom of dependent choice for completeness. This is a question within first-order logic. Hilbert and Ackermann formulated first-order predicate calculus well, in their *Grundzüge der theoretischen Logik* (1928). But they also stated: "It is still an unsolved problem if the axiom system is at least complete in the sense that all logical formulas which hold in every structure, can be deduced. It can only be said empirically that this axiom system has sufficed in every application."[330] Gödel showed this in his Completeness Theorem. I hold that it is also a proof of the importance of Reality and reality-in-particular in deciding primitives and axioms, i.e. the place of reality in theory- and system-building.

Further, the Incompleteness Theorem deals with the deductive and axiomatic completeness of systems. With the background of the above discussion it is possible to point out that every mathematical system needs an endless train of metamathematics for justification through justification of each of the justifying metamathematics. It is therefore easy to conclude that success[331] of a theory in a system implies further pushing of primitive notions and axioms by a series of

---

326 Takeuti, "Work of Paul Bernays and Kurt Gödel", 77.
327 Takeuti, "Work of Paul Bernays and Kurt Gödel", 78.
328 Takeuti, "Work of Paul Bernays and Kurt Gödel", 79.
329 *Encyclopaedic Dictionary of Mathematics*, s.v. "Gödel's Completeness Theorem".
330 Takeuti, "Work of Paul Bernays and Kurt Gödel", 80.
331 In footnote no. 44 of Gödel's unpublished work "Is Mathematics Syntax of Language? II" he summarises the concept of success of mathematical theory: "'Success', within mathematics, of some new mathematical axiom would mean that many of its consequences could be verified on the basis of the former axioms, the proofs, however, being more difficult, and moreover, it would solve important problems not solvable before. Note also that the consistency of some new axiom, provided it yields a substantially stronger system, is indemonstrable on the basis of the preceding axioms and their consistency. But the undecidability, from the former axiom,

systems, which yield greater truth-probability for propositions of systems more fundamental than later ones, due to the ontological dependency of primitive notions for their choice on ever more fundamental ones. Primitive notions could can on objects / processes only via connotative universals that reflect their relevant ontological universals. So, definitions of primitive notions should, propositionally, set the semantic sense of notions.

Propositions and their truth and falsity are bound up directly with connotative universals, because these latter are the conceptually related forms of ontological universals that are ways of being of processes. So, *pushing axioms backwards implies pushing the connotative universals behind their constitutive primitive notions, and pushing the ontological universals behind such connotatives into the more and more general realms.* Exactly this is the process implied by the ontology of couching the best explanation of the universals pertaining to particulars in the broadest of them, i.e. Reality-in-general. *This shows that the probabilistic (i.e. not-fully-defined-from-within) nature of connotatively circumscribed universals, such universals of universals etc. is the real* **explanation for the truth-probabilities** *yielded by pushing the fundamental axioms to ever more general senses by pushing the definitions of primitive notions.* In this case, the question of incompleteness of mathematical systems is an issue of logic and of the epistemology of formation, in mind, of abstract objects and their connotatives, which take their ontological relevance from their related ontological universals.

The above fact may be further brought out in the following manner. Things metalogical or metamathematical are not totally unconnected to logic or mathematics. Instead, they work at a level more *a priori*, or theoretically preparatory, to logic or mathematics. Even so, the sense in which an ontological or connotative universal is taken today is probabilizable by future questions of systemic completeness of the system in which it works. Moreover, systemically, even the terms 'epistemology', 'ontology' etc. in our context should be used to mean 'apriorily / preparatorily theoretical to positive-scientific and exact-scientific epistemologies and local ontologies'. The metaepistemology to which the connotative Reality-in-general belongs is of this probabilizable variety.

We approach herefrom the cosmology behind our epistemology of probabilistic connotative universals. *If Reality-in-general is implicit in the inductive-deductive process of pushing primitive notions and axioms, then the category of Reality-in-total too is implicit in it.* Just like there are universals and class terms,

---

of certain questions decidable by it might be so demonstrable." Gödel, *Unpublished Philosophical Essays*, 210.

there are realities-in-particular in the backdrop of perceptions behind propositions. Particularistic systems do not go deep and broad enough by pushing the definitions of primitive notions and axioms. The only important difference that we perceive from now on in system-building is this: any system where Reality and Reality-in-general are applicable is an inductive-deductive one, and hence it yields ever better truth-probabilities by reason of pushing primitive notions and axioms. Reality and Reality-in-general are themselves not pure universals. The former is the highest token of all thought, and so it is not a universal but *the* broadest entity. The latter is the generality of all connotative universals, never fully definable, but only better and better truth-probable, due to the impact of Gödel's Theorem on any system that attempts to define the terms that naturally involve connotatives. Hence, *Reality-in-general is only the dimension par excellence of universals.* There is no final system, but ever better systems can exist with their own ever better generalities of truths, achieved by reason of primitive notions of ever-greater generalities achievable. These primitive notions, in the projected dimension of ever-greater tenability, are called Eniaic categories. The connotativity of categories commits in the ultimately *dimensional* manner to both things / processes and ontological universals. *The probabilism active upon categories is therefore epistemological probabilism, resulting in ever better truth-probabilities with objectual causality in the zone of discourse of the objects / processes ontologically committed to.*

The cosmological-ontological connection between infinite and infinitesimal objectual causality and epistemological probabilism may be conceived as follows. What is objectually causal is Reality. Its conceptualisation always involves ontological universals, such universals of universals etc., up to the ideal, To Be – but all of them connotatively at the conceptual realm. That is, pushing the senses of the primitive terms, which goes to contribute to advancements in the primitive definitions that work in axioms, is the method by which we approach connotative generalities of generalities in systems and the presupposed connection of ontological universals to others of varying ontological universalities. So, pushing the definitions of primitive notions and axioms in theory yields always greater and greater generalizations in terms of Reality-in-general and greater and greater totalities at the realm of ontological universals and the To Be of Reality. The objectual is, thus, better and better grasped in its natural-law level of connotative generalities and its causal connectivity unto Reality. This is in fact epistemological probabilism regarding ontological processes, implying the lack of absolute circumscriptive connection of connotative universals with indirectly representative processual totalities that are productive of ontological universals. What is perceived and theorized about is determined by an ontological commitment that goes up to the most ontological, not merely by proofs. *This means that human knowledge about Reality is irreducibly probabilistic, whereas Reality goes its ways absolutely causally, which may be ontologically presupposed by way of the broadest ontological commitment to infinite and infinitesimal objectual causality.*

Here a passing comment is in place on the possible comparison of the contribution of postmodern (anti-)epistemology with Einaic Epistemology. The mutually probabilizing nature of both ontological and connotative universals points the way beyond postmodern ontologies and epistemologies of substitution of presence with absence. The "totally" unsettling methodology of postmodernist thoughts is thus partially unsettled in favour of the *dimension* of the ever more probabilistically universal in epistemology about ontological processes, by means of the probabilizing nature of connotative universals that bring ontological universals into consciousness. Thus, Einaic Ontology is a synthesis of the simultaneous need for staticity and change in discourse.

2.2.3 Transcendental Aspect of Particularist Probability-Makers

Chapter 1 argued for the infinite and infinitesimal objectual causality of all processes that are the case and their ontological universals; and Chapter 2 for the probabilistic nature of human knowledge and concepts and the possible worlds that arise at the interface of ontological and connotative universals. We know that some sort of ontological commitment to objects of some or other particularistic sort within the context of ontological commitment to Reality is a must for the movement to ever broader objectual ontological commitment in concepts, truths and conceptual possible worlds.

Cognition which is always truth-probabilistic has its own irreducible objectual aspect. Paul K. Moser calls it epistemologically as 'unconditional probability-maker', instead of Armstrong's 'truth-maker'. It is the purely particularistic objectual element present in perception and cognition. He thus presents what I shall call the cosmological and ontological connection of the epistemology of particularistic and probabilistic causality involved in cognition. If this could further be cosmologized and ontologized, we may have a sure objectual-probabilistic foundation for truth claims. Truth claims may further be shown to possess a transcendental and a Transcendental nature. Let us begin with Moser's proposal that nonoccurrent probability does connect the epistemic with the ontological:

> Not all evidential probability is occurrent. Some evidential probability-makers are nonoccurrent in the sense that they were present to awareness for a person, but are not now [occurrent] [....] The rejection of nonoccurrent probability-makers entails

an implausible justification [of] solipsism of the moment, and thereby raises serious problems for the possibility of justified belief in persisting physical objects.³³²

Such a stand implies Moderate Internalism, which in turn implies an epistemic foundationalism, which he claims is based on "nonpropositional experience":

> On this view some propositions can have evidential probability independently of evidential relations to any other propositions. Such a view involves a two-tier evidential probability [....] [T]here can be not only unconditional, nonpropositional probability-makers, but also derivative, propositional probability-makers [....] [S]ome propositional probability-makers are basic relative to all other probable propositions, insofar as their probability derives solely from nonpropositional experience [....] [O]ther propositional probability-makers are nonbasic insofar as their probability depends in part on other propositional items.³³³

In short, the internal element of knowledge is somehow related to evidential relations, via propositional probability-makers, which in turn irreducibly relate the internal to the external via some unconditional external elements called nonpropositional probability-makers. Moderate Internalism holds that evidential probabilities have an aspect that is non-external to the mind. This may be considered to be the universal-bounded character of theory. It is connected to foundationalism and coherentism. Foundationalism, which founds truth-probability on ontologically basic particularistic *sine qua non*s called 'nonpropositional experience', may be juxtaposed with coherentism, which believes that truth-probability issues from theoretical coherence.³³⁴ An impasse arises here, which makes synthesis of moderate internalism, foundationalism and coherentism difficult. Hence, Moser offers a synthesis through his (particularistic) theory of unconditional probability-makers based on nonconceptual, nonpropositional, experience and its object, nonconceptual object.³³⁵ "So such an experience is an ideal candidate for an unconditional probability-maker, i.e., a truth indicator in and of itself."³³⁶ Moser continues: "[...] [T]he only sense in which a conceptual perceptual experience can provide unconditional evidence is the sense in which its *non*conceptual component can provide such evidence."³³⁷ Being nonconceptual, it does not need another probability-maker.³³⁸

---

332   Moser, *Knowledge and Evidence*, 7. Square brackets mine.
333   Moser, *Knowledge and Evidence*, 7.
334   Moser, *Knowledge and Evidence*, 7–8.
335   Moser, *Knowledge and Evidence*, 35–36, 88ff.
336   Moser, *Knowledge and Evidence*, 88.
337   Moser, *Knowledge and Evidence*, 88.
338   Moser, *Knowledge and Evidence*, 89.

We reflect further on Moser's claims just presented. Insofar as the nonconceptual component results in an ontological commitment as presupposed in acts of knowing of the direct variety, the nonconceptual component is evidently an element external to the mind. But a particular ontological commitment may be commitment to an external physical element / object / process, or to an internal conceptual / abstract object (with its connotative ingredient), or to both. The nonconceptual element is not that what is propositionally concluded, but that which is presupposed in acts of knowing of the direct variety. In truths of conceptual experience, a nonconceptual component is only indirectly presupposed. But the indirect or direct element that works to communicate the nonconceptual or propositional truth to the experience or proposition is nonconceptual. So, the particular nonconceptual component involved directly or indirectly is really something external. Nevertheless, the nonconceptual component is not alone in its appearance at the level of the particular ontological commitment. On the side of the nonconceptual element, it is bound up with the ontological universals that are specific of the other physical objects involved in those ontological universals. By reason of the nature of ontological universals, the external and particular-objectual causation that encompasses the one external object is also involved in the other objects.

The identity of these two cases of occurrence is not of the physical object or process, but of the forms / universals involved. It is subject to physical interpretation, but always in terms of the identity of the ontological universal. Here rests the difficulty in identifying the nonconceptual object / process, which may be circumvented only by admitting that the identification is bound up by connotative universals, which are imperfect reflections of ontological universals, which, in turn, are part of the ontological commitment presupposed in the identification of the nonconceptual experience and the nonconceptual element in the process. Therefore, this connection is epistemological by mere imperfect mirroring – which may be perfected continuously by probabilistic pushing of axioms and perfecting of methods and actual processes of inquiry. It is ontological only by the foundational nature of the constant ontological commitment involved. These conclusions may be accepted only if the background support of ontological universals behind the nonconceptual aspect of theory makes consciousness's access to connotative universals makes theory perfectibly relative and probabilistic. Such an ontological background belongs to all conceptual universals that are perfectible reflections of ontological universals. I leave the question of the ontological universals and their background Universal for Chapter 3. The perfectible background world of conceptual universals is what I call the category of

Reality-in-general in the present chapter. It is not a purely conceptual affair, since it is something epistemic with ontological reflexive origin. To the extent that it is of ontological origin based on Reality, it has the Transcendent quality; and to the extent that it is a perfectible but foundational background world, a conceptual universal of universals of ... *ad libitum*, it has a Transcendental quality.

Moser concludes merely the play of the nonconceptual element in the act of knowing, which is merely particularistic. He does not ask after the more total ontological nexus (Reality) and more general epistemological nexus (Reality-in-general) behind the play of the nonconceptual element. That Reality-in-general is the most generalized of all the transcendently transcendental epistemic bases and the fact that Reality is the most totalised of all transcendentally transcendent ontological bases show also that they are the final epistemological (subjective) and cosmologically ontological (objectual-causal) foundations for there being nonconceptual connectives behind specific and direct or indirect acts of cognition.

Reality-in-general is the basis for what has been discussed as the probabilistic aspect of ontological discourse that we have derived from Gödel's Theorem. Pushing the definitions of primitive notions and axioms is possible only if one is after the more and more generally applicable forms of theory while pushing axioms in theory by connecting the nonconceptual element with other such elements via the causally nomic ontological universals that encompass them all in a given mode of connotative connecting. This is further generalisable, and it can end in the application of Reality-in-general as the final Transcendently Transcendental connotative primitive *dimensional* notion needed in assuring greater and greater truth-probability in all ontological discourse. Thus, Moser's probability-makers are only the particularistic ontological elements of a more generalized ontology and its epistemology.

2.3 Reality-in-general: Synthesis of Categories of Knowing

Chapter 1 dealt with the fundamental categorial dimension of all classificational cosmological categories. Chapter 2 has worked towards extending the purely epistemological categorial dimension of the confluence of the concepts of causality, laws of nature and ontological and connotative universals into the connection of knowing with being, by positing the category of Reality-in-general. The basis for there being an epistemic category in scientific ontology is that there always is some central, but probabilistically idealized, core in all ideals in knowledge. This core is what I have called Reality-in-general. Ontological universals culminate in To Be, but their conscious counterparts that are instantiated in concepts culminate in Reality-in-general, without the benefaction of which no ontology has so far been done. These universals are not merely linguistic referents, but general, epistemologically objectual and nomic abstractions that strain from Reality, and do not merely occur in language from nowhere. So, even this category may be taken as an ontological category, but with the epistemological aspect of ontology defined.

Ontological categories are maximal realizations of the modalities available in thought in that they are simultaneously descriptive of the totality of beings and of specifics that are the case. The proposition 'It is necessary that Reality is', is posterior to the objectual fact of the To Be of Reality, expressible in the statement, 'Reality is', because it expresses the fullness of modalities, i.e. of possibility and necessity, in their relation with the givenness of all that is and the processes of theorizing. For this reason, ontological categories are always instantiators of all that are possibly the case. Hence, the ultimate epistemological category of ontology also has for its most general nature the generality of all generalities implied in all epistemic activities. A realistic exercise of modal ontology needs the Einaic categories to accrue sense in the discourse. Thus, e.g., Reality-in-general can be taken as the totality of all possible worlds thinkable. To Be is the possibilizer of realization of any counterfactual that stands the test of Reality. And Reality is the original ontological source of formation of modal notions out of counterfactual propositions. For this reason, ontological universals cannot be merely instantiators and potentialities of specific things, but the nomic / acausal entities (universals in the ordinary sense and nomically as laws of nature) active in consciousness and active from the bosom of the highest causal possibilizer of beings / events, i.e. Reality. To the extent that we have discussed in these two chapters, they are simultaneously two-pronged in action: Chapter 1 dealt with what is cosmologically ontological and Chapter 2 with what is epistemologically ontological.

The epistemological realities are Reality-in-general and connotative universals in thought. The cosmological ones are Reality, reality-in-particular, ontological universals and To Be. These ontological universals are what tend to be connoted (noted together from particular instances, as representative of the ontological universal instantiative in any group of processes) in consciousness and give rise to connotative universals. Connotative universals are not in things, but are reflections of ontological universals. Similarly, Reality-in-general is not in Reality, but in thought. It is at the same time instantiative of Reality, but it is so in thought. There shall be an ontological category that is instantiative of Reality in itself. This we call To Be. So, it should be emphasized in this very chapter that

Reality-in-general is not To Be, because the former occurs in thought, and the latter in Reality. On a purely ontological footing, To Be is the very total processual category of Reality, is instantiative of Reality in itself, and also instantiative of Reality-in-general in thought. To Be as the purely ontological category is yet to be studied, in Chapter 3. It suffices for us now to say that Reality-in-general is the epistemological category proper to Reality. Epistemologically, i.e. to thought, it is the ideal of Reality. The contrary is the case with universals and classificational categories that we find in ontological and epistemological particularism. They in the limited horizon of their semantic attitude do not accommodate categories that totalise or generalize unto the maximum limit, as shown throughout these two chapters. The confluence of the concepts of causality, laws of nature and universals has been the method used to transcend such particularist categories and to posit Reality-in-general. Such confluence is part of collusion.

Reality-in-general, as the conceptual ideal of Reality, is the epistemic Transcendental *par excellence*. The important qualification implied about it is that it pertains to thought, and so, is the highest Transcendental in thought. It is at the same time of the Transcendent – Reality, the causally objectual Entity – but in an acausally objectual manner. The objectual aspect of Reality is Transcendently present in Reality-in-general. Hence, let us agree to call Reality-in-general as the Transcendently Transcendental category.

Conclusion

One usually speaks of things in the several either epistemically, giving purely epistemological definitions of individuation, or at other times one encounters purely physical definitions of it emphasizing merely the ontological processes involved and not showing how and why the ontological processes are reflected in the epistemic activity. With the admittance of Reality-in-general as the Transcendently Transcendental and epistemic category, the question of individuation has an epistemological solution that is at the same time ontologically connected with ontological processes and ontological universals. We have seen that in consciousness the process of individuation has connotative but highly partial reflection of ontological universals. Ontological universals are not had in consciousness, but they, by reason of their procession through elements of actual and specific entities into consciousness, are reflected connotatively (i.e. by noting together) in consciousness. This is not merely a set of particularist epistemic processes, but of ones that have their base rooted in the category of Reality-in-general, which is the conceptual universalization and idealization of the concept of Reality in its To Be. The process of system-building, by synthesizing the cosmological and the epistemological with the ontological, requires that the categorial orientation in all the three realms be applied to the process. This is epistemically made possible by the category of Reality-in-general.

To Be, the category analogous to Reality-in-general on an ontological footing, involves Reality-in-total, Reality-in-general and also reality-in-particular simultaneously, in the best manner possible. But it does so in a purely ontological manner and makes even discourse of particulars possible. But it is purely ontological in that it is the highest and deepest ontological universal. Hence, To Be is not a Transcendently Transcendental category. It is the purely Transcendental category, which possibilizes all others theoretically. The conceptual transcendental aspect (Reality-in-general) is not presupposed for To Be's tenability; instead, the latter implies the former and makes it relevant, which points to a major ontological flaw in particularist, analytic, hermeneutic and metaphysical ontologies. It shows also that ontology is prior to epistemology at the order of existence and possibilization. It is true that the conceptual, the epistemological, is prior with respect to the process of knowing. Reality, the thing that is most objectual-causal by reason of its total involvement of To Be and universals, is prior even to things epistemological – a fact forgotten by all non-ontologically semantic, anthropological and particularist analyses. To Be is ontologically prior, instantiative and potential of both Reality and Reality-in-general in theory. Hence, these two categories need the support of what possibilizes them: To Be. Without the thinking of To Be, the thinking of Reality-in-general and Reality is incomplete. It is the conceptually instantiative connotative universal of universals that this chapter has been attempting to clarify. I think that the absence of operation of the epistemological ideal in the epistemology of ontologies in the history of philosophy has done much harm to philosophy. The category of Reality-in-general redresses the same at the source of the theoretical enterprise. It also serves to distinguish epistemology from ontology by differentiation of the epistemological concepts of Reality-in-general and connotatives from the ontological concept of To Be.

Accordingly, truth may be defined as justified true belief in terms of ontologically and epistemologically compatible logical results of any or the entirety of theories about what there are and what is the case, through *a priori* commitment to objects / processes, empirically cognitive approach to processes and regard to the horizon of Reality, within the background of the knowledge-level ideal of Reality-in-general and the process-level ideal of ontological commitment to Reality and the ideal of To Be.

Occam's razor wouldn't be raised against the connotative universals that I have proposed. Connotative universals are not absolute and reifying universals, which latter, supposedly, have a definiteness and existence proper to entities or to Reality. They are conceived as pure entities that could only be pure and absolute *vacua*, but, paradoxically, in existence too. Truth-probabilistic connotative universals have their immediate foundation in the prehension of states of affairs in consciousness, which are in their turn veiled mirrorings of ontological universals noted together in consciousness. Thus, though they do not hold correspondence with anything, they do connotatively correspond best to the ontological universals at work in states of affairs, i.e. truth-probabilistically in consciousness's way of noting together, by means of the systemic coherence, pragmatic fixations and relativistic differentiations that are part of systems constructed by the mutually most collusive, synthetic and Einaic of categories, in consciousness. This fact will be clearer as we are through the next chapter.

Chapter 3. Ontolgoical Categorial Transcendental of Reality-in-total

Introduction

The mode of procedure here differs from Chapters 1 and 2. First I put down the ontological conclusions facilitated by those chapters and proceed to critically overcome the semblances of objectual ontology by going beyond particularistic ontological commitment in the work of Willard van Orman Quine, who has produced a typically particularist analytic ontology. I take for granted that most analytic and scientific ontologies and the objectual and nonconceptual counterpart of probability-makers attainable in them are particularistic. I take up Quine's unique concept of ontological commitment and its implications for a more adequate ontology in order to overcome its particularism by Einaic Ontology. This makes our results extendable to other particularist ontologies. Then I move to the two categories derived in Chapters 1 and 2 and connect them with the To Be of Reality for ontological discourse.

Though Russell, Carnap, Putnam and others are a must for an ontological discourse taking shape from the analytical, such references are minimal. Nevertheless, the ontological implications (existence, particularism, truth, causation, measurement etc. based on the nearest available qualia) of their work are presupposed, from which one can proceed to juxtapose them via Quine with the Einaic objectual-causal ontology. I begin with ontological commitment in Quine and extend it beyond its particularistic limits. It will be done in such a way that the inevitability and merits of involving To Be even in ontologies of the particular will be stressed. I do not discuss possible worlds here, though it is connected to Reality-in-general.

Developing a system of Einaic Ontology where the particular is sufficiently stressed is beyond the scope of this chapter. It stresses the need of things Einaic as most essential in discourse of the particular and shows why particularist "to be" is too thin to be the ontological To Be. A *most universal* To Be to stand in lieu of the traditional Being and as the objectual universal proper of Reality is our subject matter. I stress the possibility and necessity of To Be as the inevitable Transcendental element even of particularist ontological discourse, as *implicitly* admitted by all systems and non-systems of philosophy. Einaic Ontology offers a scientific ontology that is not based merely in forms of doing positive science.

3.1 To Be: The Einaic Ontological Transcendental *Par Excellence*

The verbal To Be (the gerundive-nominal Being) is not a higher or the highest genus in Aristotle. Each of the ten categories is the *genus generalissimum* of its species. The ultimacy of each of these categories is challenged by the ultimacy of the Transcendently and/or Transcendentally unique domains of definition of Reality-in-total, Reality-in-general and To Be. Their uniqueness makes them ideals beyond the classificational categories of Aristotle. Our To Be is the *that* (selfsameness, identity, thus-ness) of Reality which is the highest natural kind, and the deepest ontological *condition* for there being reality-in-particular. To Be is the ontological universal without particularity, so it is ontologically purely general. Reality-in-total is the particular and total token *par excellence*, defined uniquely against itself as the domain of definition. To Be does not contain the ontological totality of Reality-in-total, nor the ontological particularity of reality-in-particular and ontological universals, nor the epistemological generality of Reality-in-general, nor the epistemological particularity of connotative universals. Instead, To Be is the ontologically proper and general *that* of both Reality and Reality-in-general. Thus, To Be is ontologically the *genus generalissimum* of all classificational and collusive categories. Cosmologically and epistemologically it is not a genus at all, but an *a priori* in its own nature – put differently, in the purely ontologically manner it is the condition for possibility of all.

Reality is the highest and only token / entity / process that scientific ontology, cosmology and epistemology can find as the highest Transcendent category. If matter-energy in general has been a category, even better is Reality. It is objectually, by maximal ontological commitment, not referable or related to anything else outside of Reality. It may be conceptualised in ever better ways by reason of the aspect of the probabilizing origin and nature of ontological and connotative universals and Reality-in-general. The objectual selfsameness of Reality is an Einaic Ontological equivalent of the 'absolutely analytic truth' expressed in the unique tautology 'Reality is Reality'. The truth of ontological selfsameness too is probabilizing by nature, because its epistemology allows continuous betterment of definition and explication of the terms. Commitment to the objectual identity of Reality is far broader than particularistic ontological commitment that allows particulars only in a language-bounded manner,[339] and not in the manner of

---

339 For example, Quine says: "To whatever extent the adoption of any system of scientific theory may be said to be a matter of language, the same – but no more – may be said of the adoption of an ontology." Quine, *From a Logical Point of View*, 17. This

an objectual-causal ontological commitment based on To Be. In Quine ontological commitment is epistemologically particularistic,[340] whereas the objectual categorial identity of Reality ranges over itself and its ontological (the way of being of a set of entities) and connotative (the imperfect mirrorings of ontological universals in consciousness) universal aspects – To Be and Reality-in-general, respectively – for validation, not over human "to be" or human thought or particulars. These latter particulars have only one objectual aspect in Quine – reality-in-particular.

There is also an element of ever greater certainty about the to be of the self, if we are seeking ever greater truth-probabilities for the pragmatic judgments derived from the specific concepts involved. But the very foundations of Einaic Ontology of truth expectancy even regarding the self require connexity unto the broadest extent of Reality as the *sine qua non* for grasp of the genuine dimension of truth-probability in theoretical endeavours. Hence, the To Be of Reality, rather than the to be of the self, is the most self-evident intuition. This is an Einaic Ontological improvement of the Cartesian-Husserlian Cogito and the particularistic traditions of ontology of to be from Plato onwards. The generalities involved in thinking have it that the To Be of all that there exist, in their mutual connexity, is the final categorial Fundament.

This is the case of the intentional referentiality of thought to its transcendental object and to the transcendentally justified actual objects and processes in Husserl,[341] and also in the intensional-logical certainties in logic. The concept of

---

claim of Quine will be further substantiated under the various sub-sections 3.2 and 3.3.

340 This claim too will be substantiated in the various sub-sections under 3.2 and 3.3.
341 Husserl takes for granted the thinghood of things. These are not what are spoken of in merely physical-thematic imagination. "On the side where the results lie, we have as the theme of logic the manifold forms of judgment-formations and cognitional formations, which accrue to cognitive subjects during the performance of their thinking activities and do so, moreover, in the particular manner characteristic of 'theme'. The formations accruing on the particular occasion are indeed what the thinking subject is aiming at and intends to make his abiding acquisition; while at the same time they are meant to serve him as means to serve him as means for gaining similar new acquisitions. At any particular time, something has come into being, not just somehow or other, but rather as the thing aimed at in his thinking action: in a particular manner the thinking subject *'directs himself' to it*; he has it before him *'Objectively'*. In their higher forms, to be sure, these formations transcend the current sphere of presence to consciousness." Husserl, *Formal and Transcendental Logic*, 33. [Italics in the original.] He acknowledges bare presence to consciousness

the actual object is not objectual-causal in Husserl. This is clear from his work towards a philosophical methodology (wherein the study of abstract objects he might call ontology) from the very inception.[342] I hold that the objectual-causal deficiency in Husserl is due to his over-insistence on use of a semblance of the

---

as an elemental aspect. It is essentially connected to consciousness in thinking in the Objectivising manner.

What he means by 'Objective' is clear from the following: "These Objective affairs all have more than the fleeting factual existence of what comes and goes as a formation actually present in the thematic field. They have also the being-sense of abiding validity; nay, even that of Objective validity in the special sense, reaching beyond the subjectivity now actually cognising and its acts. They remain identical affairs when repeated, are recognized again in the manner suitable to abiding existents; in documented form they have Objective factual existence, just like the other objectivities of the cultural world: Thus they can be found in an Objective duration by everyone, can be regeneratively understood in the same sense by everyone, are intersubjectively identifiable, are factually existent even when no one is thinking them." Husserl, *Formal and Transcendental Logic*, 34–35. This two-sidedness is the reason for the fact that logic has not yet begun to develop in a steadily ontological manner.

342 The physical object as such is given in subjective determinations, not in Objective science which is a productive formation comprising different non-physical formations. "A purely Objective science aims at a theoretical cognising of Objects, not in respect of such subjectively relative determinations as can be drawn from direct sensuous experience, but rather in respect of strictly and purely Objective determinations: determinations that obtain for everyone and at all times, or in respect of which, according to a method that everyone can use, there arise theoretical truths having the character of "truths in themselves" – in contrast to mere subjectively relative truths." Husserl, *Formal and Transcendental Logic*, 38. In short, what we call theoretical / mathematical / abstract objects are not separate entities, but connective formations in Objective discourse. This stand may be derived from Quine too, *Logic*, 38. We posit such objects as ontological connectives in connotative and conscious thought.

In what is said above on Husserl, he was influenced by Bolzano. He quotes from Bolzano's youthful essay, in Heinrich Fel's new edition of his 1810 essay: *Beiträge zu einer begründeteren Darstellung der Mathematik* (*Contributions to a More Grounded Exposition of Mathematics*), § 8 (page 17) as a promise of a definition of formal ontology: "I think that one might define mathematics as a science that treats of the universal laws (forms) with which things must accord in their existence [*Dasein*]. Under the word thing I comprehend here not merely such things as possess Objective existence, existence independent of our consciousness, but also such things as exist only in our presentation [*Vorstellung*] and do so, more particularly, either as individuals (that is to say intuitions) or as mere universal concepts; in a word then:

Cartesian Cogito for methodological certainty. From a transcendental self locked up epistemically (to be sure, somewhat solipsistically, at least from the methodological considerations) within phenomena (the effects of which Heidegger has converted into the "Showing-itself process of To Be"), Husserl passes in his last phase over to the *Lebenswelt* as the really real stuff for methodological mooring.

To maintain the objectual-causal categorial identity of Reality in thought, as ranging over every possible definition or elaboration of 'Reality', its selfsameness is brought under the objectually ontological universal of Reality, i.e. To Be. This universal is not merely Reality-in-general, since the latter is the acausally considered theoretical generality of all the causal strains (propositional truths in laws; nomic concepts, and universals) of Reality, to the extent that causal strains are active in thought by objectual intermissions from reality-in-particular in conscious activities. For the above identity we need something nomic of objectual-causal totality, pervading not merely thought but also Reality. It must be ontological in an all-inclusively *dimensional* fashion. A dimension is not fully fixed. It should be primarily the Way of beings, and only as a result the Way of thought via the generalizing connotative, Reality-in-general. Such dimensional category is To Be, an ontological universal, properly Einaic *nomos*, not merely of acausal nomic generalities but also of the objectually causal Reality. Nothing obstructs us from considering conscious generalities as instantiations of various values of conceptual probability of ontological potentialities in Reality and reality-in-particular, all communicating via reality-in-particular. *This safeguards the ontological status of the total, the general and the particular at one go.*

Universals are attributes instantiated in names of sets of natural kinds and qualities. Even verbal-processual concepts may be ontological universals, connotable in mind, e.g., 'doing' can have a verbal-processual ontological universal, and parallel to it, 'to do' also. Similarly, we may classify a set of objects under the universal 'being' or 'beingness', or call the universal attribute of that level as 'to be'. This allows us also to speak of the universal verbal attribute of Reality as To Be. These are (1) ontological when they are mainly the connectives of processes, and (2) epistemological when they are instantiated or had access to in consciousness. Beyond their purely ontological status as objectually and causally instantiating potentials, these universals or essences have a purely intentional, acausal activity in consciousness. These too are in parts of Reality and in the final analysis characterized by To Be and Reality-in-general. This is a fact

---

everything that can be at all an object of our faculty of presentation." Husserl, *Formal and Transcendental Logic*, 85.

that most analytic-, phenomenological- and scientific realist researchers do not pay attention to. We have called this ontologically objectual-causal aspect of processes as ontological universals / essences, and the epistemic activity of the same in thought as connotative universals / essences.[343]

Ontological universals are a world of objectual-causal potentialities, always steeped in *the actualities* of Reality that have *varying degrees of particularity*. The relevant universals of commonality ingress into token processes from within the bosom of ontological universals active in processes that have been. That is, they are the lawlike ways of being of beings transferring themselves always through causal processes in various intensities into actualities and further processes. The world of universals comprises the To Be of Reality, the to be's of realities-in-particular and the processual attributes of beings (both Reality and reality-in-particular). The way of ingression[344] of To Be, anthropic to be's and other universals in consciousness may be called connotation: a selective, one-against-many "noting together" of nomic possibilities caused by processes via consciousness.

---

343 Lack of distinction between ontological and epistemological universals has very much affected Heidegger's work. For example, just as Hegel, whom he studies extensively, Heidegger too similarly gets muddle-headed while drawing his favourite conclusions from Hegel, due to such a lack while discussing Hegel's epistemology: "Consciousness means being conscious of something; that something is in the state of being known. But what is known exists in knowledge, and exists as knowledge. What is known is that to which consciousness relates in the mode of knowing. What stands in this relation is what is known. It is in that it is 'for' consciousness. What so is, is in the mode of 'being for […]'. But, 'being for' is a mode of knowing. In this mode something is 'for consciousness' from which it is 'at the same time distinguished' insofar as it is known. But, in general, what is known is not merely represented in knowing; rather, the representing intends that which is known as something real that *is* in itself, hence, something that truly is. This being-in-itself is what is known as truth. Truth, too, is one thing (something represented) and at the same time another (something that is in itself) 'for the same consciousness'. The two determinations of consciousness, knowledge and truth, are distinguished as 'being for' and 'being-in-itself'." Heidegger, *Hegel's Concept of Experience*, 89. Expressions like 'at the same time distinguished' begin to muddle between the ontological and the epistemological, and no point of conceptual transition is identified between them.

344 This term is from Whitehead, and its explanation is similar to his. *His concept of eternal objects are bifurcated in Einaic Ontology as ontological and connotative universals.* Thus, ingression occurs when "[…] an eternal object can be described only in terms of its potentiality for 'ingression' into the becoming of actual entities; and […] its analysis […] discloses other eternal objects." Whitehead, *Process and Reality*, 23.

These ways are modelled after the connotative manner of ingression of To Be in thought in terms of Reality-in-general, but always sheltered by the constant objectual-causal ingression of To Be in Reality.[345] To Be is thus the existential quality of Reality.

An important distinction to be made here is as follows. Though To Be may be consciously instantiated in thought, it is also possible for it to be instantiated and represented as 'To Be' in it, as a denotative[346] of the To Be of Reality. This representation works in thought as a nomic universal connotatively (Reality-in-general), and as a denotative (a term) of the Way of being of Reality (the term, 'To Be'). To the extent that it is connotatively conceptualised, it is not To Be. Being a denotative representation ('To Be'), it is of the To Be of Reality-in-general. This follows the rule that even the ontological universals of beings, ways of being of beings and attributes are denotable by conceptual universals. These are denotables of ontological universals, but instantiated connotatively in consciousness.

The category of Reality-in-total is cosmological, even if it may include the Divine: (1) the cosmic and the divine *need not be* mutually truncated while studying any one of them, since such mutual alienation has led to the more than two millennia of mis-identification of the purely ontological Transcendental To Be with the Divine Transcendent (the Divine is not To Be, but the infinitely conscious site of connotative realization of To Be within Reality); and (2) it is an illusion about ontology that any talk about the ontology of To Be is a recurrence of essentialistically[347] reifying metaphysics, which (a) epistemologically reifies the epistemic, transcendently transcendental concepts (connotative universals) as *tokens* that exemplify Divine conscious activities and qualities (e.g. universals as *some things* graced by the Divine), (b) ontologically mistakes To Be as the epistemological and Transcendently Transcendental Reality-in-general in full realization in the

---

345 This arouses the question of the mode of correlation of ingression of universals as causal roots of beings and ingression of connotative universals – which is nothing but the question of derivation of values of truth and falsity. This may be discussed only within the context of all the three Einaic categories. As concentration on our theme of categories may not do justice to this aspect of the epistemology of ingression of universals in consciousness, I do not work on the logic and epistemology of possible modes of correlation of causal universals and acausal connotative universals.

346 *Anything to which a connotative applies is a denotable.* For example, a token entity is a denotable, and also a universal when it is made to represent ways of being.

347 Essentialism is a theory of reality (not existence) of individual and general essential properties. It has various expressions. The extreme case of it is reifying essentialism, wherein essences are taken as actually existent.

Divine (To Be is realized in full only in Reality, but realised in the Divine in the best way possible in *an* entity, i.e. infinitely consciously), and in consequence, (c) cosmologically reifies To Be as instantiated as the Divine itself (a mistaken identification of the Transcendental with the Divine Transcendent). Reality-in-general is in fact the purely epistemic aspect proper to Reality, exemplifiable in consciousness, both human and Divine. To Be is not properly (ontologically) exemplified anywhere but in Reality.

Thus, just like Reality has its ontological aspect proper in To Be, so too Reality-in-general has its ontological aspect To Be by reason of which there is instantiation of 'Reality' through reality-in-particular, and also representation of Reality and Reality-in-general in consciousness. To Be is not the same as Reality-in-general, because the latter is the universal of all universals working at the epistemological level. Objectually and at the level of Reality, the purely ontological aspect of Reality-in-general (i.e. the ontological universal) is To Be. By reason of the nature of To Be, consciousness can instantiate the two others in itself connotatively.

'To Be' as a term does not give anything to man or to beings, because it is an abstract / processual (occurring *as* the processual relationalities of processes) entity based on the ontological entity, Reality. As far as the conceptualisation of To Be is concerned, our concept of Reality-in-general would seem, to insufficiently differentiating consciousness, to be the exact equivalent of the concept of Being, as we find often in Heidegger. A solution for this is Einaic Ontology's distinguished way of differentiating between To Be and Reality-in-general. The present chapter builds ontology (1) beyond the Platonic-Aristotelian metaphysics, and (2) beyond analytic, especially Quinean[348] particularist scientific ontology, both

---

348 In Quine a biconditional is always analytic of the terms involved. "Sentences are synonymous if and only if their biconditional (formed by joining them with 'if and only if') is analytic, and a sentence is analytic if and only if synonymous with self-conditionals ('If *p*, then *p*')." Quine, *Word and Object*, 65. In the context of ontological thinking, analytic and synthetic sentences need to be conditional upon Reality-in-total and the two universals To Be and Reality-in-general for their full meaning. That is, analytic statements (directly implying essential statements) are not purely analytic, and synthetic statements (directly meaning inductive statements) are not purely synthetic *sub specie Realitatis*, "under the species (sight / mirroring) of Reality (-in-total)". The above makes concepts of the analytic and synthetic mutually complementary in varying degrees of incurrence of both in one statement.

Strawson holds a similar position. He says: "In speaking of entailment-rules we make use of the distinction between analytic and synthetic (or contingent) statements. In speaking of type-rules we make use of the distinction between the literal and the figurative use of language. But we must not imagine these distinctions to

of which differentiate neither between Reality-in-general, Reality and To Be, nor between ontological universals and connotative universals.

Thus we have *a special set of theorems of the present work*: (1) that Reality-in-general can be an inductive-generalizational[349] instantiation of To Be through ontological universals and to be's, and (2) that 'Reality-in-general' can be an "inductive-generalizational" instantiation of Reality, through reality-in-particular and their universals and to be's – only in thought, i.e. merely connotative-universally, and not ontological-universally, i.e. not in Reality, and (3) that To Be is instantiated properly (ontologically) as To Be only by Reality, and is only *analogously* instantiated by thought as Reality-in-general, or as the denotative 'To Be'. This is a set of doctrines that go diametrically opposite to all epistemic idealisms in cosmology and ontology, which, by their substitution of the cosmic concept of Reality by epistemic constructions, result in sophistic inversions of the priority of the objectual with that of the epistemically objective. I propose that what is theorized objectually, within the purview of the To Be of Reality, bears the highest chance of yielding adequate ontological truth.

Einaic Ontology's essentialist idealism is two-fold: (1) In Einaic Ontology any property involves universals, and none of those universals is purely of one entity,

---

be very sharp ones, any more than we must imagine our linguistic rules to be very rigid [...] [W]e may very often hesitate to say whether a given sentence is analytic or synthetic; and the imprecision of this distinction, as applied to ordinary speech, reflects an imprecision in the application of the notion of entailment of ordinary speech [...] [I]f we realize that we are at best describing only the standard and typical uses of certain kinds of expression, we shall be less disconcerted by untypical cases." Strawson, *Introduction to Logical Theory*, 230–31.

349 Inductive generalization follows the axiom of infinity defined thus: "If *n* be any inductive cardinal number, there is at least one class of individuals having *n* terms." Russell, *Introduction to Mathematical Philosophy*, 131. Russell defines a particular as "[...] the objects that can be named by proper names," proper names being "[...] terms which can only occur as *subjects* in propositions [...]". His explanation of the axiom may as well be applied to abstract objects: "It is, of course, possible that there is an endless regress: that whatever appears as a particular is really, on closer scrutiny, a class or some kind of complex. If this be the case, the axiom of infinity must of course be true. But if it be not the case, it must be theoretically possible for analysis to reach ultimate subjects, and it is these that give the meaning of "particulars" or "individuals." If it is true of them, it is true of a class of them, and classes of classes of them, and so on; similarly if it is false of them, it is false throughout this hierarchy. Hence it is natural to enunciate the axiom concerning any other stage in the hierarchy. But whether the axiom is true or false, there seems no known method of discovering." Russell, *Introduction to Mathematical Philosophy*, 142–43.

except in the case of Reality-in-total (which has To Be as the only most fitting and co-extensive universal). By reason of the epistemology of ontological activity, the To Be, to be's and other ways of being (properties and types) of anything – when they are in conscious instantiation of objectual essences – *are universals based on and originating from the processes of certain reality-in-particular (and not by To Be)*. But the respective causal reality-in-particular has broader or other parts of Reality for cause. The irreducible idealism here is epistemological, based on epistemic activity, and shows that universals are a necessary ingredient in reality-talk. (2) By reason of the ontology of Reality, the To Be, to be's and ways of being are actually in beings including Reality: whereby essentialism and acausal idealism are ontologically real (as against epistemologically real) in the sense that ontological universals involved in Reality and reality-in-particular are the synaptic, unifying, qualities in things, and not distantiable from them. Ontologically To Be, to be's and ways of being are not processes. They are real non-entified (non-token) ontological connectives in processes. *This is an admissible ontological idealism devoid of unnecessary reifications.*

Such a need to admit non-reifying essences is a state of affairs that no ontology – not even phenomenological or analytic – has been able to dispense with. In general, any ontological commitment to particulars and finally to tokens involves ontological universals / abstract objects at the realm of processual relationality of things. One analytic philosopher of consequence who admits this is Quine – though he steeps his admission in particularism regarding objects, which is incompatible with the nature of particulars as co-involving ontological universals in nature. Particularist statements like 'There exists x such that', even by indicating ontological commitment, are epistemological and particularist ontological statements, not tenably and sufficiently ontological at the *dimension* of To Be. As a result of the co-implication of 'Being' in different philosophies as non-entifying, Being may imperfectly be substituted by 'Reality', which (the latter) alone can *"shelter" anything and "give" anything*. Hence, the particularist concept of 'existence' has to be enhanced by To Be. The sort of idealism allowable in the epistemic level of conception in ontology is that of the place of Reality-in-general and other connotative universals, which are consciously and physically singular and aggregate finite constructs in instantiation of, but not exhaustive of, ontological universals.

Unfortunately, from Plato, Aristotle and Aquinas to Husserl, Heidegger, Wittgenstein and to the whole of phenomenological, analytic, pragmatic, postmodern, ethico-ontological and other trends, the aforesaid clarity is not to be seen due to lack of the Einaic differentiation between (1) To Be, (2) Reality-in-total, (3) Reality-in-general, (4) reality-in-particular, (5) ontological universals

/ essences that relationally based in the processes of Reality and reality-in-particular, and (6) connotative universals / essences that are epistemic ingressions of ontological universals in concepts of consciousness via connotative universals / essences, under the shelter of the implicitly objectualizing concept of Reality-in-general. In any final analysis, this clarity is based on To Be. Making the merits of Reality-in-total and Reality-in-general as categories within the reach of universals, To Be is their highest Transcendental categorial possibilizer.

Insofar as Reality is the highest token available to ontology, another *entity* under the name 'To Be' is not needed. *To Be is an ontological nicety*[350] – the best[351] at

---

350 In the context of discussing the difficulties in differentiating between the being of abstract objects and the existence of concrete objects (Quine, *Word and Object*, 241–42), Quine expresses his objection to conceiving abstract entities simply as theoretical niceties, but he condones the practice, in a footnote that refers also to Putnam: "But the familiar vague notion that the assumption of abstract entities is somehow a purely formal expedient, as against the more factual character of the assumption of physical objects, may still not be wholly beyond making sense of; see Putnam, "Mathematics and the existence of abstract entities". [*Philosophical Studies*, 7 (1956): 81–88.]" Quine, *Word and Object*, 242. He adds: "The distinction between concrete and abstract object, as well as that between general and singular term, is independent of stimulus meaning." Quine, *Word and Object*, 52. That is, without consideration of any possibility or impossibility of synonymy of types of terms and abstract objects we may translate a sentence in one language into another language, depending solely on stimulus meaning. "Synonymy of 'Gavagai' and 'Rabbit' as sentences turns on considerations of prompted assent; not so synonymy of them as terms. We are right to write 'Rabbit', instead of 'rabbit', as a signal that we are considering it in relation to what is synonymous with it as a sentence and not in relation to what is synonymous with it as a term." Quine, *Word and Object*, 52. This too hints at the possibility of conceptual independence of abstract and concrete objects and general and singular terms from the stimulus meaning accorded in sentences: "We cannot even say what native locutions to count as analogues of terms as we know them, much less equate them with ours term for term, except as we have also decided what native devices to view as doing in their devious ways the work of our own various auxiliaries to objective reference: our articles and pronouns, our singular and plural, our copula, our identity predicate [....] The native may achieve the same net effects through linguistic structures so different that any eventual construing of our devices in the native language and vice versa can prove unnatural and largely arbitrary. Yet the net effects, the occasion sentences and not the terms, can match up in point of stimulus meanings as well as ever for all that. Occasion sentences and stimulus meaning are general coin; terms and reference are local to our conceptual scheme." Quine, *Word and Object*, 53.
351 The nature of the difference of senses of abstract and concrete objects is the problem that compels us to say that this is the best at that. Quine suggests a solution

that – within Reality and in thought, based ontologically on actual processes and *lying at the acme of instantiation of all connotative entities as capable of founding all discourse*. Opponents forget this fact when they attack ontology fearing reifying metaphysics, but incur the same variety of nicety (universals) by implication of abstract but vacuous entities[352] in their entity-talk of the particularist variety. That variety of *probabilistic ontology in which reifying metaphysics evaporates and To Be guides all entity-talk from the realms of Reality and Reality-in-general is Einaic Ontology*.

I have opined that from Plato to Wittgenstein and in other trends we lack the objectual-causal clarity yielded by the Einaic point of view. Such lack of ontological commitment in contemporary 20th century and later thought has had its colossally prohibitive inspiration from Kant's phenomena-noumena distinction. In the contemporary world of phenomenology it was Brentano followed by Husserl, and in linguistic analysis it was Quine, who have brought in some ontological commitment using purely intentionality-based or language-based discussions, respectively. These have not served to bring what we call the objectual-causal aspect of ontological commitment to the forefront of ontologizing. Instead, the Quinean and Husserlian commitments to the 'objective' and to the 'object' have remained overly intentionality-based or language-based. Husserl's lack of an objectual-causal ontology led him to give an almost objectual content to abstract objects.

---

for the problem of the difference between the "existences" of abstract and concrete objects by terming it a compromise: "For predicative set theory […] substitutional quantification is both feasible and attractive. It is attractive because abstract objects seem to be parasitical on language in a way […] that concrete ones are not. This is a point that has been urged by Charles Parsons. It is not to say that substitutional quantification over abstract objects simply eliminates them from the ontological inventory. Expressions themselves are abstract […] but they are less wildly so than the denizens of higher set theory. Substitutional quantification may be viewed as according the values of its variables a tenuous grade of existence distinct from the robust existence imputed to concrete objects by objectual quantification. It is a compromise with militant nominalism." Quine, *Quiddities*, 35.

352   Quine gives examples for such incurrence: "We find philosophers allowing themselves not only abstract terms but even pretty unmistakable quantifications over abstract objects ("There are concepts with which …," "… some of which propositions …," "… there is something that he doubts or believes"), and still blandly disavowing, within the paragraph, any claims that there are such objects." Quine, *Word and Object*, 241. Thereafter he alludes also to illustrative texts from Ayer and Ryle, mentioned by Church in his "Ontological Commitment".

Quine's merit is that he has clearly made universals objectually empty agents of entity-to-entity connectivity. For him 'object' means existents and 'objectal' means pertaining to existent objects. I shall characterize also Quine's objectually empty universals as really objectual, which yields the existent and real (related to existents) forms of objects, including the To Be of Reality and to be of processes. Thus I extend his objective ontology into the Einaic realm within the context of the categories of Reality-in-total and Reality-in-general. To Be possesses the status of the only supra-categorial category, since it is no classificational category. It makes reality-in-particular possible within Einaic Realism, and for this 'objectual' means all that processually exist and all that are real as universals in their interface. Hence, To Be which is the highest and deepest universal can be a category for thought in the purely ontological sense. I propose to accord to To Be a singular status among categories. This may be done by enhancing Quine's particularistic ontological commitment and necessitating an objectual and universalistic ontological commitment.

So far we drew up certain important ontological conclusions facilitated by Chapters 1 and 2. With the vantage obtained thus in the foregoing paragraphs we advance some of their purely ontological consequences and derive the category of the most generally instantiating ontological universal, To Be. I also show that this alone is the most proper way of studying reality-in-particular to yield the best of truth-probabilities. Study of Quine's particularism helps advance the cause of ontology from the matrix of analytical ontology. We further search into the semantic implications of these advances in order to fix the *dimensional* concept of To Be firm enough in the horizon of scientific ontology, more than in other traditional forms of ontology.

3.2 Einaic Objectual Ontology beyond Quine

3.2.1 Relevance of Quine's Ontological Commitment and Holism

There is a purposeful absence of systemic and objectual-causal commitment in ontology in Wittgenstein[353] and in most analytic thinkers. This lack is partially

---

353 Unlike Kant, who gave a metaphysics of moral action for realization of noumena, Wittgenstein hints at the realm of the "mystical", i.e. that of value, moral and less clearly the divine, without ever making it possible, without entering upon the latter realm, since meaningful propositions are possible only in science. He mystifies these realms in the *Tractatus*, by keeping them infinitely aloof from all scientific propositions. "The right method of philosophy would be this. To say nothing except what can be said, *i.e.*, the propositions of natural science, *i.e.*, something that has

filled by Quine in his particularist ontology that vouches for linguistic holism.[354] In preparation for working out the Einaic Ontological implications of To Be as the highest Transcendental, we study one of the most pronounced renderings of systemic mixture of thought, abstract objects and token entities in analytic philosophy: Quine's concept of ontological commitment and linguistic holism by taking for granted the concepts of quantification, reference, etc. We do not have to begin with Quine's arguments for stretching analyticity and syntheticity via synonymy, for us to attain ontological commitment and the ontogenesis of reference. A study of his overcoming the traditional analytic-synthetic distinction is meant for another context.

3.2.2 Objectual Ontology vs. Particularist Ontological Commitment

Some parts of the study of Quine here is partially a repetition but unavoidable. While responding to Saul Kripke's evaluation of Quine's ontological reduction as Kripke studies his ontological commitment and ontological criterion, Quine puts down explicitly what he means to accomplish in his ontological reduction:

> One of Kripke's moral precepts deplores 'the tendency to propose technical criteria with the aim of excluding approaches that one dislikes'. He notes in illustration that I adopted a criterion of ontological reduction for no other reason than that it 'includes well-known

---

nothing to do with philosophy: and then always, when someone else wished to say something metaphysical, to demonstrate to him that he had given no meaning to certain signs in his propositions. This method would be unsatisfying to the other – he would not have the feeling that we were teaching him philosophy – but it would be the only strictly correct method." Wittgenstein, *Tractatus Logico-Philosophicus*, 6.53. Thus, philosophy ends up as discourse on certain particularistically allowable and justifiable aspects of science. Insofar as it is just such a discourse, no generalization is admissible, and so no systemist metaphysical attitude to all values and universals is possible. The only possible way of metaphysicising is silence. *Tractatus Logico-Philosophicus*, 7. The position that only this-worldly propositions are within our reach is further elaborated by the later phase of his thought, with more emphasis on the dos and don'ts of language games, despite the fact that he had dealt a boomerang on his own justification of scientific propositions by reducing reason to mere beliefs (with foundation in ostensive definition, teaching and learning), as in "What I need is *certainty* not wisdom, dreams, or speculation and this certainty is faith." Wittgenstein, *Notebooks: 1914-16*, 73. The difficulties involved in having ostensive definition, teaching and learning of values, morals and things divine leads him to take them all as metaphysical. Metaphysics tends to be systemic. Pure ostension resists systemism. Systemism is meaningless for Wittgenstein.

354 Quine, *From a Logical Point of View*, 41–43.

cases and excludes undesired cases'. I protest that mine was expressly a quest for an objective criterion agreeing with our intuitive sorting of cases. This is a proper and characteristically philosophical sort of quest, so long as one knows and says what one is doing.[355]

A major difference in my arguments in earlier chapters on the connection between ontology and epistemology has been to base objectivity on objectuality. A fuller picture is given in Chapter 3 here, which divests Quine's presuppositions of ontological commitment of its particularism. His ontological commitment stood as ontological support for his ontological reductions. In ontological reduction he attempts "[…] an objective criterion agreeing with our intuitive sorting of cases." This shows that Quine was after epistemic objectivity in ontological reduction. Whether he was successful in it is an unsettled question. In his own words:

> Not to discriminate between elimination and explication, there remains an important sense in which the physicalism contemplated above may be said to be less clearly *reductive* than Frege's version of number. [Here he acknowledges his debt to Davidson and Feigl.] When Frege explains numbers as classes of classes, or eliminates them in favor of classes and classes, he paraphrases the standard contexts of numerical expressions into antecedently significant contexts of the corresponding expressions for classes; thus 'has-…-members' gives way to 'ϵ', and arithmetical operators such as '+' give way to appropriately definable class-theoretic operators. But when we explain mental states as bodily states, or eliminate them in favor of bodily states, in the easy fashion here envisaged, we do not paraphrase the standard contexts of the mental terms into independently explained contexts of physical terms. Thus the 'Jones is in' of 'Jones is in pain', the 'Jones is' of 'Jones is angry', remain unchanged, but merely come to be thought of as taking physicalistic rather than mentalistic complements. The radical reduction that would resolve the mental states into the independently recognized elements of physiological theory is a separate and far more ambitious program.[356]

As a solution for the question of objectivity of knowledge are given observation sentences that translate mental qualities to truths or not in expressions in language:

> To philosophers, 'observation sentence' suggests the datum sentences of science. On this score our version is not amiss; for the observation sentences as we have identified them are just the occasion sentences on which there is pretty sure to be firm agreement on the part of well-placed observers. Thus they are just the sentences on which a scientist will tend to fall back when pressed by doubting colleagues. Moreover, the philosophical doctrine of infallibility of observation sentences is sustained under our version. For there is scope for error and dispute only insofar as the connections with experience whereby

---

355 Quine, *Theories and Things*, 175.
356 Quine, *Word and Object*, 265–66. Square brackets mine.

sentences are appraised are multifarious and indirect, mediated through time by theory in conflicting ways; there is none insofar as verdicts to a sentence are directly keyed to present stimulation. (This immunity to error is, however, like observationality itself, for us a matter of degree.) Our version of observation sentences departs from a philosophical tradition in allowing the sentences to be about ordinary things instead of requiring them to report sense data, but this departure has not lacked proponents.[357]

As close to observation sentences and as a result of the particularist thrust in observation sentences he suggests more remote theoretical sentences.[358] Observation sentences and theoretical sentences constitute Quine's ontological commitment. By 'ontological commitment' he means a non-empty class[359] of abstract

---

357 Quine, *Word and Object*, 44. Observation sentences point to a duality between concept (knowing what a sentence means) and doctrine (knowing whether it is true). The latter is the repository of evidence for scientific hypotheses. The former is the arena of entry into language for the neophyte and the field linguist alike, by observing what correlate to utterance. The observation sentence, as "[...] situated at the sensory periphery of the body scientific, is the minimal verifiable aggregate; it has an empirical content all its own and wears it on its sleeve." Quine, "Epistemology Naturalized", 29.

358 The particularist (atomistic) thrust in observation sentences is still clear from his response to M. J. Cresswell's criticism of the concept by not agreeing to liken it with Russell's logical atomism: "Cresswell compares my view with Russell's logical atomism and rightly finds them incompatible. 'He certainly has no sympathy', he writes of me, 'with any theory which would make the atomic facts simple facts about our experience, each logically independent of all others.' True, but still it is instructive to compare my observation sentences with this doctrine. They are not about experience, but they are fair naturalistic analogues of sentences about experience, in that their use is acquired or can be acquired by direct conditioning to the stimulation of sensory receptors. Moreover, simple observation sentences are in most cases independent of one another. The profound difference between my view and Russell's atomism is rather that the rest of the truths are not compounded somehow of the observation sentences, in my view, or implied by them. Their connection with the observation sentences is more tenuous and complex." Quine, *Theories and Things*, 180–81.

359 "A class [...] is simply a property in the everyday sense of the word, minus any discrimination between coextensive ones." After the word entered taxonomical discussions and is taken for 'a specific level of classification', it proliferated into 'property' and 'attribute', "[...] except [...] for the extensionality constraint". "The reasoning behind Russell's Paradox applies to properties precisely as to classes, and shatters likewise the platitude that whatever is said about a thing ascribes a property. Whatever set-theoretic restraints may be imposed on the existence of classes, in order to preserve consistency, would need to be imposed *pari passu* on properties if we

objects of talk common to and necessary for there being the ontology that fulfils some regimented theory.[360] More explicitly, "[...] a theory is committed to those and only those entities to which the bound variables of the theory must be capable of referring in order that the affirmations made in the theory be true."[361]

Concepts are for him primarily stimulus meanings[362] *denoting* objects. When the objects assumed ("[...] general and singular terms, singular and plural

---

were so perverse as to continue to recognize properties in lieu of or in addition to classes [...] [T]he notion of property or its reasonable facsimile is needed for technical purposes in scientific theory, especially mathematics, and in these contexts classes are the reasonable facsimile that takes over, since these contexts never hinge on distinguishing coextensive properties. One instance among many of the use of classes in mathematics [... is...] in the definition of number. For science it is classes *si*, properties *no*." Quine, *Quiddities*, 23–24. In Section 1.1.1 we have discussed how Chisholm has shown that classes are reducible into properties.

360 "Ordinarily interpreted scientific discourse is as irredeemably committed to abstract objects – to nations, species, numbers, functions, sets – as it is to apples and other bodies. All these things figure as values of the variables in our overall system of the world. The numbers and functions contribute just as genuinely to physical theory as do hypothetical particles." Quine, *Theories and Things*, 149–50.

361 Quine, *From a Logical Point of View*, 13–14.

362 "[...] [T]he *affirmative stimulus meaning* of a sentence such as 'Gavagai', for a given speaker", is "the class of all the stimulations (hence evolving ocular tradition patters between properly timed blindfoldings) that would prompt his assent. More explicitly [...] a stimulation σ belongs to the affirmative stimulus meaning of a sentence S for a given speaker if and only if there is a stimulation σ' such that if the speaker were given σ', then were asked S, then were given σ, and then were asked S again, he would dissent the first time and assent the second. We may define the *negative* stimulus meaning similarly with 'assent' and 'dissent' interchanged, and then define the *stimulus meaning* as the ordered pair of the two. We could refine the notion of stimulus meaning by distinguishing degrees of doubtfulness of assent and dissent, say by reaction time; but for the sake of fluent exposition let us forbear. The imagined equating of 'Gavagai' and 'Rabbit' can now be stated thus: they have the same stimulus meaning." Quine, *Word and Object*, 32–33. Moreover, "[...] [o]ccasion sentences and stimulus meaning are general coin; terms and reference are local to our conceptual scheme." Quine, *Word and Object*, 53. This he shows by arguing as follows: "We cannot even say what native locutions to count as analogues of terms as we know them, much less equate them with ours term for term, except as we have also decided what native devices to view as doing in their devious ways the work of our own various auxiliaries to objective reference: our articles and pronouns, our singular and plural, our copula, our identity predicate. The whole apparatus is interdependent, and the very notion of term is as provincial to our culture as are those

predication, truth functions, and the machinery of relative clauses; or, equivalently and more artificially, instead of plural predication and relative clauses we can admit quantification"[363]) are taken as values of variables or of pronouns, then in such quantification various useless theoretical entities (plural predication and relative clauses) disappear and new ontic[364] commitments emerge. According to Quine this results in the choice of new and relevant objects and also in simplicity in one's overall system of the world.[365] Note that simplicity due to ontic commitment and the possibility of choice of new objects is due mainly to the fact that "[...] the objects assumed are the values of the variables, or of the pronouns. Various turns of phrase in ordinary language that seemed to invoke novel sorts of objects may disappear under such regimentation."[366] New objects needed for the ontological edifice are always welcome, not novel sorts of them. Novel sorts are untenable objects unlike spatio-temporally (for us, extension-motion) referential objects (general and singular terms, singular and plural predication, truth functions etc.) yielded by ordinary quantification.

Before proceeding further with Quine we make a distinction. Granting that spatio-temporal (extension-motion) reference is the touchstone for actual existence, there are two problems hard to crack within Quine's thought, due mainly to the lack of just the sufficient differentiation and appropriate connection between

---

associated devices. The native may achieve the same net effects through linguistic structures so different that any eventual construing of our devices in the native language and vice versa can prove unnatural and largely arbitrary." Quine, *Word and Object*, 53.

363 Quine, *Theories and Things*, 9–10. "The distinction between concrete and abstract object, as well as that between general and singular term, is independent of stimulus meaning." This is because 'Gavagai' is "[...] a singular term naming a recurring universal, rabbithood". For example, "[...] the singular term 'Bernard J. Ortcutt': differs none in stimulus meaning from a general term true of each of the good dean's temporal segments, and none from a general term true of each of his spatial parts." Quine, *Word and Object*, 52.

364 Of the use of 'ontic', Quine says in a footnote: "Of the three evident advantages of 'ontic' over 'ontological', in the special sense of 'as to what there is', brevity is the least. In thus reforming my usage I follow Williams." Quine, *Word and Object*, 120. In *From a Logical Point of View* ["On What There Is," 8–13], he uses both 'presuppose' and 'commit' to denote the same. In *Word and Object* he uses 'ontic commitment' instead of 'ontological commitment'. In most of his later works and in the tradition that has accepted the concept, it appears as 'ontological commitment'. See also Hodes, "Ontological Commitment Thick and Thin", 257, Note 3.

365 Quine, *Theories and Things*, 9–10.

366 Quine, *Theories and Things*, 10.

ontological universals and connotative-epistemological universals: (1) whether the connotative universal, the denotative concept of an existent thing and the same of a non-existent thing can have the same sort of extension-motion reference, and (2) whether only a term that denotes an actual existent can have any sort of reference. Ontological universals are actual in and essential for general ontological commitment in the act of reference, in propositions etc. and thus also in connotative universals ingredient in concepts that refer; and this bases ontic commitments (to immediate objects of all kinds) on ontological commitments (effects of the wider connexity between terms and their linguistic and natural contexts even in Quine).

Such connexity need never be limited to one or more layers, but only by the final *dimension* of their continuity. Hence, the final objects of ontological commitment are *necessary dimensional posits* beyond the interconnectedness of ontological and connotative universals. We cannot end the final objects of ontological commitment with an actual entity circumscribed in an extension-motion or its representative abstract object. The abstract objects that have led to this commitment *refer* beyond the extension-motion object's identity, into the realm of other entities in objectually causal and extension-motion roots of the same, and nomically idealized in the pertinent universals. Hence, conceptual and propositional reference to any state of affairs is not the final word in ontological commitment, but instead, objectual-causal ontological commitment over the Einaic categories.

That is, in Quine ontological commitment is not purely ontological but very much ontic or just more than ontic, meaning it is based on identifying tokens behind particulars. What at the most is practically presupposed in the commitment is the 'objectivity' of theories based on facts / states of affairs. This is thoroughgoing particularism. The particularist presupposition behind Quine's ontological commitment is insufficient for there being a full-fledged maximal ontology. Instead, we need objectual necessitation of at least some form of further objects and their totalities in the causal roots of the objects in ontological commitment. This is possible only via other wider universals of actuals. Such universals allow real reference to actual (past, present or future) existents of natural processes within the context of actual and non-actual possibilities in a regimented theory, and disallows non-existents of past and present and impossibles of future from interacting with theory. Therefore, we need entities with ultimate nomic connections as the final result of ontological commitment. Just 'entities within respective contexts' is not adequate to the notion of the dimension of the final in the implications of ontological commitment. Without the notion of the final to intervene, ontic commitments are not justified and well defined. We need the widest possible context of the concepts of 'entity' (Reality, the highest token) and ontological and connotative 'universals' (To Be and Reality-in-general) to make ontology justified in its entirety. The merit of such maximising of ontology with Reality-in-total, Reality-in-general and To Be is that it allows progress beyond the stipulations we might derive from ontic orientations.

Quine speaks of ontologizing science in terms of abstract (general) objects, which then shrink into abstract objects within limited quantificational contexts:

> [...] [T]here are others who, [...] making light of the distinction between abstract singular and concrete general terms, decide against abstract objects. Apparently these thinkers have appreciated, for whatever reasons, that concrete general terms carry no commitment to attributes or classes, and then have concluded the same for the corresponding abstract singular terms, by dint of drawing no distinction. This line of thought derives wishful vigor from a distaste for abstract objects coupled with a taste for their systematic efficacy. The motivation has proved sufficient to induce remarkable extremes. We find philosophers allowing themselves not only abstract terms but even pretty unmistakable quantifications over abstract objects ("There are concepts with which ...," "... some of which propositions ...," "... there is something that he doubts or believes"), and still blandly disavowing, within the paragraph, any claim that there are such objects.[367]

Quine goes relentless over the need to admit abstract objects just for theoretical purposes, even while apparently opposing them. If quantification over abstract objects is admissible on condition of referring them not only to epistemically spatio-temporal but also to physical-ontologically extension-motion objects, we require not merely objective ontological commitment, but an objectual-causal one – by reason of the Ontological Principle of Excluded Vacuous Middle – attuned to the totality of all that are in infinite and infinitesimal causal process. What follows in the passages below is a proposal for objectual ontological commitment in lieu of objective.

With a view to facilitating abstract objects in theoretical expressions, Quine presupposes the confusion between 'existence' and 'subsistence' as the odd ground of the two problems of spatio-temporal (extension-motion) reference. Quine's solution for the obfuscation is to prefer 'is', a form of 'to be', instead of 'exist', which latter might end in positing existence of theoretical or purely imagined entities beyond the possibilities yielded by ordinary quantification.[368] He is

---

367 Quine, *Word and Object*, 241.
368 "Wyman [a hypothetical subtle mind] [...] is one of those philosophers who have united in ruining the good old word 'exist'. Despite his espousal of unactualized possibles, he limits the word 'existence' to actuality – thus preserving an illusion of ontological agreement between himself and us who repudiate the rest of his bloated

ready to equate 'exist' with 'is', and would like not to take merely 'exist' as meaning spatio-temporal (better, extension-motion) concreteness as against the proposedly different 'is' or 'to be'.[369] The particularist question of 'is' in Quine does not account for the 'is' of Reality through the very abstract objects that are formed in thought via what I have called ontological universals of classes of entities and processes. The 'is' of particulars is in fact steeped in the 'is' of Reality – a fact that Quine overlooks in his rush to avoid essentialist metaphysical reification of abstract objects. This preoccupation makes him deal with the question of existence merely by means of the 'is' of particulars and tokens. The 'is' of particulars is in fact embedded in Reality via the purely Transcendental To Be. This justifies the phrase 'actuality in possibility' as qualification for To Be. Allowing actuality in possibility makes ontology objectual-causal.

The confusion between existence and subsistence is found by the tradition as the hitch behind extension-motion reference,[370] whereas this does not exhaust it. 'Is' has three senses to convey,[371] of which 'exist' is only the one that involves the

---

universe. We have all been prone to say, in our common-sense usage of 'exist', that Pegasus does not exist, meaning simply that there is no such entity at all. If Pegasus existed he would indeed be in space and time, but only because the word 'Pegasus' has spatio-temporal connotations, and not because 'exists' has spatio-temporal connotations. If spatio-temporal reference is lacking when we affirm the existence of the cube root of 27, this is simply because a cube root is not a spatio-temporal kind of thing, and not because we are being ambiguous in our use of 'exist'. However, Wyman, in an ill-conceived effort to appear agreeable, genially grants us the nonexistence of Pegasus, and then, contrary to what *we* meant by nonexistence of Pegasus, insists that Pegasus *is*. Existence is one thing, he says, and subsistence is another. The only way I know of coping with this obfuscation of issues is to *give* Wyman the word 'exist'." Quine, *From a Logical Point of View*, 3. Latin *Pegasus*, Greek *pēgasos*, the winged horse supposed to have sprung from the blood of slain Medusa. This word is from *pēgē*, "spring", "fount".

369 "I shall find no use for the narrow sense which some philosophers have given to 'existence', as against 'being'; viz., concreteness in space-time. If any such special connotation threatens in the present pages, imagine 'exists' replaced by 'is'. When the Parthenon and the number 7 are said to be, no distinction in the sense of 'be' need be intended. The Parthenon is indeed a placed and dated object in space-time while the number 7 (if such there be) is another sort of thing; but this is a difference between the objects concerned and not between senses of 'be'." Quine, *Methods of Logic*, 198.

370 Quine, *From a Logical Point of View*, 3.

371 In the scholastic sense, a given essence is a universal. It is not the 'is' in the sense merely of an existential instance (instantiation), identity (sameness), or predicate

existential quantifier 'there is', indicating often (not necessarily) particularism of a formalist-objective variety without objectual necessities, whereas 'subsist'[372] indicates a particular instance of a given essence, i.e. the concretion of *a* universal / essence of many existents or of an existent. Many essences – if we take them for ontological universals – can subsist in a substance / existent, which is, by Aristotelian and Scholastic systems, made of proper form and the matter. The form proper to an existent is a mixture of many essences. Subsistence is of ontological universals, not merely of an entity. Subsistence of many ontological universals within the totality of an entity or a particular is the existence or the quantifying 'is' / 'there is' of that entity / existent.

If *subsistence is the concretion of ontological universals* (not directly of things), if connotative universal is the qualitative (ontological universal) dimension of entities or classes in general in consciousness, and if subsistence depends on

---

(predication). *The Cambridge Dictionary of Philosophy*, second edition, s.v. "Is". (We discuss these three senses of 'is' under the next sub-section.) Essence is not an existential instance of a universal, making the term not amenable to fads like "individual essence". Essence is a universal-particular instance of a universal greater than anything processual of which it is predicated; it can be identical with itself in abstraction; and it can be a predicate to many. There is no singular essence of a thing, but only a few essences / universals together confluencing in one thing / process.

Subsistence is the actual instance where universals / essences of a many levels of ontological instantiation are instantiated in a process or many, which can in fact be real only as a coalescence of many universals in a process or many. An essence is a universal or universals; subsistence is the process or processes in which many essences / universals / qualities "stand under". Insofar as subsistence is a specific instance of *an* essence or more, it has within it also the essence in instantiation.

What is considered in the case of subsistence is not the whole object / process, i.e. the substance – which is the instantiation of all the essences that subsist in the being / process – but only the instantiation of the many specific essences in a substance. Thus, for example, a certain essence subsists in a substance, and the substance is made up of the matter and all possible forms that have gone into the formation of the substance. Substance is the underlying concrete stuff, considered in absolute union with all the essences that it instantiates. Muller, *Dictionary of Latin and Greek Theological Terms*, 290. It is interesting to note here that the whole theory of essences / universals is a refined repetition of medieval ontological theories.

372   The exact equivalent of 'subsistence' may not be the Latin *persona*, but the Greek *hypostasis*. *Hypostasis* is a sort of existence, especially of conceptual entities, because they "stand under" (Greek, *stasis*, "standing", "placing", "setting"; Latin, *stāre*, *sistere*, "to stand"; Greek, *hupó* and Latin, *sub-*, "under") actual entities for their sort of existence.

concretion of many ontological universals in an entity / process, then subsistent entity is not merely the crass spatio-temporally measurable and extension-motion-referable entity. Quoting from Russell's *Analysis of Mind* [191, 194], Quine argues against Russell in *Theories and Things*:

> [...] Russell commonly uses the word 'meaning' in the sense of 'reference'; thus "'Napoleon' means a certain individual" and "'Man' means a whole class of such particulars as have proper names". [...] What matters more than terminology is that Russell seldom seems heedful, under any head, of a subsistent entity such as *we* might call the meaning, over and above the existent object of reference. He tends, as in the 1905 paper "On Denoting," to blur that entity with the expression itself. Such was his general tendency with subsistents.[373]

Quine holds that, "[...] for want of distinctions, Russell tended to blur meaninglessness with failure of reference."[374] This means that, though Quine holds that meaning is the subsistent entity, and the referential / denotative entity is the object (abstract and concrete, or only concrete?) or part (abstract and concrete, or only abstract?) of the actual entity, he does also lack the appeal to something more than just what is "objective". This additional area is in fact what contributes the objectual-causal element to ontology. Quine does not realize its Einaic implications. To substantiate this claim I quote a passage where Quine stresses the about-ness of attributes with respect to reality-in-particular: "The ontology of attributes [...] allows an attribute corresponding to any sentence [...] that we can formulate about a thing. Complex singular terms for attributes commonly take the form of gerundive clauses (e.g., 'bearing spines in clusters of five'), preceded or not by 'the attribute (or quality or property) of.'"[375] Quine is aware of the metaphysical misuses of abstract terms.[376] He speaks also of the origin of abstract ontology, but clarifies the tenability of abstract singular and general terms:

> [...] [T]he disreputability of origins is of itself no argument against preserving and prizing the abstract ontology. This conceptual scheme may well be, however accidental, a

---

373 Quine, *Theories and Things*, 80.
374 Further, "This was why he could not banish the king of France without first inventing the theory of descriptions." Quine, *Theories and Things*, 80.
375 Quine, *Word and Object*, 122.
376 For example, he stipulates: "[...] we can use general terms, for example, predicates, without conceding them to be names of abstract entities [...] we can view utterances as significant, and as synonymous or heteronymous with one another, without countenancing a realm of entities called meanings." Quine, *From a Logical Point of View*, 12.

happy accident, just as the theory of electrons would be none the worse for having first occurred to its originator in the course of some absurd dream.[377]

He distinguishes abstract singular terms as necessary for ontological talk:

> Devices conceived in error have had survival value, and are to be assessed on present utility. But we stand to increase our gains by clearing away confusions that continue to surround them; for clarity is more fruitful on the average than confusion, even though the fruits of neither are to be despised. Hence we do well to distinguish abstract singular terms from concrete general ones by faithful uses of '-ness', '-hood', and '-ity', at least in contexts of philosophical analysis, despite the fact that the inception of abstract singular terms probably depended on the absence of a distinctive mark.[378]

That is, universals are predicable of relationality in objects and processes. Objects and processes are not only 'objective' but also 'objectual'. Objectual objects need not presuppose what he fears: merely metaphysically multiplied abstract entities. They can be conceived as ways of process of beings. The only condition is that abstract objects should be used in acceptable ways:

> The interlocked conceptual scheme of physical objects, identity, and divided reference is part of the ship which, in Neurath's figure, we cannot remodel and save as we stay afloat in it. The ontology of abstract objects is part of the ship too, if only a less fundamental part. The ship may owe its structure partly to blundering predecessors who missed scuttling it only by fools' luck. But we are not in a position to jettison any part of it, except as we have substitute devices ready to hand that will serve the same essential purposes.[379]

Nothing obstructs Quine from talking of the wider Reality in addition to reality-in-particular, using a conceptual scheme of physical and abstract objects. There is no unbridgeable distance for him to Reality via reality-in-particular, from the point of view of the kind of universals he admits. Abstract objects interject between particulars too. Abstracta are part of talk of the very theoretical and actual objectual reality of Reality and its parts.

---

377 Quine, *Word and Object*, 123.
378 Quine, *Word and Object*, 123.
379 Quine, *Word and Object*, 123–24. Otto Neurath holds that the work of organized conceptual scheme involves "the simplicity and straightforwardness of scientific empiricism", and "generalizations and predictions" made therefrom. Yet, "[…] the integration of science is an inevitable part of man's scientific activities." Neurath, "Unified Science as Encyclopedic Integration", 22–23. In the process of using non-metaphysical conceptual schemes, one should, as Neurath admits, use generalizations. These are abstract objects, different from but connected to physical objects. In short, even Neurath would not object if one said that these two sorts of objects are interlinked in discourse. Quine underlines exactly this point.

Reality is the entity identifiable as actualisation of subsistence of all ontological universals. What then of the 'is' of abstract objects (ontological universals) that subsist in actual existents: Reality and reality-in-particular? It gives the predicational sense of 'is', but not the identity-sense. The predicational sense yields the subsistence-sense of universals inhering in actual existents, which alone can give the identity-sense – though the identity-sense may be had in a relation from a universal to itself, say from 1 to itself, in analogy to existents.

Quine has not taken seriously the status of Reality as the highest token. Hence he was not in a position to allow To Be as the highest theoretical entity in a regimented theory of Reality. *Hence the confusion between existence and subsistence in Quine. Existence is for him of tokens; and subsistence of ontological universals in processes allows us to talk, due to which universals of connotative type are to be accorded the status of theoretical entities.* As subsistence is interwoven in existence to shape an 'is', and as subsistence carries existence beyond the 'is' of tokens of less than total stature, the 'is' of Reality is excluded, and the ultimate implication of 'is' differs drastically.

The 'is' of Reality points to the climactic entities: actual (Reality), ontologically abstract (To Be) and conceptually abstract / connotative (Reality-in-general), without the maximised senses of which the particularistic commitments of regimented theory in terms of subsistence and existence are incomplete. Ontologically by reason of the maximising effect of the Einaic categories it is apt to claim that Quine's particularism confuses the concepts of existence and subsistence, due to the absence of the categorial notions of ultimacy. The moment these are involved in regimented theory the whole shape of the ontology of objectivity is transformed into objectual-causal ontology, wherein the infinite and infinitesimal objectual causality (by reason of the Ontological Principle of Excluded Vacuous Middle discussed in Chapter 1) within Reality is brought into full effect through the ontological universals that causality produces, for formation of the connotatives that go into the conceptual framework of Einaic Ontology.

3.2.3 The 'Is' of Abstract Objects: Quine-Carnap Dialogue

In the context of the question of the Einaic Ontological connection and difference between the 'is' of abstract objects and physical objects, Quine's earlier stand and later change of mind concerning the possibility of the 'is' of abstract objects is of interest. Penelope Maddy quotes what in 1947 Quine and Goodman wrote about the ontological status of abstract entities of the realism of abstract (mathematical) entities, in the opening sentence of an article:

We do not believe in abstract entities. No one supposes that abstract entities – classes, relations, properties, etc. – exist in space-time; but we mean more than this. We renounce them altogether [....] Why do we refuse to admit the abstract objects that mathematics needs? Fundamentally this refusal is based on a philosophical intuition that cannot be justified by appeal to anything more ultimate.[380]

For Quine of the 1940s, abstract objects did not exist, nor did they subsist in entities.[381] Later in *From a Logical Point of View* (1980) in a parenthesis within the bibliographical entry on the afore-said article Quine retracted it: "Lest the reader be led to misconstrue passages in the present book by trying to reconcile them with the appealingly forthright opening sentence of the cited paper, let me say that I should now prefer to treat that sentence as a hypothetical statement of the conditions for the construction in hand."[382] A clearer passage may be quoted from his 1990 remark on Putnam's and Parsons' economizing on abstract objects by using a modal[383] operator:

> Hilary Putnam and Charles Parsons have both remarked on ways of economizing on abstract objects by recourse to a modal operator of possibility. We have [...] the other side of the coin: the positing of objects can serve to reinforce the weak truth functions without recourse to modal operators. Where there are such trade-offs to choose between,

---

380 Goodman and Quine, "Steps toward a Constructive Nominalism", cite in Maddy, *Naturalism in Mathematics*, 95.
381 It is queer that Quine, a doctoral student of Whitehead, who worked on a topic in connection with Russell and Whitehead's *Principia Mathematica*, was not influenced by Whitehead's later ontological work on eternal objects, and had to wait for many more years for ontological insistence on abstract objects. Perhaps, Quine was too pragmatic and analytic to admit for universals any more than a mere functional space in discourse.
382 Quine, *From a Logical Point of View*, 173–74. (The said comment is found in this work as a short passage in explanation of the former position, in the bibliographical entry on Goodman and Quine, "Steps toward a Constructive Nominalism".)
383 The possibility of a modal version of second and higher order Einaic Logic and resultant Einaic Semantics are mooted in the General Conclusion of this essay (throughout sub-heading no. 3), as a way of dealing with ontological universals and connotatives that appear to be classes of classes of... This is suggested against the background of what I call the ways-of-being interpretation of ontological universals and connotatives. The connection between ontological universals and connotatives requires some manner of reflecting the former in the latter. Hence, we need higher-order logics and semantics to explicate tis possibility.

> I am for positing the objects. I posit abstract ones grudgingly on the whole, but gratefully where the alternative course would call for modal operators.[384]

Thus, for Quine of at least the close of the 1970s and much thereafter, there 'are' abstract objects having a sense of existence different from the 'is' of actual, particular, physical objects. His grudge is not to abstract objects, but about the philosophical state of affairs where universals / essences are still vaguely defined. The ways-of-being interpretation proposed earlier and will be elaborated upon in the present chapter promises a fresh look at it with a better explanation without incurring particularism, relativism and absolutism.

Famously, the said conceptual change in Quine is due to Carnap. Concerning the Quine-Carnap dialogue[385] on the connection and distinction between pragmatic and theoretical languages, we may make the following conclusions. The dialogue concluded in Quine's realization that the language of scientific inquiry and the mathematical / theoretical language are both adopted on pragmatic grounds, the main reason being that even abstract entities take the status of values in quantification. But according to him the Carnapian absolute distinction between the thing-language (of physical existence) and the number-, set- and other languages of theory / mathematics cannot stand.

> Ontological questions [...] are on a par with questions of natural science. Consider the question whether to countenance classes as entities. This [...] is the question whether to quantify with respect to variables which take classes as values. Now Carnap has maintained that this is a question not of matters of fact but of choosing a convenient conceptual scheme or framework for science. With this I agree, but only on the proviso that the same be conceded regarding scientific hypotheses generally. Carnap has recognized that he is able to preserve a double standard for ontological questions and scientific

---

384 Quine, *Quintessence*, 114. "Operator is a one-place sentential connective, i.e., an expression that may be prefixed to an open or closed sentence to produce, respectively, a new open or closed sentence." *The Cambridge Dictionary of Philosophy*, second edition, s.v. "Operator". 'Modal' is a logical term "[d]esignating or pertaining to a proposition involving the affirmation of possibility, impossibility, necessity, or contingency or in which the predicate is affirmed or denied of the subject with a qualification [...]", or "[...] (of an argument) containing a modal proposition as a premiss." *The New Shorter Oxford English Dictionary on Historical Principles*, s.v. "Modal". The counterfactual variety of modalities yield the pure cases of possible worlds necessitated by ontological discourse, where systems and sets of our ontological universals and truths therefrom can belong, provided such possible worlds admittedly have origin and conceptual foundation in past realizations of universals.

385 For a succinct study of it, see Maddy, *Naturalism in Mathematics*, 95–101.

hypotheses only by assuming an absolute distinction between the analytic and the synthetic; and I need not say again that this is a distinction which I reject.[386]

Even without discussing the rationales for Quine's praiseworthy lifelong effort to keep away from the centuries-long absolute analytic-synthetic distinction, we can understand his position on the distinction between the two languages:[387] (1) What apply to mathematical abstract objects apply also to other abstract objects, since both are intuitive generalizations over sets of qualities / ways of being of entities / processes,[388] never absolutely axiomatizable and are subject to the process of connection between tokens and abstract objects / ways, and (2) the two aspects of the languages of science and of mathematics / theory are but two poles of the same language, which is in effect one important fact that Quine has attempted to show in his relativization of analytic and synthetic propositions.[389]

---

386   Quine, *From a Logical Point of View*, 45–46.
387   Maddy makes her own summary statement of the Quine-Carnap dialogue: "Aside from their disagreement over whether it is better to say 'The pragmatic virtues of the linguistic framework of *x*s confirms the existence of *x*s' or 'The pragmatic virtues of the linguistic framework of *x*s make it advisable to adopt the *x* language', Quine and Carnap are not so very far apart at this point: they see the adoption of an overall scientific language / theory as justified by similar considerations. But there is a further disagreement between them. To see this, recall that to accept the thing language is 'to accept rules for forming statements and for testing, accepting or rejecting them. The acceptance of the thing language leads, on the basis of observations made, also to the acceptance, belief and assertion of certain statements'." Maddy, *Naturalism in Mathematics*, 98.
388   Gödel has attempted to bring up this aspect in an important passage in the context of correlating mathematical intuition with sense-intuition: "It should be noted that mathematical intuition need not be conceived of as a faculty giving an *immediate* knowledge of the objects concerned. Rather it seems that, as in the case of physical experience, we *form* our ideas also of those objects on the basis of something else which *is* immediately given. Only this something else here is *not*, or not primarily, the sensations. That something besides the sensations actually is immediately given follows (independently of mathematics) from the fact that even our ideas referring to physical objects contain constituents qualitatively different from sensations or mere combinations of sensations, e.g., the idea of object itself, whereas, on the other hand, by our thinking we cannot create any qualitatively new elements, but only reproduce and combine those that are given. Evidently the 'given' underlying mathematics is closely related to abstract elements contained in our empirical ideas." Gödel, "What Is Cantor's Continuum Problem?", cited in Maddy, *Naturalism in Mathematics*, 91.
389   Quine has relativised analytic (taken as a purely linguistic device independent of experience!) and synthetic (taken to be based only on experience!) propositions by

The one sort of entity or language continues upon and is made possible by the other. Scientific pragmatism, being adopted by empirical-theoretical exigencies, employs also the "purely" theoretical / mathematical / ideal language; and the vice versa is also a must since the universals / essences need thing- or process language for their formulation. This is also a result of Gödelian realism of pushing the definitions of primitive notions and axioms, discussed in Chapter 2. The Frege-Russell and Russell-Whitehead programmes to construct a logico-mathematical integration were somewhat undermined by Gödel's result. This necessitates a probabilistic and ontological synthesis in science. But Quine is committed to empiricism projecting particularist ontology as the only possible shape of thought. Particularist ontology falters on the question of the extent of access to infinitesimal theoretical- and thing-concepts for definition of the concept of the particular. If theoretical- (mathematical-) and thing- (empirical-) languages are mutually continuous, then even 'Reality', a term in ordinary and scientific languages, denotes the most total particular (empirical) token connecting all sorts of tokens by way of ontological universals, especially To Be.

Therefore, the probabilistic and ontological synthesis in science should be based on coalescence of the ontology of science in general that integrates both

---

use of synonymy and by means of relativising reductionism, and has held thus: "[…] [I]n general the truth of statements does obviously depend both upon language and upon extralinguistic fact; and […] this obvious circumstance carries in its train, not logically but all too naturally, a feeling that the truth of a statement is somehow analyzable into a linguistic component and a factual component. The factual component must, if we are empiricists, boil down to a range of confirmatory experiences. In the extreme case where the linguistic component is all that matters, a true statement is analytic. But I hope we are now impressed with how stubbornly the distinction between analytic and synthetic has resisted any straightforward drawing. I am impressed also, apart from prefabricated examples of black and white balls in an urn, with how baffling the problem has always been of arriving at any explicit theory of the empirical confirmation of a synthetic statement. My present suggestion is that it is nonsense, and the root of much nonsense, to speak of a linguistic component and a factual component in the truth of any individual statement. Taken collectively, science has its double dependence upon language and experience; but this duality is not significantly traceable into the statements of science taken one by one." Quine, *From a Logical Point of View*, 41–42. Quine puts the connection between science and mathematics thus in a famous passage: "[…] [A] self-contained theory which we can check with experience includes, in point of fact, not only its various theoretical hypotheses of so-called natural science but also such portions of logic and mathematics as it makes use of." Quine, "Carnap and Logical Truth", cited in Maddy, *Naturalism in Mathematics*, 101.

the scientific (empirical) and the mathematical (theoretical / abstract) aspects of language. Ontological universals / abstract entities have no actualty or particularity; they are no entities or tokens; they are connective realities of the interworld of actual entities. Their nature is connectivity at actualisation. Beyond this they lack a different individual essence; they cannot constitute an "individual essence". *Their essence is subsistence in actual entities in their 'many'.* Such subsistence is their only essence; hence, they are ways of being of processes that *always* have causal connections from the infinite extension-motion past and into the future without end, in an infinitesimally determining manner. Essences, as ways, cannot exist merely with particular extension-motion actualities (because they do not exist, but only subsist), but can be the principles of ontological connectivity severally in the many that causally connect unto other such abstracta for their own subsistence. Even when essences / universals are connotative singular terms denoting unique classes or beings and are values in quantification, they have connotative ingredients that apply their related ontological universals ontologically to the many. Though there are singular terms, abstract universals contributing to them are not limited to particulars. They always have connections proper to other past processes and lead to future processes. *Therefore, processual and objectual-causal connectivity is the essence, the form of the 'is', of all possible abstract entities, whether singular or plural.*

Abstracta / universals are subsistents; they are capable of future subsistence as ways of processes via objectual-causal instantiation. They subsist now in actual particulars, and will do so in the future as ways of processes tending towards other such ways that relativise and probabilise the former. Thus, the ontology of inherence of abstract objects in particulars and tokens is such that particulars (actual objects / their classes) have extension-motion reference in themselves. They are not particulars or tokens. *They have a sort of 'is' that makes them ontologically subsist in the many and connotatively subsist in consciousness in ways derived from their ontological occurrence.*[390]

---

390   This allowance of subsistence to abstract "objects" does not mean that any sort of existence is granted them. But they are related more to real processes, states of affairs and matters of fact than are connotative universals. I distance this sort of ontology – that accepts existence-related primacy for ontological universals over connotative ones – from Aristotle's ontology of overly realistic and overly abstract pure universals and from Husserl's phenomenological ontology with its queer judgment-entities originating purely from within (or the external connection is insufficiently mentioned). Evidently, Aristotle is an extreme case. To quote Husserl on this problem: "In the first place, Aristotle's establishment of analytics as apophantics,

To connect the 'is' of abstract objects with other senses of it, let us study the three possible senses of 'is', contrast them with a possible objectual ontological extension of the same and argue for a trans-particularist extension of Quine's

> as a logic of the predicative statement and, correlatively, the predicative judgment, proved itself a hindrance. However necessary that was a beginning, it involved a deeply rooted difficulty [...] of abstracting thematically from the judging activity and, while remaining consistent in so doing, regarding the judgment-sphere theoretically as a specific Objective field of apriori ideality, just as the geometer regards the sphere of pure geometrical shapes and the arithmetician regards the sphere of numbers." Husserl, *Formal and Transcendental Logic*, 81.
>
> Husserl continues to present his variety of ontology of universals: "It is because of the intrinsic nature of the affairs themselves that the ideal Objectivity of judgment-formations could not gain recognition, and that even in recent times – after having been brought out systematically and vindicated by a critical refutation of empiricisitic psychologism – it has not yet won universal acceptance. Judgments are there for us originally in judicative activities. Every work of cognition is a multiple and unitary psychic activity in which cognitional formations originate. Now, [...] external Objects too are originally there for us only in our subjective experiencing. But they present themselves in it as Objects already factually existent beforehand (Objects "on hand") and only entering into our experiencing. They are not there for us, like thought-formations (judgments, proofs, and so forth), as coming from our own thinking activity and fashioned by it purely (not, perchance, out of materials already on hand and external to it). In other words: Physical things are given beforehand to active living as objects originally other than the Ego's own; they are given from outside. Contrariwise, the formations with which logic is concerned are given *exclusively from inside*, exclusively by means of spontaneous activities and *in* them. On the other hand, [...] after having in fact been generated they are still taken to be existent; one "returns to them" as the same formations, and does so repeatedly at will; one employs them in a sort of practice, connects them (perhaps as premises) and generates something now: arguments, proofs, or the like. Thus one does actually deal with them as with real physical things, even though they are far from being realities. And so they float obscurely between subjectivity and Objectivity. To accept them seriously as irreal Objects, to do justice to the evidences *on both sides* (which it may well have been illegitimate to play off against each other), and to fix one's eye on what is seriously problematic here and take it seriously as problematic – that is something one does not venture, old inherited fears of Platonism having made one blind to the doctrine's purifiable sense and the genuine problem implicit in this sense." Husserl, *Formal and Transcendental Logic*, 81.
>
> Einaic Ontology attempts to have abstract objects *cum fundamentum in processe* ("with foundation in process"), not merely *in re* ("in thing"). Hence, Einaic Ontology posits ontological universals *in processe* ("in process") and their objectual-causal counter-forms *in mente* ("in mind").

idea of the priority and greater breadth of 'is' in comparison with 'exists', so to harvest the full meaning of our broadened notion of ontological commitment beyond his particularistic one. The three possible senses of 'is' are categorised thus: "The 'is' of existence (*There is a unicorn in the garden: $\exists x\ (Ux \wedge Gx)$*) uses the existential quantifier. The 'is' of identity (*Hesperus is Phosphorus: $j = k$*) employs the predicate of identity. The 'is' of predication (*Samson is strong: Sj*) merely juxtaposes predicate symbol and proper name."[391] Existence is token-existence or particular-existence as a variable. Identity is equality or identity of direct reference / denotation. Predication is expressing an essential (attributive: say, 'is red') or other (predicative: say, 'is human') quality of an existent. The third is the sense in which abstract objects are used. 'Is' implies all these. The reason why 'is' is used statically (not processually) for all these three senses is that somehow actual existence and identity are connected to the ideality of predication without each time realizing that these predicable abstract objects are not denotatives denotative of the ontological universals of processes but ways of processes that are flexible.

Connotation is such by reason of the ontological universals in processes connoted by connotatives in consciousness (it is impossible to denote by connotatives) through formation of denotation by concepts in which connotatives are ingredients. Hence, we should have a philosophy of logic that amplifies the statically, non-extension-motion[392] and acausally nomic object-nature of abstract

---

391 *The Cambridge Dictionary of Philosophy*, second edition, s.v. "Is".
392 Husserl presents the whole dilemma and proposes to solve it by accepting abstract objects / essences as atemporal: "We answer: Certainly essences are "concepts" – if by concepts one understands, in so far as that ambiguous word allows, precisely essences. Only let one make clear to himself that *then* it is nonsense to talk about them as psychical products and likewise as concept-*formations*, provided the latter is to be understood strictly and properly. One occasionally reads in a treatise that the series of cardinal numbers is a series of concepts and then, a little further on, that concepts are *products* of thinking. At first cardinal numbers themselves, the essences, were thus designated as concepts. But are not cardinal numbers, we ask, what they are regardless of whether we "form" or do not form them? Certainly, I frame *[vollziehe]* my numbers, form my numerical objectivations adding "one plus one." These numerical objectivations are now these and when I then form them a second time in an identical way, they are different. In this sense, at one time there are no numerical objectivations of one and the same number, at another time there are many, as many numerical objectivations as we please of one and the same number. But just with that we have made (and how can we avoid making) the distinction; the numerical objectivation is not the number itself, it is not the number two, this single member

objects and makes them accompany processes in extension-motion, as ways of being of processes; and remain as ideals of extension-motion-, causal and objectual processes in their connotative reflections that serve the purpose of denotation by concepts by connotatives' being ingredient in concepts. The 'is' as 'existence' is idealized token-existence without reference to its ways of process. It should bring in Reality as the highest token, with the support of which alone the 'is' of existence has meaning through other ways of processes. The predicative meaning of 'is' has to bring in the highest ontological universal To Be for predication for meaningfulness of 'is' in Reality. Connotation should be through the ideal of thought, i.e. Reality-in-general which vaguely reflects To Be and Reality.

Meinong has held that 'is' is broader than 'exists', because 'is' produces truths when combined with 'deer' and 'unicorn', and 'exists' produces truths only when combined with 'deer' and not with 'unicorn'. This is in line with the fact that Aquinas and others use *esse* to denote the very essential activity of the 'is-dimension' or 'being' of every existent, implying that "[…] with 'is' they attribute more to an object than we do with 'exists.'"[393] In Brentano, for a presentation to have an object it is not necessary that the object exists. "According to him, to be an object and to exit are widely different predicates. The object as such is beyond being and non-being; it is to be defined simply as that which can be grasped by an act."[394] So, it is possible to speak of positive and negative states of affairs and assign to every judgment its own corresponding state of affairs.[395] This necessitates that even

---

of the numerical series which, like all members, is an atemporal being. To designate it as a psychical formation is thus countersense, an offence against the sense of arithmetical speech which is perfectly clear, discernible at any time and therefore which *precedes* all theory. If concepts are psychical formations then those affairs, such as pure numbers, are not psychical formations. As a consequence, one *needs* new terms if only to resolve ambiguities as dangerous as these." Husserl, *Ideas*, 42.

393 *The Cambridge Dictionary of Philosophy*, second edition, s.v. "Is".
394 Schuhmann, "Brentano's Impact on Twentieth-century Philosophy", 285. "Thus to judge 'God exists' includes, in addition to the act of judging, the presentation 'God' as its matter, but in addition has God's existence as its content. On the other hand, the judgment 'God does not exist' is about exactly the same presentation or judgmental matter 'God" but its content is God's non-existence." Schuhmann, "Brentano's Impact on Twentieth-century Philosophy", 287.
395 Husserl's student Adolf Reinach developed the most comprehensive phenomenological theory of states of affairs, and has held so. Schuhmann, "Brentano's Impact on Twentieth-century Philosophy", 288.

impossible beings have a truthmaker value.[396] Quine opposes this view, though he vouches for 'is' as the term that is broader than 'exists'. To make 'is' possible as broader than 'exists', Quine takes recourse to abstract objects in the third sense of 'is', i.e. the predicative. This shows that the 'is' of abstract objects is for him merely subsistence at the consciously connotative realms of talk of sets of entities or processes, and not at the connective and ontological realm of processes themselves. Again, the lack of the distinction between ontological and connotative universals in processes is what causes this inconsistency.

To posit the actuality of things Quine has brought in the notion of ontological commitment, in which the Quine-Carnap dialogue has made a mark on Quine by making abstract entities possible at least as subsistent entities. But the actuality of things and the ideality of abstract objects are not connected well within the meanings of 'is' as is done in the ways-of-being interpretation. I propose towards the end of this chapter that taking abstract objects as ways of being of actual, past and future processes can unify the three meanings of 'is' into To Be and give rise to Einaic Ontology.

3.2.4 From 'Is' to Language-Ladenness and To Be

Quinean ontological commitment is particularistic because its fundamental tokens are necessarily reality-in-particular (at the realm of the bound variable 'some', meaning 'at least one'), and never does he use any equivalent of Reality as one of the tokens or as the totality of tokens. It is possible to de-particularize 'is' in favour of ontological holism by extending Quine's holism by the cosmological and epistemological arguments of Chapters 1 and 2. *This implies leaving the nominal, particularist and singular status of 'is'* and accepting its infinitive background, the verbal-processual 'to be', which works at the fundamental attributive and processual-verbal levels of entities.

This too implies some sort of particularism, since 'to be' is the 'to be' of an entity / process. This 'to be' is the first sense of 'is', i.e. not merely as in an existential quantifier, but in addition, as it is rendered in the infinitive in order to

---

396 "Russell took exception to the fact that impossible objects violated the principle of non-contradiction. Meinong agreed, but could not see a problem here. This principle applies, after all, only to objects that can exist or obtain. Impossible objects by definition cannot, and therefore are exempt from the principle. Still, they function as truthmakers of true and wrong statements: it is correct to state that the round quadrangle is round, but it is wrong to state that it is elliptic." Schuhmann, "Brentano's Impact on Twentieth-century Philosophy", 297.

conceptualise it and make it processually (i.e. verbally, as any verb *denotes* processes) attributive. Reality is the Transcendentally Transcendent category. The verbal 'is' of Reality is 'To Be', rendered in the processual-attributive infinitive. It implies not merely the direct ontological commitments of first order predicates, but also the 'is'es of all levels of logic.

We now analyse the nature of the 'is' of ontological commitment in Quine within the quantificational framework, with a view to facilitating our extension of his holism in the next sub-section. For Quine ontological commitment is of the province of the theory of reference (semantics in the loose or broad sense), not directly of the province of the theory of meaning (semantics proper).[397] According to Hylton it is in and through semantics that there are the direct counterparts of Quine's ontological commitment, given in first order logic:

> [...] [W]hen we have a body of theory cast in the notation of first-order logic, its ontological commitments, Quine claims, are apparent. For every existentially quantified sentence that the theory contains or implies, there must be an object of which the corresponding open sentence is true; such an object must exist if the theory is to be true. (The "corresponding open sentence" here we would obtain by simply deleting the quantifier.) This much seems to be implied by the explanation of quantification...[398]

Here Hylton speaks of tokens proper and nothing else as the yield of ontological commitment in first-order logic. This is emphasized in the context where Quine agrees with Kripke's remark that the connection between ontology and referential quantification "[...] is trivially assured by the very explanation of referential

---

397 For Quine, semantics is the theory of meaning (though Tarski has done much correspondence-level theory of reference in the name of semantics), and involves concepts like synonymy (sameness of meaning), significance (possession of meaning), analyticity (truth by virtue of meaning alone) and entailment (analyticity of the conditional). The theory of reference involves the concepts of naming, truth, denotation (truth-of), extension and values of variables. Quine, "Notes on the Theory of Reference", 130. For the purpose of this chapter I use the term 'semantics' in a general sense, as implying both theories of meaning and of reference by them, as the latter is the product of the former.

398 Hylton, "Quine on Reference and Ontology", 123. First-order logic is predicate calculus (lower functional calculus, elementary quantification theory), which is the study of valid inference in first-order languages. First-order languages are languages "[...] built up from an expressively complete set of connectives, first-order universal or existential quantifiers, individual variables, names, predicates (relational symbols), and perhaps function symbols." *The Cambridge Dictionary of Philosophy*, second edition, s.v. "Formal Logic".

quantification."[399] Quine adds: "The solemnity of my terms 'ontological commitment' and 'ontological criterion' has led my readers to suppose that there is more afoot than meets the eye, despite my protests [...]."[400] This shows clearly that the objects he means by the objective counterparts of his ontological commitment in first-order logic are possible extension-motion (measurementally "spatio-temporal") tokens and natural kinds with direct physical access for thought to verify the actuality or specificity of. But Quine almost contradicts the direct meaning of this self-justification while attempting to vindicate himself against Kripke's criticism:

> The expressions 'five', 'twelve', and 'five plus twelve' differ from 'apple' in not denoting bodies, but this is no cause for disinterpretation; the same can be said of such unmathematical terms as 'nation' or 'species'. Ordinary interpreted scientific discourse is as irredeemably committed to abstract objects – to nations, species, numbers, functions, sets – as it is to apples and other bodies. All these things figure as values of the variables in our overall system of the world. The numbers and functions contribute just as genuinely to physical theory as do hypothetical particles.[401]

The actualities which Quine is ontologically committed to are now also theoretical entities that theoretically make tokens and particulars possible. Singular and plural abstract objects (ways of being) are thus also implied in ontological commitment. In itself, such commitment does not refer in the broad sense to actual entities. Hylton elucidates by giving an example:

> We need to understand [...] why Quine does not see names as a source of ontological commitment on a par with quantified variables [...] His claim is that when we are concerned with exposing the ontological commitments of a given language, we should reformulate it so that the names are all eliminated. The easiest way to see why [...] is to consider the fact that there are names that do not actually name any object [....] The name 'Vulcan', for example, was introduced to name a tenth planet whose existence was postulated – wrongly [...] – to explain certain astronomical phenomena [....] Now someone who asserts, say, 'Vulcan is a small planet' *is* thereby committed to the existence of Vulcan. No such object actually exists: Any criterion of ontological commitment will tell us what a given body of theory is committed to but will not tell us what there really is unless we add the claim that the body of theory is true. Difficulties of this sort convinced Quine that the ontological commitments of a language are most clearly displayed when the names of the language have been eliminated, along the lines suggested by Russell's application of his theory of descriptions. Suppose we have a name [...] 'Socrates', and [...] a predicate, 'is human' [...] We introduce the predicate 'S' that we take to apply to

---

399   Quine, *Theories and Things*, 174.
400   Quine, *Theories and Things*, 175.
401   Quine, *Theories and Things*, 149–50.

Socrates and to no one else [....] Then we can say, there is an object that is S, and that object is human.[402]

When names are eliminated, there remain possible particulars under bound variables, always universal-laden and language-laden. So, in general, ontological commitment is never referentially specific of the token entity (object) at issue, but it is of a (natural) kind which is, again, universal-based, given the basic mode of definition of ontological commitment over a domain (natural kind): "For to say that a given existential quantification presupposes objects of a given kind is to say simply that the open sentence which follows the quantifier is true of some object of that kind and none not of that kind."[403]

This explicitly excludes connotative universals; at the same time, his universals are not fully ontological too. This is the source of the worst of confusions concerning universals in his thought and elsewhere, say, in most analytic thinkers, phenomenologists, hermeneuticians, pragmatists, scientific ontologists etc. To add, the fact remains that Quine's ontology reels in linguistic-particularistic ontological commitment, without the natural objectual-causal placing of linguistic expressions and their nomic acausal commitments (connotatives ingredient in concepts) in ontological universals and tokens, and finally in the ultimate referable / denotable Reality and its Transcendental (ontological universal) To Be. So, his *abstracta* are overly particularistic and language-bounded, is aimed at objectivity, and is least objectual-causal. For these reasons, his ontological commitment is unable to possibilize concluding the token-existence of beings and the particular-subsistence of abstract objects in the *really out-there sense*. This state of affairs excludes concluding the To Be of Reality as the foundation of ontological commitment, without which a final possibilization of actual objectual tokens within the framework of the highest possible Transcendental is not feasible.

Paradoxically, he admits the place of system in the language-ladenness of ontological discourse, not realizing that systematicity may be extended to the whole of Reality if ontological universals are admitted. The particularism of Quine's ontological commitment is all the more evident in what I consider as his Scientific Criterion of Ontological Commitment:

> Our acceptance of an ontology is, I think, similar in principle to our acceptance of a scientific theory, say a system of physics: we adopt, at least insofar as we are reasonable, the simplest conceptual scheme into which the disordered fragments of raw experience can be fitted and arranged. Our ontology is determined once we have fixed upon the

---

402 Hylton, "Quine on Reference and Ontology", 124–25.
403 Quine, "Notes on the Theory of Reference", 131.

over-all conceptual scheme which is to accommodate science in the broadest sense; and the considerations which determine a reasonable construction of any part of that conceptual scheme, for example, the biological or the physical part, are not different in kind from the considerations which determine a reasonable construction of the whole. To whatever extent the adoption of any system of scientific theory may be said to be a matter of language, the same – but no more – may be said of the adoption of an ontology.[404]

Why exactly should we limit with the particular science or science in general? What is at stake here is the realm of generalities that pertain to Reality, which is also a token like reality-in-particular. Why should bound variables not be made to mean cases that pertain to Reality too? An answer to this betrays the crux of particularism in Quine, and in analytic, linguistic and pragmatic traditions in general. This lack is clear in the following passage from Quine, where he deals with the inseparability of tokens from universals:

> […] [W]hat are classes? Consider the bottom layer, the classes of physical objects. Every relative clause or other general term determines a class, the class of those physical objects of which the term can be truly predicated. Two terms determine the same class of physical objects just in case the terms are true of just the same physical objects. Still, compatibly with all this we could reconstrue every class systematically as its complement and then compensate for the switch by reinterpreting the dyadic general term 'member of' to mean what had been meant by 'not a member of'. The effects would cancel and one would never know. We thus seem to see a profound difference between abstract objects and concrete ones. A physical object, one feels, can be pinned down by pointing – in many cases, anyway, and to a fair degree. But I am persuaded that this contrast is illusory.[405]

He argues from the implied background framework of linguistic holism, by which anything in propositions is referable to other propositions and finally to the whole of a language. This shows it is impossible to isolate extension-motion objects from class-names and the presupposed ontological universals. One must wonder why the inseparability of particular tokens from universals does not make him extend the possibility of systematicity beyond that in language, into that in Reality. As a result, the acausal level of his abstract nomic "objects" remains, and tokens are not philosophically connected intimately to their processual universals beyond particularistic universals, which latter is not defined clearly, either as ontological or as connotative or both. To solve this predicament we extend the concept of tokens and natural kinds to totalities and to totalities of totalities and finally to Reality, which is the unique token and natural kind

---

404 Quine, *From a Logical Point of View*, 16–17.
405 Quine, *Theories and Things*, 16.

allowable. The manner in which any token is universal-bounded is also the manner in which Reality is universal-bounded.

That is, the language- and universal-ladenness or theory-ladenness of ontological commitment resists ostensive isolation of specific entities as absolutely 'this' or 'that', without connecting them to their ontological-processual and theoretical counterparts: i.e., other entities and finally Reality in their possible ontological connexity with the entity / process in question, and also the senses and ontological universals of ever-more general varieties, by means of the connotatives ingredient in concepts. Quine may be judged as abstaining from making relevant conclusions from ontological commitment in combination with the language-ladenness of the tokens that result from commitment, if he does not inductively generalize abstract objects unto the ontological To Be of the inductive totalization of tokens, i.e. of Reality. The moment ontology is recognized as both language-bounded and system-bounded, it is straightforward to conclude that the universals that language-boundedness entails are also system-bounded within the context of the highest token entity, Reality. The universal involved here is To Be. System-boundness in Reality implies ontological universals. This is justified by postulated physical-conceptual schemes for ontology in terms of the universals involved in particulars (natural kinds) and tokens. This is stressed by Quine too – though not as comprehensively as we have – in terms of his rule of simplicity with respect to sense data, but not with respect to the connotative ingredients of sense data and the ontological universals of objects / processes:

> The rule of simplicity is indeed our guiding maxim in assigning sense data to objects: we associate an earlier and a later round sensum with the same so-called penny, or with two different so-called pennies, in obedience to the demands of maximum simplicity in our total world-picture [....] The physical conceptual scheme simplifies our account of experience because of the way myriad scattered sense events come to be associated with single so-called objects; still there is no likelihood that each sentence about physical objects can actually be translated, however deviously and complexly, into the phenomenalistic language. Physical objects are postulated entities which round out and simplify our account of the flux of experience, just as the introduction of irrational numbers simplifies laws of arithmetic. From the point of view of the conceptual scheme of the elementary arithmetic of rational numbers alone, the broader arithmetic of rational and irrational numbers would have the status of a convenient myth, simpler than the literal truth (namely, the arithmetic of rational) and yet containing that literal truth as a scattered part. Similarly, from a phenomenalistic point of view, the conceptual scheme of physical objects is a convenient myth, simpler than the literal truth and yet containing that literal truth as a scattered part.[406]

---

406 Quine, *From a Logical Point of View*, 17–18.

What he stresses are physical-conceptual schemes in ontology. Quine's ontological commitment is particularistic, language-bounded and universal-bounded. In general, Quine's ontological commitment is not *referentially* specific of any token entity (object), but of a (natural) kind, which is, again, universal-based, given the basic mode of definition of ontological commitment over a domain (natural kind). The elementary commitment to physical objects is only a special case of commitment over the domain of natural kinds.

Universals must then be not merely connotative / mind-based, but also ontological, i.e. always process-based. The processes implied by a universal are not merely a few particular ones but the many (probably infinite) processes of the kind denoted by the connotative universal in question. The natural kind denoted by the connotative universal implies the connotative as derived not merely and purely from the capability of the mind to create denotative terms by referring to the "seeming" aspect of similarities (recall Ockham's concept of naming and Hume's concept of causality). Connotatives are derived from the members of the natural kind that is in direct or indirect connotative relation with the mind. The members of natural kinds are encompassed by a connotative concept, but these members have an ontological aspect and rationale for the occurrence of the connotative universal in question. This ontological aspect is called the ontological universal / essence with respect to the connotative. In short, it is not only the case that terms, words, sentences and languages are connotatively universal-laden; but these, along with connotative universals, are all simultaneously ontologically universal-laden. Whitehead acknowledged this fact less than a century ago as he discussed his organismic-ontological concept of 'social order':

> Thus a society is more than a set of entities to which the same class-name applies: that is to say, it involves more than a merely mathematical conception of 'order'. To constitute a society, the class-name has got to apply to each member, by reason of genetic derivation from other members of that same society. The members of the society are alike because, by reason of their common character, they impose on other members of the society the conditions which lead to that likeness.
>
> This likeness consists in the fact that (i) a certain element of 'form' is a contributory component to the individual satisfaction of each member of the society; and that (ii) the contribution by the element to the objectification of any one member of the society for prehension by other members promotes its analogous reproduction in the satisfactions of those other members. Thus a set of entities is a society (i) in virtue of a 'defining characteristic' shared by its members, and (ii) in virtue of the presence of the defining characteristic being due to the environment provided by the society itself.[407]

---

407 Whitehead, *Process and Reality*, 89.

A member or members of a society can also be defined as a society by following the same criteria in the miniature. Whitehead's footnote to this passage refers also to p. 34 of the same book, where he recalls also that

> [t]he notion of 'defining characteristic' is allied to the Aristotelian notion of 'substantial form'. The common element of form is simply a complex eternal object exemplified in each member of the nexus. But the social order of the nexus is not the mere fact of this common form exhibited by all its members. The reproduction of the common form throughout the nexus is due to the genetic relations of the members of the nexus among each other, and to the additional fact that genetic relations include feelings of the common form. Thus the defining characteristic is inherited throughout the nexus, each member deriving it from those other members of the nexus which are antecedent to its own concrescence.[408]

His 'defining characteristics' as yielded by 'eternal objects' are analogous to our 'connotative universals'. Our 'ontological universals' are nothing but his eternal objects in creative ontological (not merely consious) instantiation in processes. This fact is clear in Whitehead's concept of the public aspect of things, though he has not made the ontological-connotative distinction in eternal objects. His concept of the privacy of things is not the same as our connotative universals:

> An eternal object considered in reference to the privacy of things is a 'quality' or 'characteristic': namely, in its own nature, as exemplified in any actuality, it constitutes an element in the private definiteness of that actuality. It refers itself publicly, but it is enjoyed privately.
> The theory of prehensions is founded upon the doctrine that there are no concrete facts which are merely public, or merely private. The destinction between publicity and privacy is a distinction of reason, and is not a distinction between mutually exclusive concrete facts. The sole concrete facts, in terms of which actualities can be analysed, are prehensions; and every prehension has its public side and its private side. Its public side is constituted by the subjective form through which a private quality is imposed on the public datum. The separateions of perceptual fact from emotional fact; and of causal fact from emotional fact, and from perceptual fact; and of causal fact from emotional fact, and causal fact, from purposive fact; have constituted a complex of bifurcations, fatal to a satisfactory cosmology. The factsof nature are actualities; and the facts into which the actualities are divisible are their prehensions, with their public origins, their private forms, and their private aims. But the actualities are moments of passage into a novel stage of publicity; and the coordination of prehensions expresses the publicity of the world, so far as it can be considered in abstraction from private genesis. Prehensions have public careers, but they are born privately.[409]

---

408   Whitehead, *Process and Reality*, 34.
409   Whitehead, *Process and Reality*, 290.

Whitehead speaks of the universal-ladennes (the public nature) of things and processes, but Quine limits himself to the universal-laden nature of language. The Quinean theory-laden or universal-laden characteristic of language demonstrates the need to posit connotative universals; and the particular (private) instantiative characteristic of eternal objects in their publicity and their derivation from the formal aspect of infinite (public, i.e. ontologically universal) sets of processes demonstrates the need to posit ontological universals.

If all talk is universal-laden, what is the universal aspect of the natural kind 'being/s', whose fundamental attributive level is that of the 'to be's'? This is generality called the To Be of Reality. It is a *sine qua non* in ontological talk, provided Reality-in-total and Reality-in-general are categories. Why is particularistic ontology naïvely considered as simultaneously language-laden and universal-laden, and not as laden by ontological and connotative transcendentals? This is because the language-boundedness of discourse makes one, rightly, to admit that discourse is universal-laden, but persuades one, incorrectly, to identify ontological universals with the connotative universals involved in the concepts that are part of the language-boundedness of discourse. *This, I believe, has been* **an unconscious mistake in Quine and in all particularist ontologists**.

The above fact points to the only way to making ontology possess objectual-causal ontological commitment, i.e. the way of generalizing unto the ontological Transcendental character of To Be and the ontological transcendental character of to be's and qualities as the natural sequel of the connotative transcendental character of the universals involved in the concept of any referable / denotable particular. Consequently, my contention is that theoretical products of semantics, and especially the semantic theory of reference, are not broad enough to accommodate the whole ontological referential milieu of any term defined within the theory of meaning, because ever after our explication of the Einaic Ontological implications of Gödel's result, no primordial, intermediate or final ontological universal is free from definitional relativization and no connotative universal is free from pushing of axioms proper to the system within which meanings of terms may further be confirmed by Gödelian definitions. Hence, ontological quantification should be placed in the wider conceptual setting of the universal, To Be, of which we may further have only better Gödelian explications, because beyond the concept of To Be we may at the most think of more and more adequate explications of the same. Placing quantification in the wider setting of ever better explications of To Be is possible only if, by reason of Gödel's result, the conceptual and axiomatic foundations of the logic of quantification may be pushed sufficiently backwards to its primitive-conceptual backyard of

Reality, Reality-in-general and To Be. The above will result in Einaic Semantics that begins with the Einaic Ontological categories for determination of meaning. Naturally, we may have also an Einaic Theory of Reference issuing from such semantics.

Quine is explicit on the presupposed particularist nature of implications of truth of statements. I cite his famous Criterion for Ontological Commitment, by first noting that bound variables are 'something', 'nothing' and 'everything', and by presupposing that mathematical objects are for him a variety of the objects of set theory, based for their justification in the ontology of abstract objects:

> Classical mathematics, as the example of primes larger than a million clearly illustrates, is up to its neck in commitments to an ontology of abstract entities. Thus it is that the great medieval controversy over universals has flared up anew in the modern philosophy of mathematics. The issue is clearer now than of old, because we now have a more explicit standard whereby to decide what ontology a given theory or form of discourse is committed to: a theory is committed to those and only those entities to which the bound variables of the theory must be capable of referring in order that the affirmations made in the theory be true.[410]

In view of Einaic Ontology, no objectual-causal commitment is in fact without involvement of the nomic universals concerned. These are the ways of being of beings / processes. To Be is the highest such; and it is the bound or unbound set *par excellence*, taken in the purely verbal sense of this abstract object. Hence, an ontological commitment based on the bound variables 'something', 'nothing' and 'everything' is not theoretically self-sufficient when it has to do with ontology. It automatically points beyond the causal aspect of them all, to the acausal-processual aspect ontologically committed to even in mathematical *abstracta*, i.e. to the To Be of all processes and ways of being taken together. *This allows a horizon of discourse that connects something, everything and nothing to Reality.* The Einaic criterion of connecting all unto To Be for truth-probability is purely ontological, and the Quinean criterion is particularistic. Theory must be ever better true only if the set of bound variables absorbs To Be as its limiting *dimensional* unbound variable.

Without any foundation on the Einaic Ontological presuppositions – the categories: Transcendentally Transcendent Reality, the Transcendently Transcendental Reality-in-general and the Transcendental To Be – Quine bases the particularism of ontological commitment on the reification implied in observation sentences, which is the (ordinary) first step of reification of meanings: "Incipient

---

410   Quine, *From a Logical Point of View*, 13–14.

reification can already be sensed in the predicational observation sentences. That mode of combination favors, as components, observation sentences that focus on conspicuously limited portions of the scene; for the compound expresses co-incidences of such foci."[411]

The second step of reification beyond ordinary observation sentences is had in focal observation categoricals of the type: 'whenever this, that'. The focal observation categorical requires two given features, e.g. 'Raven' and 'black', to fuse together in the scene. But ordinary predicational observation sentences simply predicate. The child perceives the difference between predicational and focal observation sentences gradually.[412] The secondary reification here is a prerequisite of ontological commitment in the first step of reification in first-order logic. Without pursuing the ontological universal-level categorial implications of these steps and the others presupposed in them, Quine further bases ontological commitment on first-step observation sentences at the realm of first-order logic, by appeal to a possible way of positing objects through intersubjective agreement and implying that this avoids subjectivism:

> There is generally no subjectivity in the phrasing of observation sentences, as we are now conceiving them; they will usually be about bodies. Since the distinguishing trait of an observation sentence is intersubjective agreement under agreeing stimulation, a corporeal subject matter is likelier than not. The old tendency to associate observation sentences with a subjective sensory subject matter is rather an irony when we reflect that observation sentences are also meant to be the intersubjective tribunal of scientific hypotheses. The old tendency was due to the drive to base science on something firmer and prior in the subject's experience [...].[413]

Though intersubjective agreement avoids pure subjectivism and empiricism,[414] it does not guarantee the best available objectivity, since objectivity is genuine

---

411 Quine, *Quintessence*, 109.
412 Quine, *Quintessence*, 109.
413 Quine, "Epistemology Naturalized", 28.
414 A similar view by Dingguo is noteworthy: "The ontic reality is the reality which is external and independent of human beings, having strong objectivity. One of its essential characteristics is its transcendentality [in Einaic Ontology we call it transcendent-ness] [...] [H]olders of [...] ontic realism [...] think that the notion of ontic reality possesses an epistemological implication. They insist that although the ontic reality is beyond experience, it is within reach of mind [....] The empirical reality is the reality which depends on subjects and cannot be separated from experience so as to have the weak objectivity characterized by intersubjectivity [...] [E]mpirical realism [...] insists that science (and ordinary knowledge as well) is indissolubly linked with human experience, so that the task of science is but to

only in objectual-causal positing of the maximal categories, and not in language-bounded object-orientation through intersubjective agreement. The hindrance is the analytic-ontologist (logical-positivist) forgetfulness to explicate bodies as through and through causal extension-motion processes.

In Einaic Ontology the language-bounded object-orientation through intersubjective agreement above – of first-order logic – is sadly naïve in that Quine limits the implications of observation sentences to the simplistic 'corporeal subject matter' (in other words, tokens devoid of the higher and higher ontological universals explicative of their process nature), and never unto their totality which, by reason of the place of infinite and infinitesimal causal pasts of corporeal subject matter, presupposes nomic universals within To Be. The *ultimate nomic / universal* aspect of language in observation sentences is thus overlooked. Theoretically, without them in concepts and in the relational aspect of processes it is impossible to make observation sentences oriented to corporeal subject matter in the objectual-causal manner. Otherwise it is only particularistically objective. Once we admit the connotative nomic-ladenness and universal-ladenness of observation sentences, it is straightforward to connotatively involve objectually ontological universals in such sentences. That is, if the concept of natural kinds and ultimately that of tokens are language-laden, and naturally also bounded by what we call (epistemic) connotative universals (for him, 'universals' / 'abstract objects'[415]), then, for the sake of possibilizing objectually ontological

---

describe the phenomena constructed by the collective experience of human beings [....] To offer an image, empiricism is a boat that is sailing between ontic realism and pure instrumentalism, and completes the instrumentalism with ideas about empirical reality as natural as possible [....] The ontic reality and empirical reality are two complementary notions existing side by side with their own implications and functions respectively. On the one hand [...] intersubjectivity is the foundation of all empirical sciences, and the notion of empirical reality is of paramount importance to practical sciences. On the other hand [...] the ontic reality potentially dominates the activities of mankind and the progress of sciences. Although ontic reality is always hidden, its various surface structures can be perceived by human beings in the form of empirical reality. And its temptation for human beings is the eternal motivation of the development of sciences." Dingguo, "On the Neutral Status of QM in the Dispute of Realism vs. Anti-realism", 307–8.

415 Quine on universals as stimulations: "Yet a stimulation must be conceived for these purposes not as a dated particular event but as a universal, a repeatable event form. We are to say not that two like stimulations have occurred, but that the same stimulation has recurred. Such an attitude is implied the moment we speak of sameness of stimulus meaning for two speakers. We could indeed overrule this consideration,

commitment why not the ontological universals of relationality and the *naturally widest possible "placing"-background of 'corporeal subject matter'* (i.e. Reality), be placed under the possible ontological Transcendental proper to it (i.e. To Be)?

To Be is an ontological universal. It cannot be equated with its epistemically connotative counterpart, Reality-in-general. To make Reality connotatively theory-laden and ontologically universal-laden (by implication from the language-ladenness of natural kinds and tokens), *we need the ontological Transcendental, To Be – neither merely the connotative concept Reality-in-general, nor merely any other connotative universal/s*. Moreover, to make the 'corporeal subject matter' objectual-causal, too, we need To Be, the universal of relationality of Reality; and without making it objectual-causal we are not justified in any entity-talk.

The methodological burden in Quine's position of particularistic and "objectively" ontological commitment is to explain the status of the immediate connotative universals of abstract terms of observation statements as the tools (1) to justify observation statements with the presupposed ultimacy of their theory-laden empirical consequences (unconsciously considered dear by him and by the whole analytic tradition) with respect to their truth-probabilities, (2) to justify the truth-probabilities of 'sentences higher up in theories' – which are, evidently, sets of more theory-laden and empirically less consequential statements (of higher-order logic) – and (3) to explain the subsequent (but presupposed) more extensively applicable individual or maximal sets of counterfactual truths, laws of nature and universals as either purely connotative or as connotatively represented ontological universal sets proper to 'sentences higher up in theories'. This predicament is evident in the theoretical *discontinuity of connotative universals* between observation statements and highly theoretical statements, implied in his

---

if we liked, by readjusting our terminology. But there would be no point, for there remains elsewhere a compelling reason for taking the stimulations as universals; viz., the strong conditional in the definition of stimulus meaning." Quine, *Word and Object*, 34.

Quine on universals as unrealised particulars: "[…] [C]onsider again the affirmative stimulus meaning of a sentence S: the class Σ of all those stimulations that *would* prompt assent to S. If the stimulations were taken as events rather than event forms, then Σ would have to be a class of events which largely did not and will not happen, but which would prompt assent to S if they were to happen. Whenever Σ contained one realized or unrealised particular stimulatory event σ, it would have to contain all other unrealised duplicates of σ; and how many are there of *these*? Certainly it is hopeless nonsense to talk thus of unrealised particulars and try to assemble them into classes. Unrealised entities have to be construed as universals." Quine, *Word and Object*, 34.

statement on the importance of observation statements as the tools for learning meaning in a language, as distinguished from purely theoretical statements: "The observation sentence is the cornerstone of semantics. For it is [...] fundamental to the learning of meaning. Also, it is where meaning is firmest. Sentences higher up in theories have no empirical consequences they can call their own; they confront the tribunal of sensory evidence only in more or less inclusive aggregates."[416]

Even in Quine epistemic, connotative abstract objects are the tools of expression for observation statements and theoretical statements. If they are inalienable ingredients of both the statements in every language, and if for him the concept of objectivity of the 'fact' is not equivalent to the concept of the object / token, then one should search for the ontological universals and the limits of possible ontological universals at play in the object of ontological commitment in 'facts' and inquire after their place in determining the objectivity of objects and states of affairs.

The part of ontological universals in concepts and formulations of truths leads to what I call the 'connotative universal-ladenness' of theoretical facts and facts about objects and ultimately to the ontological universal-ladenness of objects and the final objectual-causal token, Reality.[417] But tokens up to Reality are processes with ontological ways of being. This necessitates recognition, in theory, of the 'ontological universal-ladenness' of token processes up to Reality – the latter of which is to be understood as ontologically most universal-laden and at the same time ontologically most objectual within its extension-motion processual nature. This necessitates the inclusion of ontological universals in terms and propositions by their Einaic Ontological commitment to the three Einaic categories, which commitment is implied in terms and propositions. In short, connotative universals presuppose ontological universals. Connotative universals are laden by ontological universals, especially by To Be, and yield genuine objectual realism beyond mere factual-particularistic realism in object- / token-language. The theory-ladenness of connotatives and the shyness of Reality and reality-in-particular to lend to consciousness make realisms probabilistic.

Unless there is communication between connotative universals and Reality, we cannot explain their confluence in theory about particulars (reality-in-particular) and token objects / processes (including Reality, the most inclusive

---

416 Quine, "Epistemology Naturalized", 29.
417 Insofar as the denotable Reality-in-total (not the denotative idea of it, i.e. 'Reality-in-total', nor its connotative concept 'Reality-in-general') is not a term, it directly denotes a token.

token). Their theoretical confluence is made possible only by ontological universals issuing from causal routes in Reality. These are couched in Reality, not merely in ontological universals considered as end-objects. The latter, again, is possible only if the ontological Transcendental, To Be, is the active Way of objectual-causal routes in Reality. In short, *unless the theory-ladenness and the consequent connotative universal-ladenness of observation statements* and higher theoretical statements are continuous upon each other by means of objectual-ontological universal-ladenness of higher and higher extensions of ontological universals / ways and tokens justified by the extension-motion nature of tokens of any extent, and *finally by means of ladenness by To Be, theory about theory-ladenness of facts and observation statements is meaningless.* Quine does, though insufficiently, hint at this possibility in the following, which treats not about ontological universals but about nomic generalities in the sciences, which was written in the context of arguing for regimenting ordinary language for scientific and ontological purposes: "The terms that play a leading role in a good conceptual apparatus are terms that promise to play a leading role in causal explanation; and causal explanation is polarized. Causal explanations of psychology are to be sought in physiology, of physiology in biology, of biology in chemistry, and of chemistry in physics – in the elementary physical states."[418] No more can observation sentences of the immediate language-level be the cornerstone of the ontology of semantics – instead, greater and greater reaches of extension of ontological universals obtained in ontological commitments and instantiated in connotative universals can be. It is also foundational to learning of language by the child and the neophyte. What is prior in time in learning process need not be what is prior in ontological primordiality. Hence, Quine's limiting of ontology to language, meaning and reference is fallacious by reason of the natural extension of his own convictions regarding regimentation by means of greater and greater ontological universality.

In short, genuine, objectual ontological commitment is possible only in a theory of meaning and a theory of reference suffused with the Einaic categories, which ontological commitment implies by presupposition. Dependence on the immediate connotative universals implied in observation sentences for firmness of meaning and for the truth-probability that issues in observation statements, is not justified by the Einaic presuppositions of ontological commitment. We need to go for ontological universals that depict the causal roots and routes of any object/s yielded by observation statement/s. It involves going for the presupposed laws of nature, nomic universals and categories within the

---

418   Quine, "Facts of the Matter", 168–69.

ontological commitments of the extension-motion contour/s of the object/s in question (which, again, is ontologically universal-laden). Thus, we need to inductively generalize unto the Transcendental To Be; inductively totalise unto the Transcendental Transcendent Reality; and inductively generalize unto the Transcendent Transcendental Reality-in-general.

We have thus transcended the limited concepts of 'is', the particularistic ontological commitment of the 'is' and justified the ontological universal To Be using the thought of Quine. The fact that we have thus justified the most total token, the most general connotative ultimate and the most general ontological ultimate shows the path to ontologically justifying the place of reality-in-particular and discourse. Without the particular there is no ontology. This endeavour follows, through a preliminary attempt to sense the semantic sense that should be attached to reality-in-particular.

3.3 Einaic and Semantic Ontology of Actualizing Possibility

3.3.1 The Ontological Transcendental vs. Semantic Holism

We need now an Einaic variety of semantic holism that fits well with the three categories I have proposed, and also fits with the sub-category of reality-in-particular, without sacrificing the implications of the widest context of objectual-causal ontology. Einaic holism is viable only at the level of connotative universals and terms – since these are what actually occur as theoretical concepts and structures are formed; and towards this we must keep the semantics of such universals within the purview of ontological universals and To Be as best as we can.

Variables that represent general (universal) terms are of (1) a multitude of extensions by way of external boundedness and (2) a multitude of intensions by way of inner compression of sense. The term 'extension' here in connection with intension is the linguistic-logical and not the cosmological one that appears in connection with motion. Universal-boundedness shows into how many (extension) entities / processes a connotative ingression has consciously occurred or can occur, thus developing certain objectual meaning/s (intension), by reason of the ontological universal/s involved. Similarly, compression shows the sort of shades of connotatives compressed or noted together internally into the term (intension). Connotation is not intension; it is productive of the extensive and intensive aspects in consciously occurrent concepts. That is, every term in consciousness that results from an ingression of an ontological universal into the form of a connotative and then a denotative manner has a certain extension and intension. Both are mutually related and presupposed in the concepts.

Each connoted concept is bounded through the extension of capability of each ontological essence to permit conceptual intension in meaning/s of the concept with the given extension in a certain number of the type – both through denotative (in terms) and denotable (in things) forms of extension – thus defining also the extent of external extension of the ontological commitment to actual token processes.

Now we are in a position to define Einaic Ontologically the extension and intension of terms. *The boundedness of semantic extension of each value (term) of a variable co-defines its conceptual sense / intension by reason of its intrinsic partial connection with the intension.* **Extension** is the dimension of ontological externality (quantitative and spatiotemporally measured extension-motion aspect of objects and states of affairs referred to) of term-sense, phrase-sense and sentence-sense. In short, *the extension of terms is the denotation that refers to the denotables (existent or generally real objects and processes) of the set.* **Intension** is the dimension of internality or number of connotative senses (and their togetherness) of the terms, phrases and sentential truths propositionally defined in a manner mutually connective of the ontological universals (not the existent processes) – made generally representational by ontological commitment to abstract objects (implied in existent processes), wherein the number of connotative senses are contrasted coherently and constructively in propositions. Connotative senses are not the same as connotative universals, but presuppose these universals. In short, *intension (of a term or a phrase or a sentence) is the sum total of the differences and connections the term or phrase or sentence has with other similar ones.*

In contrast: Intension is commonly defined as "[…] the meaning or connotation of an expression" and "[…] extension or denotation […] consists of those things signified by expression."[419] These two common definitions may be purely semantic, but not sufficiently ontological and holistic. Hence the complexity of the Einaic definitions, which vow to be more holistic. Though Quine is a semantic holist, his holism is thoroughly particularistic, not Einaically Ontological. The major defect of Quine's semantic holism is particularism, which may be discussed here, so that a more viable, Einaic and actuality-imbued holism may be spelt out here. Some of the key texts of Quine on holism, with their contexts, are the following (of which the key portions I have italicized). As we shall see further, this variety of holism may be converted into a tool against his own ontological

---

419 *The Cambridge Dictionary of Philosophy*, second edition, s.v. "Intension".

particularism in the ontology of semantics, by recourse to a few passages representative of his semantic holism.

(1) The first of the said passages explicates the ontology of semantic holism by citing the decidability of truth of statements solely in their union with all other statements of the theory:

> The dogma of reductionism survives in the supposition that each statement, taken in isolation from its fellows, can admit of confirmation or information at all. My counter-suggestion, issuing essentially from Carnap's doctrine of the physical world in the *Aufbau*, is that *our statements about the external world face the tribunal of sense experience not individually but only as a corporate body*.[420]

(2) The second passage is about the whole field of our knowledge or beliefs and about the possibility of re-evaluation of statements in such holism:

> The totality of our so-called knowledge or beliefs, from the most casual matters of geography and history to the profoundest laws of atomic physics or even of pure mathematics and logic, is a man-made fabric which impinges on experience only along the edges [....] A conflict with experience at the periphery occasions readjustments in the interior of the field. Truth values have to be redistributed over some of our statements. Reëvaluation of some statements entails reëvaluation of others, because of their logical interconnections – the logical laws being in turn simply certain further statements of the system, certain further elements of the field. Having reëvaluated one statement we must reëvaluate some others, which may be statements logically connected with the first or may be the statements of logical connections themselves. But the total field is so underdetermined by its boundary conditions, experience, that there is much latitude of choice as to what statements to reëvaluate in the light of any single contrary experience. *No particular experiences are linked with any particular statements in the interior of the field, except indirectly through considerations of equilibrium affecting the field as a whole.*[421]

(3) The third passage is on the revisability of statements on the basis of his famous relativization of the boundary between synthetic and analytic statements, thus paving the way for a firm ontology of semantic holism:

> If this view is right, it is misleading to speak of the empirical content of an individual statement – especially if it is a statement at all remote from the experiential periphery of the field. Furthermore it becomes folly to seek a boundary between synthetic statements, which hold contingently on experience, and analytic statements, which hold come what may. *Any statement can be held true come what may, if we make drastic enough*

---

420 Quine, *From a Logical Point of View*, 41. Italics mine.
421 Quine, *From a Logical Point of View*, 42–43. Italics mine.

adjustments elsewhere in the system [....] Conversely, by the same token, no statement is immune to revision.[422]

Semantic holism in Quine (as also in Donald Davidson, Gilbert Harman, Hartry Field and John Searle, in general) may be defined *generally* as

> [...] a metaphysical thesis about the nature of representation on which the meaning of a symbol is relative to the entire system of representations containing it. Thus, a linguistic expression can have meaning only in the context of a language; a hypothesis can have significance only in the context of a theory; a concept can have intensionality only in the context of the belief system.[423]

The need of context necessitates an ever broader ontological setting for semantics. Rightly, Quine holds that sentences face truth-test systemically, as a corporate body, not individually. *He does not give the systemic and objectual-causal extension and subjective-objective intension, which sentence-senses and word-senses can and should take in the context of their being ever broader revisable.* Is this not an Einaic Ontological mistake in Quine? Either these senses can be dealt with in pure semantic particularism, which is semantic atomism; or they can be dealt with in pure semantic holism, which would automatically be based on the dimensional categories of Einaic Ontology. There is no reason for a via media, an abrupt stoppage of the course of attaining breadth beyond atomism.

(1) Semantic particularism / atomism is impossible, since there are no pure particulars without being universal-laden even in Quine: the tokens he finalises upon as the final objects theoretically implied in ontological commitment are never reached well concretised in propositions of any sort. (2) Pure semantic holism has looked unwieldy for Quine, since it is impossible to fetch, straight into sentence-, phrase- and word-senses in consciousness, all sorts of connotative instantiations of ontological essences, which are in fact active in actual processes, i.e. in the objectually causal routes and roots of processes and beings, with the quality of nomically accompanying acausal ideals. Therefore, in effect it may be concluded that Quine and his like have had recourse to pragmatism of immediate sentential, empirical and theoretical contexts for fixing the extensions and intensions of senses. This is a pragmatic holism in semantics, and hence it is not a sufficiently ontological holism.[424]

---

422 Quine, *From a Logical Point of View*, 43. Italics mine.
423 *The Cambridge Dictionary of Philosophy*, second edition, s.v. "Semantic Holism".
424 See the first of the three numbered quotes above from Quine, *From a Logical Point of View*, 41.

All senses inherit not merely from their own and other immediately wider corporate fields of sentences in language, but also from the widest possible – the most total and the most general – of contexts, which whole dimension happen to be ontological, not merely linguistic. The pragmatic question whether it is possible to incorporate the farthest possible of all acausal essences in discourse can be posed in two contexts: (1) the conversational, linguistic and empirical / experimental, and (2) the ontological and mathematical-scientific. For the former, the unwieldy character of the purely holistic issue is difficult to grasp, since its purview is limited, however large the proper object of the discourse / science in question is. For the latter, the unwieldy and involving character of the purely holistic issue can make much sense, for these fields deal not with just one point of view, but points of view of many, nay, ideally largest and inclusive, *dimensions* of the higher and higher in points of view. This is an Einaic Semantic contribution to philosophy.

In Quine 'the totality of our [...] knowledge or beliefs' is in fact only the *objective content* of word-senses and sentence-senses, resulting from ontological commitment.[425] The final *objectual-causal content*, not yielded in Quine by way of his tokens and their theoretical / universal elements as included in particulars, and which we have further generalized and totalised into the Einaic Ontological Transcendent Reality and Transcendental To Be can fruition in ontology, if we insist on categorial inclusion of the objectual elements consistently. Holism can thus exist at its best in ontology, not so much in experimental science. If ontology caters only for particular sciences, and not for their whole in their interconnections, it can reach only up to Quine's particularism that grudgingly approves abstract objects in the hope of saving particulars (tokens, of course, remain saved already). One forgets in particularism that every particular / natural kind has its own finite or infinite set of tokens, and that Reality itself is a token.

3.3.2 To Be as Possibilizing Actuality in Possibility

How to include the objectual-causal aspects of Reality in Einaic Semantics? Will it still allow the particular to be intact and ready to hand to theory and still allow dynamism within the overbearing interference of To Be, Reality-in-general and

---

425 Ontological commitment may be connected ultimately to belief. This problem was made famous especially by the treatment of the issue by Hume and Kant. I do not study the inner realms of the transparent and opaque sorts of belief that issue from constructions. Quine, *Word and Object*, 145. A construction is a general term predicatively used. Quine, *Word and Object*, 144.

Reality-in-total? The answer lies in transforming some of Quine's work. He has been the most ardent advocate of relativization of the Kantian[426] dogma of the analytic (*a priori*) and the synthetic (*a posteriori*, empirical), and of the revisability of any statement whatever, be it empirical, mathematical or ontological. Quote (3) above from Quine on revisability[427] is but a summary statement to this effect. If even the most general of all statements is subject to revision, it is by questioning the very axioms of its system of origin. *This resonates well with our Gödelian epistemology of pushing the definitions* (and connotatives) of primitive notions and axioms for greater and greater truth-probabilities. Thus, revision of truth of statements is nothing but the direct result of revision of word-, phrase- and sentence-senses by pushing axioms of respective system/s.

This tells upon any alleged absolute status of our Einaic Cosmological, Epistemological and Ontological categories too, because there is nothing too analytic or synthetic about them to make them absolutely analytic or synthetic. They become merely relatively more absolute primitive notions, and always subject to revision of senses of all in result of revision of any one of them. Synthetic statements are already understood to be such. Our categories are synthetic by their propositional (definitional) sense and, at the same time, postulated as axiomatic and analytic in their very propositional (definitional) sense. This fact makes them simultaneously *a priori* and *a posteriori* to some extent, yielding a case of amalgamation of the analytic and synthetic. Nothing is purely analytic or synthetic. They are simultaneously mutually connected, since they are coextensive upon Reality, and one is the Transcendentally Transcendent, the other Transcendently Transcendental, and the third (by agreement to call it a category) ultimately Transcendental, dimension. It is easy to see them as mutually collusive and continuous, since none of them is self-sufficient within itself and needs the other two for genuine theory. That is, the analytic and synthetic aspects of these categories in their definitional sense contain elements of each other in varying degrees. This makes the hardest of analytic (theoretical, mathematical) statements in Einaic Ontology subject to Einaic Semantic revision. Both analytic and synthetic statements have a certain measure of dependence on actuality and on many statements of the (synthetic or analytic) nature other than deemed as its own. Statements about actuality should depend on both (empirical) actuality and theory. Propositions maximally involving any of the categories are also maximally analytic-synthetic.

---

426 Kant, *Immanuel Kant's Critique of Pure Reason*, 48–58 (A7–10 and B11–24).
427 Quine, *From a Logical Point of View*, 43.

Actuality as worthy of direct or indirect perception is always particular and culled out using connotative universals for demonstration within the extension-motion limits through spatiotemporal limiting. The whole (which is also actual), if taken as infinite, is in extension-motion realms and connotatively infinitely actual and denotable, by reason of the ontology of the axiom of infinity that allows inductive totalization and generalization of denotables. It is not meant for direct perception. Reality as the denotable *par excellence* is a whole by inductive totalization, but not particularistically. So, *objectively* the infinite objectual-causal actuality of Reality is a theoretical actuality, occurring by inductive totalization. That is, by its very positing, it involves ontological and connotative universals.

The extent to which Reality involves only ontological universals / essences may be termed its *objectual actuality* as the single-member token-natural-kind Reality, and the extent to which ontological universals are connotatively involved may be called its *objectual actuality* as Reality-in-general. The same may be said by substituting reality-in-particular and connotative universals in place of Reality and Reality-in-general respectively. The mode of Reality in which it is instantiated by a single same-extension ontological universal is To Be. This does not apply to reality-in-particular as such. Hence, the particular / natural kind in Quinean ontological commitment can still be theoretically the token plus abstract objects related to it, if considered within the inductive totalization and generalization of Einaic Ontology. This allows the free occurrence of reality-in-particular – both actual particular tokens and particular natural kinds – within the purview of Reality and in theory.

Reality-in-particular is impossible without Reality, and vice versa. The fact that they already are also tokens makes the task of objectual ontological talk relative upon the *specific fact* of Reality. We are prone to talk in terms of the particular. Particularist talk naturally takes universals directly, and To Be indirectly, by inductive generalization of ontological commitment. When To Be is involved, both the particular and the whole are talked of in their mutual possibilization and are guaranteed in each other. *This talk is not particularistic, but particular-possibilizing.* It will tend to augment particular statements of fact at the level of the to be of processes with those at the level of the To Be of Reality. Now we can remedy what has gone wrong with the semantic particularism of Quine, so that we are in a position to possibilize reality-in-particular in Reality, from within our semantic holism of To Be and Reality.

We are now sure that analytic and synthetic statements are subject to revision and mutual connection. The reference of a term is given not merely in one judgment, but in an array of them, that *tend to define ever better* the extension and intension of their senses by systemic unification of their analytic and synthetic aspects, for which reason also the sense of a term that is represented in the related sentence sense is capable of being defined ever better. Therefore, **sense is definable as** relatively judgementally settled empirical extension and intension of a term involving relatively more or less of analyticity and syntheticity. The meaning is primarily intensional; and the reference is primarily extensional. *This theory connects the reference and sense of a term and phrase with the sense of a judgment and does away with absolute distinctions between the senses of a term and phrase, the sense of a judgment and the sense of them given in theory.* They are distinct in particular, but they enhance each other. Intersubjective settlement on the basis of observation sentences and theoretical advances is only the ground in which it is done in a continuous and historical manner. So intersubjective settlement too is no final word. To that extent, the concept of To Be can always be brought in ontological talk under the condition that such talk is not particularism, but involves particular facts about reality-in-particular, Reality and Reality-in-general *sub specie* To Be.

The "*sub specie* To Be" application point can be a valid transition point between special Einaic ontologies and Einaic Ontology, at the practical realm of mutual possibilization of various universals and reality-in-particular on the one hand, and To Be and Reality on the other. But reality-in-particular or classificational particularistic categories or universals or the very fact of mutual possibilization cannot be a starting point for doing philosophy. For starting points we need all-encompassingly universal and total categories in their mutually collusive possibilization and possibilization of the particular, than mere epistemic, scientific and classificational ones which cannot make the most total and general possible. Insofar as To Be is the best starting point of all ontologies, based on Reality, Reality-in-general and reality-in-partcular – To Be is the possibilizer of actuality in possibility.

3.3.3 Dynamism of Einaic Semantic Connectivity

What is the connection and difference between the word- and sentence-senses and our mutually collusive categories?

(1) The senses of ultimate categories of ontological starting points tend closer to word-senses than to sentence-senses. Insofar as-word senses and sentence-senses are complementary, so also are senses of mutually collusive (and other) categorial senses.

(2) The representational quality of word-senses applies also to categories. But the representational structures of word-senses and category-senses vary widely.

In one sense category-senses are very broad with respect to word-senses, and in another they are the most capable word-senses available for forming true and real statements. The structures of senses are determined in their extension and intension by To Be. The Einaic representational structure of a word-sense, phrase-sense or sentence-sense is in fact the extensional and intensional connection or the boundedness of extension and intension. *The more intensely bound the word-senses are extensionally and intensionally, the more is their mutually collusive capacity as word-senses,* because this alone makes them more dynamic and receptive of overlapping of senses. The greater their mutual implication, the better are they bound up Einaically.

(3) The fact of the non-absolute boundary between extension and intension defines the probabilistic measure of determination of word- and sentence-senses. This shows that *the extensional and intensional aspects of senses never complete each other, and thus give rise to the probabilistic nature of sense determination.* According to Quine, the bound variables (something, nothing, everything) are not things in the sense that is yielded by the final objects of his ontological commitment:

> We can very easily involve ourselves in ontological commitments by saying, for example, that *there is something* (bound variable) which red houses and sunsets have in common; or that *there is something* which is a prime number larger than a million. But this is, essentially, the *only* way we can involve ourselves in ontological commitments: by our use of bound variables. The use of alleged names is no criterion, for we can repudiate their namehood at the drop of a hat unless the assumption of a corresponding entity can be spotted in the things we affirm in terms of bound variables. Names are, in fact, altogether immaterial to the ontological issue, for I have shown, in connection with 'Pegasus' and 'pegasize', that names can be converted to descriptions, and Russell has shown that descriptions can be eliminated. Whatever we say with the help of names can be said in a language which shuns names altogether. To be assumed as an entity is, purely and simply, to be reckoned as the value of a variable. In terms of the categories of traditional grammar, this amounts roughly to saying that to be is to be in the range of reference of a pronoun. Pronouns are the basic media of reference; nouns might better have been named propronouns. The variables of quantification, 'something', 'nothing', 'everything', range over our whole ontology, whatever it may be; and we are convinced of a particular ontological presupposition if, and only if, the alleged presuppositum has to be reckoned among the entities over which our variables range in order to render one of our affirmations true.[428]

The so-called boundedness of bound variables (something, nothing, everything), therefore, is partial, but close to completeness within the specific structure of

---

428   Quine, *From a Logical Point of View*, 12–13.

given concepts and statements. Nevertheless, their incompleteness as lack of absolute clarity of concepts and statements and the probabilistic nature of truths of the definitions therein, both together point to the transcendental tendency of their ingredients beyond their own extension and intension, untimately to including the Transcendental To Be. This again connects actual beings (including bound variables) with universals and ever broader totalities, and finally connects the highest total Reality with To Be. Moreover, there is nothing that stops us from talking of Reality in terms of reality-in-particular. All these show that the universal / essence To Be as the final case is implied by the bound variables, which need not be taken as the token-like actuality of Reality.

The difference between reference and sense in referential sense may be expressed as follows: Referential sense has two aspects: (1) the referential aspect, i.e. the connection of propositional truths to those abstract entities and concrete processes of the world that are objectual-causally (not merely objectively) involved in the formation of those concepts, and (2) the sense aspect, which is the connecting link between the relevant ontological universals / generalities / *qualia* and the acausal abstractiona / idealizations of these causally constituted qualities (which are in partial identity and partial difference with each other). The question as to what a sense is always hints beyond the causal (physical) objects to the acausally constituted referential aspects of entities and processes, i.e. to pertaining to ontological universals / essences / *qualia* as objects. A sense has never directly to do with entities and processes but to abstract objects. Reference by a name or concept is either to entities / processes in relation with their universals, or to individual or collective universals without entities / processes; and a sense is the connotatively expressed aspect in general, abstract terms or names to the universal objects alone.

If we can speak abstractly of the bound variables 'something', 'nothing' and 'everything' in particularistic logic and ontology without at times having actually to find anything directly referring to them for verification, and we still do refer to actual entities / processes by generalization using them directly and indirectly, then it is licit to grant that any term that generalizes, generalizes not in a vacuum. Even the term 'Not To Be' that has no direct or indirect reference has an indirect sense and meaning. Determination of the exact sense of the term 'Not To Be' will lead into the processes in actual world that lead cumulatively into the subterm 'not to be' in abstract reference to the absence or non-identity of what is determined at some extension-motion region, another determined at another extension-motion region, etc. Thus, the exact processes behind the senses of 'not to be' and 'Not To Be' yield the real internal constitution of the indirect sense (not direct sense or meaning) or lack of content of these concepts in relation to the senses and contents of the constituent words and notions which surely possess ontologically justified sense and content. The non-content of 'not to be' and 'Not To Be', being indirect, they are referential with the help of improper direct senses. This is therefore also a source of error in propositions, theory and discourse – a field we do not discuss here.

The sense aspect of 'Not To Be' and 'not to be' is related negatively to general terms or concepts of processes and their senses which have been used to formulate the concepts of 'Not To Be' and 'not to be' – which are always formulated in relation to the absence or lack or non-identity of some universals and/or particulars and/or tokens. Their sense is therefore an altered version of those of the constitutive terms and notions of 'Not To Be' and 'not to be' – the references of these constitutives possessing partial identity and partial difference due to the extension-motion nature of the processes referred to. Thus, even the concepts of 'Not To Be' and 'not to be' have relevance in ontological discourse. They too contribute to the dynamism of the study of Reality and reality-in-particular through word- and sentence-senses.[429]

The question of the connective link of actuals and universals has been a source of ontological chaos and metaphysical bewitchment from the beginning of the exercise of philosophical wisdom. Such muddling may be lightened by connecting extensional sense with denotables and intensional sense with connotatives. This will allow presentation of denotatives (names or terms) as the interplay of denotables (entities / processes along with universals as ways of being of processes) and connotatives (the conceptual occurrence of universals). *This shows the highest ontological dynamism in the connective activity of denotables and connotatives in their confluence.* This is possible only if reality-in-particular and universals are connected by the highest connectors: Reality, Reality-in-general and To Be. Ultimately, the dynamism of abstract objects and concrete tokens among themselves is due mainly to these three Einaic categories.

Einaic Ontology is the purest case in which both language and their objectual counterpart are unified, and Quine's Quinean holism is a theoretical rendering of a collective linguistic phenomenon. By Einaic Ontological extension of Quinean holism we do not find reason for squaring reality-in-particular with

---

429 The above is the sense in which Buddhism uses the term 'nullity'. It leads to the meaninglessness or unreality of the supposed absoluteness of the objects, exeriences and mental constructs about all objects and experiences, and each such construct is allowed to pass from experience without the attention getting fixated on them due to the presupposed absoluteness of them all.

Reality-in-general and universals with To Be for mutual connecting. But we know that (1) the to be and the universals (extension, motion etc.) of reality-in-particular are the ways of being of reality-in-particular, and the To Be and the universals (extension, motion etc.) of Reality are the Ways of being of the totality of reality-in-particular, and (2) all transcendents (reality-in-particular) and the Transcendent (Reality) are objectual-causally connected by ontological universals, and they are in turn connoted in consciousness by connotative universals, which are made to appear as actual denotatives in concepts. Hence, *ontological universals, of which To Be is the highest dynamic member and to be's and universals are general members, are the ultimate connectors of reality-in-particular and Reality-in-total.*

Ways and the Way of being do not make sense in isolation from beings. They subsist as dynamic ways and Ways in beings and Reality. So, reality-in-particular and Reality, as dynamic, are the connected cites of universals, to be's and To Be. To Be studied by means of itself and of Reality is Einaiology; and Reality-in-total and Reality-in-general studied by To Be is General Ontology. The dynamism of word- and sentence-senses can be defined only if universals and to be's in their particularity, and To Be in its ultimate *dimensional* totality are taken as ways of being of beings and of Reality respectively, because their dynamism consists in their connexity to other universals, these with others etc., so that, though not every universal is related to every other, every one of them has some or other connection to more general ones that possibly include these subsidiary, less general ones. So, the Einaic dynamism of word- and sentence-senses rests on taking universals as ways. Universals in Einaic Ontology are nothing but the verbal, partially moving-and-static/extended, aspect of processes of beings. They are not merely representative of the static dimension alone, because they are not acausal linguistic or mathematical entities, but relationalities in processes.

Any variable in language or logic is a general term representative of tokens and particulars. The comparative difference of extension of a variable from another is the index of the extent of specificity or generality in the variable or the general term. The process of conceptual representation of senses is particular construction by reason of connotation, and the ontological process of it is Einaic (i.e. from the broadest possible realm of ontological essences / universals). The epistemology of the process of knowledge construction is that of Einaic realist construction using connotatives. Even in Einaic construction, sense is a relatively finalized synthesis of extension and intension. However well judgment-arrays fix the sense of a term by fixing the sense of a judgment, it always remains true that sense is relatively finally delivered in arrays of judgments through the involvement of ontological essences from processes. Judgmental senses always attempt to excel in being representational of such processes. Representations are idealizing of generalities / essences of actual processes. Actuals taste of ontological universals. Hence, connotatives (ideals / essences in conscious representations) are never exclusively fixed (i.e. they are productive of probabilistic truth) in their own nature, but fully based in ontological essences active in beings. This ends up in inductively generalizing unto To Be as the highest ontological universal and Reality-in-general as the highest epistemologically probabilizing connotative essence of Reality. Hence, we may very well identify ontological essences as ways of being active in connotation and theoretical truth-probabilities.

In Einaic Semantics, the sense of every term is ultimately a result of fixing the senses of many relevant judgments. (It affects also the Einaic categories.) But every term has its antecedent, extensionally and intensionally explicative, judgments, which fact yields an ever-explicative *regressus in infinitum*. That in turn makes continuous intensional extension and extensional intension of foundational definitions, axioms and principles possible. This is where Gödel too has led us. Hence, the Einaic Ontology of Semantics should rest on continuous re-sensing of categories and terms in concurrence with the Gödelian demand for ever-higher truth-probabilities. Applicable to categories of all forms of discourse, the need of avoidance of vicious retrogression of meanings in sciences may have us fix meanings of categories, terms and sentences. But in Einaic Ontology the farther backward we push the meanings and senses of categories in theoretical simplicity of retrogression, the better will ontologies, cosmologies and epistemologies work and yield ever deeper truths. It is explicitly admissible to carry theory further into the extensional-intensional inquiry of senses of terms (in our contexts, upto the Einaic categories) and judgments and bring about their integration into one organized whole of wholes by use of the categories (and axioms). Such categories are necessarily capable of continuous collusion with each other, since they are based on continuous extension and intension under the umbrella of To Be. In short, continuous collusion of the senses of categories is the way to make ontology dynamic.

Continuous collusion stipulates that the continuous semantic clarification by Gödelian pushing of axiomatic structures of meaning of any set of fundamental categories proposed in any system is the only instrument of philosophical clarity in the field of the semantics of categories, terms and sentences. Any other version of clarity is based on particularism, and is ontologically fake or lay, due to their classificational tendency and fixing particularist categories as final. *The problem with classificational categories is that they reify instead of leading to categorial dimensions, and are incapable of mutual collusion because their senses resist continuous re-sensing of terms and sentences within the purview of ever more perfect explication of Einaic connectivities.*

If word-senses and sentence-senses, as continuous and representative extensions and intensions, are never absolutely set, we may surmise that the immediate reason why word-senses and sentence-senses are probabilistically mutually continuous is that their extensions and intensions continuously snub complete[430] settlement, i.e., they are incomplete by nature. Reality-in-particular is theoretically incomplete due to the incompleteness of their connotatively acausal universals without participation of other entities and the universals different from those of the former contributing to the wider nexus. This is the ontological reason why the epistemology of truth-probabilities through word- and sentence-senses and their respective extensions and intensions are mutually continuous, probabilistically interdependent and cumulatively circumspective at further advancing senses. Such relative theoretical incompleteness of acausal universals, based on the particularity and mutual extension-motion connexity of reality-in-particular, maintains a tension with the absoluteness of Reality and To Be. *On this tension is based the dynamism of reality-in-particular.* Hence, as the ideal of ontological universals, *To Be is the principle of individuation of reality-in-particular, making actuality possible in the total and the general.*

In short, only an Einaic Ontology (not necessarily the specific variety I have worked here), which vouches for pure cosmological, epistemological and ontological holism from within and for continuous re-sensing of major and subsidiary categories, and provides for particulars in their internal and external dynamism, can be the ideal variety of scientific metaphysics. Continuous re-sensing and use of To Be in ontology are the ways in which the objectual-causal and acausal aspects of Reality may be conceptually advanced and actualised in ontology. To

---

430 The semantic completeness of a formal logical system is possible when all semantically valid formulae are derivable as theorems. Semantic validity implies that "given a specified interpretation of logical operators", the statement is true "on any interpretation of the non-logical terms". In other words, any statement is a formula provided it is provable, and every word in it is semantically well defined and any operator (symbol or otherwise of relation) is consistent in its application in the system. "For example, (P v ~P) is semantically valid and is also derivable as a theorem in the propositional calculus." In this sense, the propositional and predicate calculi originating from this system are complete, because the senses of propositions and predicates are valid due to the fact that the senses of words and operators are set. Honderich, *The Oxford Companion to Philosophy*, 144.

Be makes actuality in its dynamism possible within the purview of Reality and Reality-in-general.

3.3.4 Einaic Logic and Ontology of Quantification

Universals, to be's and To Be are ways of being of reality-in-particular and Reality. Insofar as we speak of systemic or Einaic Ontological commitment as not merely of particularist reference to actual extension-motion objects, but also of Reality, we need to have a quantificational way of bringing abstract objects by implication in propositions. Such implication is possible by quantifying not merely over terms and objects, but over objectual-causal processes / events with proper attributes, and by quantifying over the bound variables 'something', 'nothing' and 'everything'. Objectual-causal processes are simultaneous nexuses of theoretically and spatio-temporally (measurementally) referential realities-in-particular, wherein objectual-causal elements of Reality are active via the respective abstract objects / ways of process, and in general via To Be. This may be justified by adducing to Quine's objection to thinkers who allow themselves abstract terms and quantification over them and still disavow the existence (there being in discourse) of such objects.

> Pressed, they may explain that abstract objects do not exist the way physical ones do. The difference is not, they say, just a difference in two sorts of objects, one in space-time and one not, but a difference in two senses of 'there are'; so that, in the sense in which there are concrete objects, there are no abstract ones. But then there remain two difficulties, a little one and a big one. The little one is that the philosopher who would repudiate abstract objects seems to be left saying that there are such after all, in the sense of 'there are' appropriate to them. The big one is that the distinction between there being one sense of 'there are' for concrete objects and another for abstract ones, and there being just one sense of 'there are' for both, makes no sense. Such philosophical double talk, which would repudiate an ontology while enjoying its benefits, thrives on vagaries of ordinary language. The trouble is that at best there is no simple correlation between the outward forms of ordinary affirmations and the existences implied.[431]

I think differentiation between the "there is" of abstract objects / ways of being of processes on the one hand and the "existence" of objectual processes on the other may be achieved by terming abstract existence as Reality-based theoretical potentialities from the past and objectual existence as Reality-based actuality where the potentialities are realized. Real potentiality is causally accrued and attestable in principle. Unreal or mere potentiality is non-causal and is prone to be part of

---

431 Quine, *Word and Object*, 241–42.

the acausally consciousness-reflected connotatives. In my opinion here lies the differentiation between really possible worlds and impossible worlds.

Abstract potential objects are non-entifying but actual and potential relationalities in the theoretical aspect of extension-motion. There are many such, as ontological commitment over them refers back to the actual processes that have constituted them processually, though these actual processes need not be the processes that have given rise to an inductive or deductive truth that at the moment refers to 'some' relevant actual entities and ontological universals. That is, as we quantify over some entities, their causal roots and the whole of Reality are ontologically implied in it. Causal roots confer ontological universals that pertain to other processes, but allow their superimposition over the processes at hand. Somehow these superimposed universals are connotatively but vaguely grasped in connection with the inductions and deductions at work on processes at hand. Those ways of being of processes at the causal roots of processes at hand are, at least partially, common. Clearly, these ways are potentiality in store for the present processes, insofar as the potentiality is really causally accrued. Hence, it is not objectionable to quantify over abstract objects / qualities / properties as *ways* of being of respective (present) processes; and consequently, quantification over processes (which implies also sub-quantification using ways of processes / universals / essences) should take the place of that over actual entities.

There are no measurementally spatio-temporally purely localized and purely circumscribed extension-motion objects / processes, as in the case of the fully defined atoms of Leucippus,[432] whose work is intertwined with that of

---

432 Leucippus, associated with Miletus, Elea and Abdera, wrote, according to Theophrastus and Diogenes, the books: *Great World-Systems* and a treatise *On Mind*, the latter being famous for the statement extant from it: "Nothing comes to be at random, but all things for a reason and of necessity." He promulgated his atomism about 430 BCE or a little earlier. He conceded to his theory of phenomena agreement with senses, and to objects generation and destruction. Atomism rescued the reality of the physical world from the Eleatics by a thoroughgoing pluralism. Melissus had held that if there were many things, each would have the attributes of the Eleatic One (infinity, immobility, lack of density). Leucippus took up the challenge to go against this theory in his system of the world. He argued in line with the Eleatics that what exists must still be ungenerated, imperishable, unchangeable, unable to be added to or subtracted from, homogeneous, finite, full, continuous and indivisible. With all these conditions, reality may be subject to sensation, if it is cut up into millions of a-tomic, solid particles. Voids could be outside them: 'void is not being yet exists'.

Democritus[433] and of the Epicureans. Quine's particularism seems to be a bit unnecessarily intertwined with the concept of atomicity of concepts in line with the fixed atoms of Leucippus, Democritus and the Epicureans.[434] In fact, instead

---

433   Democritus of Abdera in Thrace (b. *ca.* 460 BCE) wrote in his youth *Little World-System* that treats of man first in history under the title "microcosm". He followed Leucippus in the essentials of atomism, but worked out the theory elaborately. According to Aristotle (*Metaphysics* (985b4)), Leucippus and Democritus held that both the atomic plenum and void are elements.

   There is a logic and epistemology behind Democritus's atomism, which is important in the context of Quine's particularism. Democritus's *On Logic, or the Canon* gives the criteria of knowledge based on the relative trustworthiness of evidence from sensation and thought and their ability to establish the ontological status of the physical world. As an empiricist, he did not admit the deductive method (*apodeixis*) of axiomatic inference. Aristotle's references to Democritus do not admit that the latter gave definitions of ideas. But, *Metaphysics* 1042b9 "... recognized the essence of things in their actuality and not simply in matter or potentiality." He held that the basic differences between things are due to differences in shape, position and order. W. K. C. Guthrie, *A History of Greek Philosophy*, 382–502, especially 393 and 483–84.

   Guthrie mentions also Gregory Vlastos's re-examination of Democritus's philosophy, which concludes that "... the atoms of Democritus are indivisible for physical reasons, but infinitely divisible as portions of the three-dimensional extensive continuum, i.e. if regarded mathematically." W. K. C. Guthrie, *A History of Greek Philosophy*, 503. This amalgamation of the Being-level demands of the Eleatics and the admittance of sense knowledge and particular entities have given ample space for Democritus to develop an ontological particularism about things and particular-based, quality-based interpretation of ideas (for us, language, abstract objects with their singularity, plurality, abstractness, concreteness, relativity and absoluteness). The qualities of atoms (infinity, immobility, lack of density) were simply thrust on atoms. They were their universal qualities, based never on any sort of shape, position or order.

434   We reflect in continuation with the previous footnote on Leucippus, Democritus and the Epicureans. Quine's particularism ended up in a predicament similar to theirs. Predicable abstract objects are somehow built into the particulars, without any external reason in the shape, position or order of other particulars, and without any external reason in the infinity, immobility and lack of density of other particulars. These particulars behave like a-toms, since Quine's particularism had to base itself in some shape, position and order of particulars. But there is no criterion by which to fix the level at which a particular is really a particular or a token. Causal roots from without, from the bosom of Reality-in-total, are active in them all, in the shape, position and order of nomic universals. These are somehow brushed aside in order to fix on particulars and tokens by way of ontological commitment. This

of the fixed atoms, we have only the matter (particle) waves of de Broglie[435] and the QM energy wavicles, which are never fully localized in extension or temporalized in motion, not merely by reason of the mere probabilities achieved by the present reach of science, nor due to perspectival absolutism of current and possible realms of reach of science, but by reason of the very nature of processes that we happen to call as entities. Moreover, "[m]odern physics [...] has introduced the notion of the 'physical field'. Also the latest speculations tend to remove the sharp distinction between the 'occupied' portions of the field and the 'unoccupied' portion."[436] The concept of the physical field is applicable also to matter waves, which is not the measuremental spacetime but the cosmologically ontological extension-motion. Observation refers entities (here particles) to other occurrences (in other extension-motion regions), and hence, the graph of particles in motion is that of waves; and even the very nature of particular processes of de Broglie matter waves is connexity to other processes. There are no coalescences of them that settle on points at absolute space-times or extension-motion regions without referring to anything else via abstract universals. This is also why the fundamental geometrical elements of strings theory are cosmologically and ontologically well-grounded.

Similarly, there is no abstract object that does not refer to any other abstract object or concrete process that is connected to the immediate reference of the abstract object in question. So, there is no quantification over abstract ways alone. Quantification over abstract objects are sub-quantification along with quantification over processes where they causally originate as ways / universals. As universals are in instantiation in actual processes, and due to the theory-ladenness of discourse, there are only quantifications over the domain of nexuses of abstract and concrete ways of processes. In short, quantification should be transformed in terms of the understanding of processes as universal-laden processes. This yields a more tenable ontology, wherein entities and universals, along with Reality and To Be, are naturally brought about in bound variable quantification. In short, the quanta of quantification are no more commonsense entities, but theoretically and objectual-causally mutually overlapping processes and ways, some of which happen to be Reality and To Be. Ontologically, they are never fully set in extension-motion, yet they have localizations in their specificity, within the overarching nomic realm of connexity. Developing such a theory of quantification

---

    particularism bears analogy with the ubiquitous and ghost-like atoms of Leucippus and Democritus.
435 Barrow, *The World within the World*, 137.
436 Whitehead, *Process and Reality*, 72.

is not the task of the present work, but the realm is here mooted and made possible. In such a logic, it is possible to speak of the realm of abstract objects not in the Platonic sense but simply as the ontological ways of being of beings always related to actual processes. Hence, tokens and ontological universals have more to do with ontological commitment than was admitted by Quine. This fact may be made clearer in what follows.

Hodes distinguishes between the "thin" and "thick" varieties of ontological commitment. Quine's version may be seen as the "thin" variety, since it defines the final objects merely within regimented theory: "When I inquire into the ontological commitments of a given doctrine or body of theory, I am merely asking what, according to that theory, there is."[437] 'Thin ontological commitment' may be defined as commitment by which entities are merely obtained within regimentation into formal languages: "Regimentation clarifies logical aspects of syntactic, including quantificational, structure. Thus it can help us assess what I'll call the thin commitments of the chunk of discourse, a matter of *what is said to be*."[438]

Thin commitment behoves or is incumbent upon clarification as to what is and can be actual in it and what is real (Reality- or reality-imbued). The slippery realms of potentially realization-oriented and merely semantic-linguistically real possible worlds lie here. Thick commitment is much in line with Einaic Ontological commitment. Hodes defines 'thick ontological commitment' thus:

> The question of *what there would have to be in order for certain statements to be true* is a question of what I'll call thick ontological commitment. To answer it we must assess the alethic underpinnings for the statements in question: the semantic properties of their basic constituents and the recursive 'process' that determines their truth conditions. These underpinnings are a matter of semantic form.[439]

The open-ended epistemological and semantic stance of this definition allows room for the recursive process of ever more adequate truth-conditions. I understand this process as ontological, and it is based on what I have called 'ontological universals' active in connotative universals that are active in consciousness. To the extent that is works through connotative universals it is epistemological. This allows us to formulate an ontology and a semantic ontology based on the To Be and other universals that pertain properly to Reality and reality-in-particular, as the ontological universal furniture of Reality. Quine's ontological mistake is

---

437 Quine, *Ways of Paradox*, cited in Hodes, "Ontological Commitment Thick and Thin", 235.
438 Hodes, "Ontological Commitment Thick and Thin", 236.
439 Hodes, "Ontological Commitment Thick and Thin", 236.

in that he considered 'universals' univocally, as ontological and epistemological (connotative?) universals together and then equated 'universals' with connotative ones due to the absence of the distinction in his thought. *He automatically disregarded the connotatively formed universals while building his theory since they are not involved in the very building blocks of the actual universe, but merely in entity-talk.* The view of Einaic Ontology is that the final building blocks (parts) of Reality are actual entities imbued with ontological universals – the latter being the processual ways of beings in their extension-motion processual connexity – under the auspice of Reality and its To Be. Such actual entities are yielded in Einaic Ontological Commitment, which is more than thick. Here the "there is" of connotative universals is less real than that of ontological universals, which is less real than the "existence" of objectual processes. We need only to produce three verbs that represent the three instances (which is not expedient in a general work like this).

An interesting Einaic Ontological aspect of quantification over processes and ways-in-particular is that of the commitment-level implications of quantification over the process of Reality and the Way of process of Reality. Reality may never be understood to be of any finitely circumscribed extension-motion. The "present" presumably measurable time and space of extension-motion of Reality is presumably infinite. We may automatically spread the concept of the current extension-region and motion-region of Reality and reality-in-particular into that of the past and the present of it. We have to speak of them, and so, nomic universals of any particular entity spreads over its past and present. But nomic universals are in fact connotatively formulated concepts of real ways of many processes, spread over past and present. Hence, just as there "are" subsistent (subsistent in processes) universals of the present as connotatively conceivable, there "are" also uninstantiated / unactualized and instantiated / actualised potentials as connotatively conceivable and originating from present or past connotative conceiving of past ontological universals which, through the present, extend connotatively into possible future as actualisable and unactualizable potentials in the nexus of things. These are in fact not real but conceptually possible (not potential) abstract ways, ranging over part or whole of Reality surely conceptually, but need not be causally, over corresponding ontological universals. So, there is no explication of potentials without recourse to actuality; and possibles include both such real potentials (in the sense of having lent themselves to be formed out of corresponding denotative universals via connotatives) and real possibles (in the sense of a mixture of those formed out of corresponding denotative universals which lent themselves to such formation and those formed out of irrelevant denotative universals – both via connotatives). Formation of mere possibles via irrelevant denotative universals via connotatives is thus an epistemological source of error in human knowledge and planning. In Einaic Logic any quantification over mere possibles is a source of logical error.

The above discussion throws light into the problem of instantiation of abstract objects. *Just as we speak of the ways of being of actual processes (which are actual realizations of real causal-nomic ways), we may speak also of the ways of having-been and ways of will-be of processes, which are in fact the various realized or to-be-realized causal-nomic ways of processes.* Reality has a theoretically measurable motion (temporality) and extension (spatiality) that do not situate it for sure in the infinitesimal present of common sense. It is at the same time in the past, present and future of any given sub-process. Hence, there already are *ontological universal* relations to the past and future processes active in the present Reality at the level of ontological commitment – not merely the *connotative universal* references in consciousness at the realm of Reality-in-general. The ontological universal relations to the past and future are not too complete in reality-in-particular, since they are represented by the bound variable 'some' under the presupposition of particularized universals.

Reality in process in the totalized extension-motion regions can be logically verbalized only under the abstract processual object, To Be. There is no other universal that subsumes the total token. There is nothing wrong in conceiving To Be as the most general causal-nomic universal that is the highest Law unto Reality. Hence, To Be is the abstract, processual-verbal, nomic-nominal categorial Transcendental and the most universal instrument under which to quantify and ontologize Reality in discourse. This sort of dimensional Objectual Ontology I shall call Einaic Ontology, and the logic based on the quantification process of it as Einaic Logic.

Just as we speak of the ways of being of actual processes, ways of having-been and ways of will-be, may we speak also of the counterfactual ways of would-have-been? I am of the following opinion: These counterfactual ways of would-have-been have not only similarities but also direct relations with the real causal-nomic ways of Reality, at least in their conceptual formation in connotative universals. If some counterfactual ways do not belong to the causal-nomic ways, then they are just maximal sets of Reality-in-general, which is nothing but the conceptually connotative maximal set of all such sets. If these are the remaining possible worlds that do not belong as potentials to real causal-nomic ways of Reality, then they are just sets of imaginations without foundation in possible realization.

3.4 Trans-categorial Transcendental or Way of Being of Reality

To Be functions as the supra-categorial category: it must be simultaneously the category that makes categories possible and also the umbrella under which categories thrive. The history of ontology is replete with concepts of Being that are either a purely abstract entity, or an abstract but concretizing entity, or a universalised and reified form of God, or simply a mere universal in actual this-worldly entities, processes and laws, *or, worse still, an anthropologised*[440] *highest agent of giving of Being (better, Reality) in human thinking due to human's privileged position as the thinker of Being.* In lieu of these varieties of the concept of Being – of course without examining such traditions – I propose the universal To Be as the verbal (processual) Way of being of Reality, involved in the ways of being of beings[441] and involved acausally and connotatively also in human thought.

As the Way and acme of all sorts of ways, To Be is not able to "give" anything to humans or beings or God. No poetic, imaginative ontology should substitute the reifying essentialist metaphysics of old. Instead of making an eclectic system, the categories of Einaic Ontology project, through some of the best general motifs of different 'Isms', their own importance within the space that theories can offer, in the process stressing the concept of To Be as the verbal-processual be-ing of Reality. The analytic (and implicitly, the Husserlian phenomenological) requirements of particular beings and facts in ontology are met within the purview of To Be, Reality and Reality-in-general. The realist insistence on beings and the

---

440 See the Appendix for an elaborate discussion. The major twentieth century example for such anthropologization of the concept of Being is that by Martin Heidegger. Earlier Heidegger has overemphasized in his thought the acceptor-significance of *Dasein* – the Being-thinking, Being-thinking-"receiving" human – to the very giving of Being, so much so that later Heidegger found it arduous to wash his hands off the subjectively objective over-importance of *Dasein* in the study of the objectual-causal aspect of Reality-in-total, in which To Be is placed ontologically, not connotatively. He seems to confuse the connotative with the ontological. I believe that this has been caused by the lack of emphasis, in Heidegger, on the objectual-causal aspect of ontology. The objectual-causal can be stressed only cosmologically.

441 It noteworthy that the concepts of instantiation of To Be in Reality-in-total and partial instantiation of To Be are partially attempted by Heidegger via conscious appropriations of the same in conscious thinking of beings at the level of Reality-in-general: "Being used by gods, shattered by this heightening, in the direction of what is sheltered-concealed, we must inquire into the essential sway of be-ing *as such*. But we cannot then explain be-ing as a supposed addendum. Rather, we must grasp it as the origin that *de-cides* gods and men in the first place and *en-owns* one to the other." Heidegger, *Contributions to Philosophy*, 60.

systemist blaze for totalization and generalization are met in a categorial system based on the most universal trans-categorial concept of To Be. The idealistic tendency to involve and characterize tokens and Reality by theory is given an analytic justification and absorbed just to the extent that objectual-causal ontology demands, by use of the systemic nature of Reality and Reality-in-general, under the justification that To Be offers. In short, the Einaic variety of To Be colludes all and still retains its meaning and effect.

To Be is not a genus with sub-genuses, but the purely maximal ontological *sine qua non* of all processes. Why is To Be considered a category? *This problem seems sharp if we always tend to say that the particular alone exists in its to be with universals – which does not allow Reality to be always in its To Be.* The reality of the particular is meaningful only if Reality exists in its To Be; and the token and the particular / natural kind exist only under the purview of To Be. If 'to be' (the 'is' of beings / processes) and ways of 'to be' (universals) can instantiate in beings / processes, then To Be should also instantiate. Its instantiation happens to be unique, thus yielding a singleton set. If To Be instantiates uniquely, it has the classificational sense of particularistic categories, but minimally. Hence, although minimally, To Be is a category. It is the trans-categorial category, the category *par excellence*. It instantiates in the to be's of beings / processes, nor in the ways of to be (universals) of beings / processes through their causal inheritance from Reality, but it instantiates as such most properly in Reality (ontologically) and in Reality-in-general (conceptually, connotatively). Thus, for example, To Be is ontologically instantiated in the genus 'substance' partially, because 'substance' is ontologically explicated – by pushing primitive notions – only with reference to the ever-broader ontological universals in its theoretical backdrop. This process will extend ontologically unto To Be, cosmologically to Reality as the only total substance, and epistemologically (connotatively) to Reality-in-general.

It is good to note that the ontological holism and systematicity that analytic thinkers do not much strive for is present in Quine in the shape of linguistic holism,[442] and that I have made use of his thrust after linguistic holism to get at the conclusion that we should generalize farther than linguistic holism and take Reality under its objectual-processual and ontological To Be dimension. To Be, as the Way proper to Reality and as implied in reality-in-particular, is in fact the way of making the ideals of Staticity and Becoming simultaneous and non-different. There is no difference between the two at work in things in their to be.

---

442 This is why Quine was chosen for study as the main thrust area of this chapter for us to bear out To Be as the ultimate Transcendental category of Ontology.

But stressing To Be as the static dimension of Reality-in-total and Becoming as the fluent dimension is a misappropriation of the status of To Be as the Way of Reality and as inclusive of the static and becoming aspects of it. The Being-Becoming paradigm in lieu of the staticity-becoming paradigm is not adequate for metaphysics.

As far as Einaic Ontology is concerned, this has been the combined effect of the contributions of Heraklitus, Parmenides, idealists like Hegel, naturalist metaphysicians, later Heidegger and later Whitehead upon the history of philosophy. To Be as the objectual-causal universal of Reality is not recognized by these thinkers. As a preparation for the conclusion of this chapter, I propose that the re-interpretation of the concept of To Be as the objectual-causal and trans-categorial Transcendental and as the Way of process of Reality allows ontology a face-lift and offer a more adequate manner of doing that fundamental science. This manner, being a thorough-going objectual-causal systemism reaching up to Reality and its To Be, seems capable of being a scientific ontology that shows how to transcend the particularism of individual positive sciences and at the same time make these sciences possible – which is in fact the function of scientific ontology. The sciences have their own special ontologies, but Einaic Ontology as ontology proper does not belong to any one of them, or to all of them severally. Instead, it is the zenith of them all, possessing local ontologies (like Einaic Cosmology, Anthropology etc.) superior to their local ontologies, and is their unique guide.

Finally, let us also discuss the exact difference between the traditional concept of Being and the Einaic Ontological concept of To Be. To Be is the *ideal factuality or givenness* of all that are, in all their ways of being, which include the different ways of being of processes. Ways of being are themselves ideal factualities of beings at different layers of being. Similarly, To Be is the ideal factuality of Reality. To Be is not itself the totality of all becoming, or of all that become. To Be apriorily makes all becoming possible, as the factuality, givenness or thusness of Reality.

To Be is not 'All with respect to All (together) or all (separately)'. It is the ideality of all the ways of the All (together). Ways are always constantly self-differentiating if determining is in conjunction with the many other relevant ways of being. Ways of being are attributes, therefore they characterize not each other separately from causal processes but within processes, causally. To Be is not another way of being on par with attributes. It is the possibilizer of all ways of being together in the thusness of Reality. It is not the totality or generality of the All. Hence, genuine instantiation of To Be is only in Reality as its thusness, and not as instantiation properly in reality-in-particular. Instantiation of To Be is the act in Reality, by which the latter exists in partial becoming and partial staticity. But partial becoming and partial staticity together is not To Be; instead, To Be as the thusness of Reality is why partial becoming and partial staticity coexist in Reality. To Be "instantiates" in reality-in-particular only vaguely, i.e. in their to be and other attributes. Thus, To Be is that of Reality and to be belongs to reality-in-particular. This constitutes a major difference that Einaic Ontology makes with the Aristotelian-Thomistic, Modern and Contemporary traditions in ontology.

Conclusion

As promised in the General Introduction, Chapter 3 has attempted the final trans-scientific touch to ontology and showed it as essential. It had a distinct manner of proceeding. First it made a study of what has been allowed for ontology by the first two chapters and what is lacking in the first two categories. Thus it has concluded To Be as the *sine qua non* of all ontologies. *To Be, as the ever-advancing ontological quality of all that are*, has been inculcated in the nature of Reality and Reality-in-general as the trans-category that does a great deal to system-building by bridging between the three famous criteria of truth: (1) the analytic-realistic epistemology of *correspondence* between statement and fact, (2) the epistemological-ontological criterion of consequence of Reality upon theory, namely, *coherence* of the body of theory and the structure of Reality, and (3) the scientific- and particularistic-ontological criterion of *pragmatism*.

The bridging takes place by systemically transforming correspondence-level particularized truths by coherence-level generalized truths (by system-building) and pragmatic-level limitations (by the probabilism of connotative universals and of particularized and generalized truths by means of Gödelian pushing of primitive notional senses). To Be as the trans-categorial category of Einaic Ontology is presented here in order to make system-building possible. The synthesis implied by To Be is between the cosmological, epistemological and purely ontological demands of philosophical activity from the point of view of constructing scientific ontology.

Though Chapter 3 was a study of the place of the trans-categorial category To Be in scientific ontology, it has also treated of Reality, Reality-in-general and reality-in-particular in connection with To Be. This was in view of making a scientific-ontological synthesis possible by use of the three categories proposed. Though I have been making constant reference to the To Be of Reality in Chapters 1 and 2, it is the present chapter that formally derived To Be as the properly possibilizing sense of all the senses of 'is'. For this, I have indirectly questioned the Wittgensteinian and general linguistic-analytic position that all of philosophy is merely a linguistic analysis[443] and used the analytic and linguistic ontology and holism of Quine, which tacitly justifies – in a qualified manner – the Wittgensteinian and general linguistic position. Moreover, any complexity of expression and complication while elucidating that this process of basing everything on To Be has incurred seem to be fruitful, since these have helped in an epistemologically probabilistic universalized synthesis of the commonalities of discourse, and in providing the highes tools (the three categories) for such a synthesis. The reality of the particular is best explicable only within an Einaic framework that allows terminological and synthetic progress by pushing primitive notional definitions, which has facilitated Einaic Semantics. This very way shows also the relevance of Einaic Ontology for the sciences, scientific ontologies and local ontologies. For immediate practical purposes, this completes our project of creating the Transcendental and Transcendent categories for second generation scientific ontology.

---

443 For linguistic analysis even today: "The object of philosophy is the logical clarification of thoughts. Philosophy is not a theory but an activity. A philosophical work consists essentially of elucidations. The result of philosophy is not a number of 'philosophical positions', but to make propositions clear. Philosophy should make clear and delimit sharply the thoughts which otherwise are, as it were, opaque and blurred." Wittgenstein, *Tractatus Logico-philosophicus*, 4.112. It is common knowledge that the elucidatory theory of philosophy itself offers positions regarding the world, language, symbols etc., and so, these should have been recognized by the philosophical world at the very time when Wittgenstein proposed it as a doctrine salvific of philosophy from metaphysics. Whatever its Socratic methodological bearings via clarification of ideas, it is history that the elucidatory doctrines of philosophy held by analysts have produced too much of sophism to justify their rationale for continuance in relevance for any genuine scientific ontology.

General Conclusion. Prospects of the Transcendental-Transcendent Categorial Synthesis

1. Founding Second Generation Speculative Philosophy

The thoughts presented in the foregoing chapters have been preliminary to a possible system of scientific-ontological thought that takes into consideration the major achievements and setbacks of the history of metaphysics. The present work is exploratory of the *major philosophical areas* it covers. A comprehensive system has to do with all possible areas of philosophy. Einaic Ontology is only the foundation of Einaic Philosophy. Within the theme of Transcendental and Transcendent categories for ontology and in general for speculative scientific philosophy growing out of such ontology we made attempts to preclude the "convenience" mode of ontologizing and philosophising rampant in many thinkers for more than two millennia. The convenience mode was practiced by many metaphysicians by being circumscriptive of anything less than To Be, Reality and Reality-in-general. Thus they ended up in classificational categories. Einaic categories overtook classificational categories by being most circumspective and generic of them and the various ontological starting points, which are all accepted as sub-categories within the purview of a set of mutually colluding systemic categories. That is, local ontologies and particular sciences, with their specific classificational categories, work under the generic standpoint of Einaic Ontology. This gives rise to a second generation of speculative scientific philosophy.

Circumscriptively, thus, the broadest consequence of the present work and the guiding principle that it issues for future ontological activity may be spelt out: Reality is realistically the most absolute reality. It is nakedly absolute, and at the same time finitely and thoroughly mutually connective and event-ive in its parts in their respective extension-motion. To Be is ontologically (Einaiologically) absolute and normative of Reality, and Reality-in-general is absolute and normative of the process of exercise of human consciousness over Reality and reality-in-particular (Einaic Ontology and special Einaic ontologies). The mutual involvement or collusion of these categories in ontological activity is thus more than clear in the present work.

The collusive nature renders the three categories mutually supplementary. Only with Chapter 3 is the real import of Chapters 1 and 2 brought out. Except to some extent in the case of Whitehead's metaphysics, we do not find in the 20th and 21st centuries scientific ontologies that do also collude the various categories. Einaic Ontology colludes categories at the Transcendental and Transcendent levels of apriority in thought, and not merely at the classificational or particularistic level. Therefore, I believe that this methodology is adequate for second generation scientific ontologies. Whitehead has called his philosophy of organism as a version of speculative philosophy, with a mutually collusive speculative scheme. He holds that speculative philosophy, with its speculative scheme, is most useful for philosophy (and the sciences). He justifies it from different angles.[444] He is confident that, "[a]t the very least, men do what they can in the way of systematization, and in the event achieve something. The proper test is not that of finality, but of progress."[445] Einaic Ontology is committed to ever-better relative finality in terms of its truth-probabilistic appeal, especially due to the mutually collusive nature of its three categories. Truth-probabilism assures commitment to progress. Second generation scientific ontology, with its commitments to systemism, probabilism and progress, is best founded on collusive, Transcendental and Transcendent categories. It seems safe to call also the philosophy that issues from Einaic Ontology that works with the collusive methodology, as a version of speculative philosophy.

What if one does not accept these categories? The attempt in this essay has been to discover what is ontologically apparent but empirically only implicit in discourse. It is hoped that the rationality of this attempt will persuade anyone that these categories are most fundamental even to purely empirical endeavours. As is the case with anything theoretical, the truth-probabilistic and continually differentiating nature of all connotative universals shows that at least the definitions of these three categories – perhaps not the very idealized notions that represent even the train of revised future definitions of the same – should change and find newer avenues of explanation. This is the nature of anything that takes less than infinite time for explication. To that extent, both the discovery and the reception are bound to be partial and temporary. Still it has to be asserted that there will be many points of view that do not stress the highest, deepest and broadest dimensions from within their presumably truth-probabilistic nature. Einaic Ontology promises to be an exception. The Einaic point of view and the Einaic categories remain a knowledge-maximizing agent beckoning convinced reception. If one does not accept them, one is impelled thereby

---

444 Whitehead, *Process and Reality*, 14–17.
445 Whitehead, *Process and Reality*, 14.

to develop better-serving categories and systems – which is the inspiration that Einaic Ontology imparts.

Thinkers customarily settle *in qualitative and nominal terms* the categorial definitions of their systems' primitive, originary categorial notions. They conceive these terms as amply final and as representative of certain finality within the ambit of respective realities, forgetting that the terms are representations of notions idealized from the finite extents of the notional explication of realities achieved within the discourse. The fact that terms are pragmatically fitted to represent the notions with their available explications should persuade us to be aware that notions involve truth-probabilising universals, which are probabilistic with respect to advancements in explication and in the resulting improved definitions. Hence, it is important also in the case of Einaic categories to push primitive notions after the manner that results from the epistemological implications of Gödel's theorem. In the process of definitional advancements, even the terms that represent these categories may be replaced, but the ideal notions that these terms represented are still notions that may be explicated differently. In short, there is a sort of infinite ideality with respect to notions, although they too may be represented by terms that differ in shape by the explicative capacity of primitive notions used to define. Categories are the most basic of notions. Einaic Ontology in its general structure theoretically and practically points out the dimension of advancement and replacement of senses of categories and structres of systems. This makes it a perennial trend-setter, though in its individuality in the present work it remains one of the attempts that awaits being *built further upon, differentiated and overthrown*. As an individual system, no system is most essential. The dimension of philosophical activity that Einaic Ontology points to is that of second generation scientific ontolo*gies*, based on the three Einaic *dimensional* categories. Second generation speculaitve philosophy is thus founded on dimensional categories, not on fixed foundational definitions.

2. Nature of Einaic, Second Generation Scientific, Categories

We began Chapter 1 with Chisholm's reduction of entity-discourse into predicate-discourse and synthesizing Kant's phenomena and noumena, thus preparing to take on the problem of the category of substance in the philosophy of physics, by using concepts of Van Fraassen, Newton, Bohr, Einstein etc.

Kant has clearly thought beyond the merely epistemological status of Being and said that the concept of Being is unsuitable for being a mere predicate. "*'Being'* is obviously not a real predicate; that is, it is not a concept of something which could be added to the concept of a thing. It is merely the positing of a thing, or of certain determinations as existing in themselves."[446] By 'real' is not meant here anything that pertains to existents but anything existent as process, which may be attributed. Einaic Ontology overcomes the "overthrow" of the concept of Being that a few centuries of philosophy have perceived in the Kantian argument: and it transcended in Chapter 3 above the particularist tradition of analytic ontology and made the metaphysics of To Be possible. This we accomplished by taking up the trans-epistemological, i.e. the metaphysically possibilizing status of the To Be of Reality, with respect to both beings and thought. *To Be is not to be predicated of a particular thing or a few of them* – be they abstract or concrete objects. It is predicated *only of Reality*, of which it is always the truest predicate – a fact that ontologists including Whitehad forget in their rush to stress the primacy of the particular and the primacy of the individual's (direct) experience.

We take To Be as a criterion for ontologizing. *Simultaneously, also its actual and conceptual cites, i.e. Reality and Reality-in-general, become the criteria for ontological thinking.* These criteria are unobstructed by the classificational tradition in categories. Although Kant gave us the concept of conditions for the possibility of sensibility and understanding, his two aesthetic forms and twelve cognitive categories take the approach of classification by actual appearance in sensibility and understanding, not by mutual maximal collusion of the conditions for the possibility of there being anything (particular existence) and there being thought (sensibility and understanding). Kant's rejection of the Thomistic-Scotistic instantiation of Being by the concept of existence of God has thus had its toll through his disjuncting Being from the concept of the existence of God in the ontological argument. This, combined with his phenomenalist epistemology, has resulted in his exclusion of the possibility of knowledge-claims about the existence of things-in-themselves. His own epistemology debarred him from speaking of ontological categories that taste the existence of things. Hence, I have in Chapter 1 indirectly synthesized phenomena and noumena, so that Reality as the maximal substance will be the apriorily cosmological and ontological necessity behind all acts of phenomenal knowing and, as we approached the end of Chapter 3, the To Be of Reality is automatically implied by it.

In Chapter 1 we derived the category of Reality-in-total by making phenomena and noumena mutually continuous, without access to the concept of To Be. Now read the categories of Chapters 1 and 3 together in the context of Kant – and To Be is indirectly made possible in Kant by making phenomena and noumena mutually continuous. Kant went even against predicating 'to be'

---

446 Kant, *Immanuel Kant's Critique of Pure Reason*, 504 (A598 / B 626).

to specific beings, by making noumena rationally impossible and phenomena rationally possible in human thought. The reason behind it was the empiricism of the finality of the immediate in experience, which works with complete detriment to the ontological categorial dimensions in thought. He mistook 'to be' for 'To Be' and applied his ontological strictures on 'to be' too. I have suggested undoing this long history by synthesizing phenomena and noumena into one continuous whole, thus facilitating the ontological commitment to and positing of Reality-in-total as the cosmological-ontological Transcendent category. Upon the foundation of Reality is posited To Be as the highest Transcendental, *a priori* and supra-categorial category which, by nature of its purely ontological-universal apriority, is more than a predicate: it is the very and only Predicable of all that are in their totality, in their Way of being together. The final Predicable and the Way of being of Reality are the same.

The Einaic effort circumvents reifying metaphysics and particularist metaphysics and allows metaphysics in the *a priori* sense of meta-sciences like metalogic, metamathematics etc. Chapter 1 with its cosmological category has been the first step in creating a second generation scientific ontology. This ontology is truth-probabilistic with a difference, as we may express in the following. Perspectival absolutism holds that whatever has been formulated in accordance with today's level of knowledge and attainment is true, and absolutely so. The truth-probabilistic ontology expounded in this work synthesizes reasons for justified rational belief and justifiable rational doubt in one flexible system. This system does not found truth-probabilism merely on our ability to know, but on the very nature of matter / substance as involving infinite and infinitesimal causations that are not circumscribable into one or any number of perspectives. *This does not mean that nature is causally probabilistic* (my *Causal Ubiquity in Quantum Physics* studies specifically this problem: see Bibliography), but our knowledge of nature is probabilistic about the discovering of many causes, based on matter's inability to mediate knowledge absolutely well and our inability to have it transmitted to. Chapter 1 has hinted at this as it studied QM causal probabilism in passing. Our discussions on Gödel's work later in Chapter 2 has complemented by giving epistemological ways of treating this problem. *Ontology based on Reality is more than epistemological truth-probabilism. It is based (1) on the causal-objectual and intensity-level infinitesimality and extension-motion infinitude of Reality, (2) on the causal-objectual and intensity-level infinitesimality of reality-in-particular and token processes, and (3) on our inability to epistemically circumscribe all this infinitude and infinitesimality exhaustively.*

After facilitation of the way to Reality, Chapter 1 went into establishing a relevant objectual-causal concept of Reality using different philosophies of physics. This consolidated the position that an adequate concept of Reality is feasible within contemporary philosophy of physics. Some physicists, philosophers of physics and other thinkers were discussed from this perspective. This was unavoidable, and it has not made any simplistically eclectic effect.

In Chapter 2 we moved to the philosophy of knowledge against the backdrop of the philosophies of physics, mathematics and logic, by appealing to van Fraassen, Armstrong, Strawson, Gödel, Sfedoni-Mentzou etc. The epistemological aspect of Reality had to be shown as the comprehensive ideal of experience of Reality, i.e. Reality-in-general. Many 20th century thinkers have tended to mistake the ideal of the concept of experience in general or the ideal of the concept of Reality as To Be, since they have always been cautious against incurring the purely abstract or reifying or onto-theological notion of Being of traditional metaphysics. Hence, the second category I have proposed in Chapter 2 is a distinctive concept in its status as the highest epistemological ideal. I hold that it is not the same as the ontological-cosmological Reality or its purely ontological To Be. This category does not complete the epistemological picture of thought. It is always to be colluded with Reality and To Be, since systemically the three do not have independent existence.

In Chapter 3 we moved to the philosophy of being in a characteristically Einaic fashion by studying the concept of ontological commitment in Quine and pointing out the possibility of transcending his and the like particularisms as a test case of the array of anti-reificational particularisms on the twentieth century scene. Here the purely verbal-processual concept of To Be, which ontological particularism through the centuries has been dissociating from Reality, is taken as the realistic, objectual-causal, nomic-nominal and verbal-processual essence of Reality. This essence is further identified as the Way of processual being of Reality. To Be as the most fundamental category of Einaic Ontology establishes the need of a basis for scientific ontologies that treat Reality. Finally, this is the ultimate haul on the whole tradition of particularist ontologies. Quine's ontological commitment has been merely an instrument for the haul.

Quine's dislike for essences and his partial admittance of abstract objects as necessary for ontological talk have helped us to formulate an ontology that constantly beckons for continuous reformulation of the very senses of concepts, connotatives and ontological universals. Constant reformulation of senses in corporate judgment-wholes makes this ontology epistemologically probabilistic. Einaic Ontology, therefore, is simultaneously essentialistic and probabilistic.

The categories of Einaic Ontology that have thus resulted are essentialistic in the sense that they are Transcendental and Transcendent in nature, and work through the universals essential to being and thought. I have been defending the objectual-causal, nomic-nominal and verbal-processual concept of essences. Essences / universals are ways of being of beings / processes. Such ways are essences if they are treated as nomic entities with foundation in the ontological fact of Reality's totality and particularity. The categories are realistically essentialistic, because essences are not in vacuous suspension but are made to rest collusively on all the three categories. Thus, Einaic Ontology is a realistic ontology. The categories are basically not particularistic but holistic, facilitating in theoretical adequacy the advantages that particularism (i.e. justifying reality-in-particular) presupposes, on our part by justifying *the particular within the context of* the Transcendentally general and the Transcendently total.

These categories are also not mere class-names or names of classes of classes – in Frege categories are mere class-names[447] – in that they are not primarily denotatives but ontological universals with emphasis on their cosmological, epistemological or ontological aspect. These are yielded mutually – but not exclusively – by each other, and they are in fact the *a priori* conditions of ultimacy. Their connotative status is in fact based on their ontological status as derived in some or other manner from ontological universals. Hence, the Einaic categories base ontology in ways of being of beings in process.

'Reality' and 'Reality-in-general' are inferior to To Be by the latter's ontological-Transcendental (ultimate universal) quality. 'To Be' and 'Reality-in-general' are inferior to Reality, at the level of the latter's ontological-Transcendent (actually total) quality. 'Reality' and 'To Be' are inferior to 'Reality-in-general' against the latter's purely epistemic-Transcendental (conceptual) quality. Such subservience at various levels while simultaneously being mutually collusive distinguishes Einaic categories as suitable for system-building without absolutism, because *To Be is not a mere universal but the instantiable Way of Reality reflected probabilistically in Reality-in-general.*

The three categories possibilize specific entities / processes not merely by being aletheial ("opening the lid (of truth by Being)") or Being-historical (where history is of the giving / opening of Being to *Dasein*), but also by being abstractly (nomic-nominally) and partially particularistic, and verbal-processual, in the sense that the absolutely pure verbal connotative essence (To Be) is partially realized in specifics in their totality, and so, are simultaneously abstract and universal

---

447   Feibleman, *Assumptions of Grand Logics*, 37.

by nature. But they are not anthropic or anthropocentric. To Be, thus, is appropriated not merely in terms of *Dasein* or other specific beings including the Divine, but universally at the Transcendent (actually total) level of Reality – universality being a condition that Einaic Ontology yields in favour of sensible ontological-scientific realism. To Be gives itself properly in and to Reality. To Be is partially instantiated even in essences, so that they are the nomic-generic principles of processes. Connotatives or universals are therefore more than Husserl's essences with their subjective evidential correlates or the generalities in contemporary essentialism (Quine, Armstrong etc.),[448] in that they are ontologically present as ways of being of beings / processes. They are the very *possibilizers of entities in their Einaic generalities*, which happen to be had denotatively also by consciousness.

Thus, the concept of the highest connotative To Be in Einaic Ontology is (1) purer than in Aristotle[449] and Aquinas, and as I shall argue in the Appendix, purer than in Heidegger; (2) applicable in the philosophy of science past the specific, classificational and cosmological categories; and (3) applicable in Whitehead

---

448 One variety of essentialism proposes an absolute distance from modalities (and thus also from universals): mereological essentialism, which is the formal study of parts of particulars (Greek *méros*, "part"). Stanislaw Leśniewski and Nelson Goodman have held that reference to universals like sets, properties and other abstract entities is not a must. This sort of essentialism does not find favour with many analytic ontologists, since mereological essentialists cannot invoke the modality of necessity to assert that a whole has its parts necessarily. Goodman, *The Structure of Appearance*, 33–44. See also *The Cambridge Dictionary of Philosophy*, s.v. "Mereology". It is still mereological essentialism, since it happens to assert by the modality of necessity that "[…] every composite is necessarily constituted by a particular configuration of particular proper parts, and loses its self-identity if any parts are removed or replaced." *The Cambridge Dictionary of Philosophy*, s.v. "Haecceity". Lately, Michael Jubien and others have held that reference by terms is not at all by use of universals. Jubien, *Ontology, Modality and the Fallacy of Reference*, 22–24. Here too the predicament that is difficult to face is the absence of universals to connote essential properties. Roderick Chisholm tends to this view.

449 *Met. Γ.2* says: "There are many senses in which a thing may be said 'to be' [….]" They are not homonymous, but are related to one central meaning of 'being', i.e. 'substance'. Aristotle, *The Complete Works of Aristotle, volume 2*, 1584 (1003a30–34). See also Ando, *Metaphysics: A Critical Survey of Its Meaning*, 10. From the Einaic point of view, this shows the categorial confusion in Aristotle between the highest Transcendental (To Be, which is parallel to the Aristotelian nominal 'Being') and the highest Transcendent (Reality-in-total, which is parallel to the Aristotelian 'substance'). The Einaic categories, I hold, are purer than in Aristotle.

past the less than pure, transcendently transcendental Categories of the Ultimate: namely, one, many and Creativity. The resulting ontology inclines to truth-probabilism, as it applies the implications of Gödel's Incompleteness Theorem to all possible extensions in its ontologico-epistemological category, i.e. Reality-in-general. It is scientific-realistic, as it applies the concept of To Be in Reality and reality-in-particular. The three categories collude with each other by mutual implication in the process of the universe, of consciousness and of all that are.

In effect, but not by an explicitly protracted study of the data pertaining to the following relevant numbers, the attempt in this essay has been to render second generation scientific ontology possible by deriving the re-generalized and non-classificational categories, by transcending (1) the cosmological classificational categories and the concepts of beings and of Being in Aristotle and Aquinas,[450] (2) the categories, phenomena, noumena and the phenomenally impossible concept of Being in Kant, (3) the concept of epistemically transcendental essences in Husserl, (4) the aletheial, anthropic and Being-historical concept of Being in Heidegger, (5) the Categories of the Ultimate, i.e. one, many and Creativity in Whitehead, (6) late twentieth century categories like those of Strawson, Chisholm etc., and (7) the empirical-scientific and particularistic categories of space, time, essence, relation, causality, structure, object, number etc. that are mostly in use in physics, cosmology, epistemology and particularist ontologies. It was not possible to study Aristotle, Aquinas, Husserl, Heidegger, Whitehead, Strawson, Chisholm, etc. exclusively, due to spatial considerations. The re-generalized, simultaneously verbal-processual, nomic-nominal universal To Be is a *sine qua non* for all ontologies, by reason of which the other two Einaic categories become

---

450 Aristotle's categories or ten modes of predication in propositions are as follows: substance (subject), quantity, quality, relation, place, time, position, state, action and passion. These are predicable notions of some or other manner, and "[...] 'being' has a meaning answering to each of these." Aristotle, *The Complete Works of Aristotle*, 1605 [*Metaphysics*, Book V (Δ), 1017 a 25]. *The classificational approach* should then have led him to differentiate between substance (which is for him one of the categories) and the categories themselves, because all the ten categories are to be used as predicables of substance (subject). If one argues in favour of taking "being" as the subject of the ten predicates, one has to differentiate between a being, beings in a particular group, beings in total and Being as such. That is, he had got confused between these different meanings of being as the subjects of the categories, and God as the highest realisation of Being, due to his approach of classification of being to produce categories. His search for categories was not one for conditions for ontological talk based on what is real.

necessary. Thus, Einaic Ontology is a speculative scientific ontology that makes possible and transcends the particularistic categories of the sciences.

For the use of ideas from philosophy of logic and philosophy of mathematics, I had recourse to Gödel's Incompleteness Theorem that attempts to settle axiomatic problems in Russell and Whitehead's *Principia Mathematica* and other systems, and thus answers the Hilbert programme of re-founding mathematics, but answers in the truth-probabilistically partial negative. This treatment does justice to the logical and mathematical ingredients (*abstracta* or ways of being) that serve to abstractly found the question of categories. Are the maximal categories also subject to the method of pushing of axioms that has resulted from Gödel's theorem? Yes, this is possible by pushing sentence- and word-senses of axioms and primitive terms. As a result, one may discover other, more necessary and sufficient, categories.

3. Einaic Ontology: Blend of Einaiology and General Ontology

Ontology is defined variously by different thinkers. Aristotle defined his First Philosophy (ontology) thus: "There is a science which investigates being *qua* being and the attributes which belong to this in virtue of its own nature."[451] Aristotle was aware of the perils of isolating attributes, which included Being as the final case, from substance. He is sharp in his classification of ontologies into First Philosophy and regional ontologies, the latter of which distinguish the abstract, attributive and mathematical object from actual entities.

> They [the special sciences] cut off a part of being and investigate the attributes of this part – this is what the mathematical sciences for instance do. Now since we are seeking the first principles and the highest causes, clearly there must be some thing to which these belong in virtue of its own nature. If then our predecessors who sought the elements of existing things were seeking these same principles, it is necessary that the elements must be elements of being not by accident but just because it *is* being. Therefore it is of being as being that we also must grasp the first causes.[452]

The first 'Being' here pertains to reality-in-particular. The second 'Being', as being which (*qua*) 'beings' are studied, should have been defined as the proper object of First Philosophy. Aristotle was not cautious enough to avoid the problems of definition. First Philosophy will be safely defined by proposing transcending this concept of Being by the nomic-nominal and verbal-processual To Be, with

---

451 Aristotle, *The Complete Works of Aristotle*, 1584 [*Metaphysics, Book IV (Γ)*1003 a 20].
452 Aristotle, *The Complete Works of Aristotle*, 1584 [*Metaphysics, Book IV (Γ)*1003 a 25–31].

its proper object Reality as the foundation for any categorial scheme that aims to do First Philosophy. Nor did he treat the conceptual ideal aspect of thought, i.e. Reality-in-general, as distinct from To Be. Thus, Aristotle degraded his First Philosophy into a study of something other than the To Be of Reality by equating it simultaneously (1) to the Being of the Divine – whose Being could be neither Transcendental (because He is a being that is Transcendent in a manner higher than ordinary beings), nor Transcendent (because He is this Transcendental Being) in Aristotle – and (2) to the being of particular beings. This dichotomy is what Einaic Ontology objects to, not just any one of its aspects.

Nevertheless, Aristotle's wise definition of First Philosophy does consider it a meta-science, despite the fact that historically the prefix 'meta-' is a later formulation. He does also discuss the special sciences in which 'to be' means differently, from how it is in First Philosophy.

> There are many senses in which a thing may be said 'to be', but they are related to one central point, one definite kind of thing, and are not homonymous. Everything which is healthy is related to health […] so, too, there are many senses in which a thing is said to be, but all refer to one starting-point; some things are said to be because they are substances, others because they are affections of substance, others because they are a process towards substance, or destructions or privations or qualities of substance, or productive or generative of substance, or of things which are relative to substance, or negations of some of these things or of substance itself. It is for this reason that we say even of non-being that it *is* non-being.[453]

For Aristotle such a meta-science had to be of and in terms of things / substances, and at the same time of and in terms of *their* (!) qualifying To Be. Here Einaic Ontology substitutes the particularist 'they' with 'Reality'. Such a meta-science cannot be merely of God the highest Transcendent Being or of To Be in terms of God; nor merely of To Be in terms of To Be. In the former case it is the onto-theology of the Aristotelian and Scholastic tradition. In the second case it would be a pure science unconnected to substance.

If in accordance with the Einaic concept of the co-extensiveness of To Be and Reality we study To Be in terms of Reality (not in terms of God[454]) as the

---

453 Aristotle, *The Complete Works of Aristotle*, 1584 [*Metaphysics*, Book IV (Γ)1003 a 30–1003 b 11].

454 For Einaic Ontology, God is the highest conscious realization of To Be, and Reality is the highest ontological realization of To Be. Hence, Einaic Theology is a regional ontology. God is infinitely present and active in Reality by conscious action at the dimension of the To Be, which makes also individual entities / partial processes possible, acted upon and had in awareness by God, since To Be is the

Transcendent cite of the Transcendental To Be, and in terms of Reality-in-general as the Transcendently Transcendental cite of the same, then we have a justifiable meta-science. The study of To Be in terms of To Be is a pure meta-science that speaks nothing more than that, and nothing more. So, we *define Einaic Ontology* as the meta-science that studies (1) To Be in terms of Reality and Reality-in-general, and (2) Reality in terms of To Be and Reality-in-general. Let us call the first as *Einaiology*, since it studies the universal-verbal *Einai*, "To Be", in terms of the Transcendent Reality and the Transcendently Transcendental Reality-in-general; and the second as *General Ontology*, since it studies *ta ónta*, "beings", in their totality, via the Transcendental generality of To Be and the Transcendently Transcendental Reality-in-general. That is, Einaic Ontology is a synthesis of two sciences.

This provision, I believe, will answer the inadequacies of reifying-metaphysics and onto-theology on the one hand, and on the other the difficulties caused by particularist ontologies to the construction of metaphysics. The study of ontology is not of To Be or universals alone, or God as the Being or Reality or the totality of all beings. Ontology becomes comprehensive only if Einaiology and General Ontology are done with mutual supplementation. Einaiology works into the concept of To Be only through that of Reality. It needs the concept of Reality from General Ontology. General Ontology needs the Einaiological concept of To Be to study Reality. The two complementary sciences together constitute metaphysics in the positive sense.

In like manner, regional ontologies may as well be defined within the realm of Einaic Ontology. General Ontology, as the study of Reality, has, as one important aspect of it, the study of Einaic Cosmology. The concept of reality as such is never complete without the general cosmological understanding of the universe of matter-energy and anything that contributes to it. Einaic Cosmology studies

---

principle of conscious and unconscious unification and individuation. Moreover, the constitution of God is such that He is a Transcendent in whom To Be is realized Transcendentally, connotatively, in the most conscious and active manner. Hence, in contradistinction from Reality-in-total, which is the sole Transcendent ultimate, God is the Transcendentally Transcendent ultimate. God is no absolute vacuum, nor the same as the world or of its nature; so, God has to be the infinitely active and infinitely intensively existent stuff in every finite extension-motion region of finite existences. This is an actively absolute God with an infinite, infinitely intense and non-vacuous body that is different from this world. This allows Einaic Theology a basis to advance from – different from Aristotle, Thomas, and modern and contemporary philosophers including Whitehead.

the physical sub-totality of the world in terms of To Be, Reality and Reality-in-general. Einaic Epistemology studies connotative universalistic thought's capacity to construct coherent, practical, relatively more truth-probabilistic and thus most widely correspondence-enhancing, systems with maximum analyticity and syntheticity.

Similarly, Einaic Philosophies of God, Human, Mind, Morals and Beauty are also essential for determination of the nature of Reality. Einaiology with its concept of To Be is somehow instrumental in deepening the very nature of inquiry in Einaic Ontology. To that extent, Einaic Theology, Cosmology, Anthropology, Psychology, Ethics, Epistemology, Semantics, Aesthetics etc. together, with philosophical inheritance from the respective positive sciences, yield the unique subject matter of Einaic Ontology – every detail of which being studied via To Be and its corresponding categories *alone*. This meta-science is very much science-dependent in general and transcendent of science in particular. *Einaic Ontology is thus a version of scientific ontology*, with the qualification that it at the same time depends on Einaic categories and regional ontologies and particular sciences.

No more should ontology be the inaccurately defined as the 'science of being as being' and philosophy as that which studies all the provinces beings under the point of view of the study of being qua being. Einaic Philosophy is generally definable as the science that studies anything (any one of the three categories or the general possibilities of reality-in-particular) in terms of Reality, To Be and Reality-in-general. Its regional ontologies study only regions of reality-in-particular via To Be. Einaic Philosophy is generic of all possible philosophical sciences. Thus, definitions of philosophy, such as 'science of possibility', 'science of all possible things insofar as they are possible', 'general science of reality' etc. are vague, since they do not specify the extension of the region of Philosophy or regional ontologies and the intension of ways and instruments of study of the subject matter.

I gave for universals the "ways of being of processes" interpretation. Ways of being are always of beings / processes. From that viewpoint, all properties, class terms and universals are ways of beings. There is no class term applicable to one being / process alone. Any class term applicable to one and only one being is in fact composed of many universals with respectively imaginable connotatives, where the universals have ulterior reference to other beings / processes. To Be is of Reality. The latter is not *a* being, but the totality of beings / processes, to which belong all sorts of ways of being, out of which To Be is the highest and only exemplification of all ways of being. Hence, To Be refers partially to other ways of being but which have references to ways of processes within Reality. Processes within Reality are studied in regional ontologies and special sciences. These facts facilitate the conception of Einaic Ontology as a scientific ontology that transcends the sciences and regional ontologies and connects beings with To Be and Reality via the universals of beings and the To Be of Reality. The ways-of-being interpretation of universals allows also the study of reflections of ways of being in consciousness and consciousness as a process as such. That allows ample space for Einaic Semantics, Epistemology, Psychology etc. based on the Einaic categories – and of courses for the corresponding empirical sciences.

The verbal-processual To Be, too, may be taken as a class term of Reality in the several, provided it is constituted as an infinite regress: class of class of … sets *ad infinitum*. In this case, other ways of being (ontological universals that appear for consciousness as connotatives) also will be class terms: either classes, or classes of classes etc., but never reaching the infinity of the highest dimension of generalization as in To Be. This is also why To Be is taken as the supra-category of all categories, symbolizing the ideal of the *ad infinitum* at the limits of Cantor's highest cardinal, 'The Continuum'. Cosmologically, the Continuum has the size of all possibles and reals in physical reality. And if (and only if) there is infinitely continuous creation of matter-energy out of nothing is the ontological Continuum of Reality an ever-increasing cardinal. In this sense, future-possible real and existent worlds, future possible ontological universals related to the additional future-realizable ontological existent processes and future possible worlds of connotative abstract entities encompassed by Reality-in-general – they together define and determine the size of the total ontological Continuum of really possible worlds.

Description of real possibilities within the class of all possible classes in To Be are modal in nature. The future of Einaic Ontology should therefore see attempts to build Einaic Epistemology, Logic and Semantics of a modal[455] variety,

---

455   Modal logic and its semantics argue in terms of necessity and possibility, but its ontology functions under various meanings of necessity, possibility and contingency. "[…] [A] proposition is *necessary* if it holds at all possible worlds, *possible* if it holds at some." Chellas, *Modal Logic*, 3. David Lewis makes a realistic but fantastic statement regarding possible worlds: "I believe that there are possible worlds other than the one we happen to inhabit. […] I emphatically do not identify possible worlds with respectable linguistic entities; I take them to be respectable entities in their own right. When I profess realism about possible worlds, I mean to be taken literally. Possible worlds are what they are, and not some other thing. If asked what sort of thing they are, I cannot give the sort of reply my questioner probably expects: that is,

epitomized in the working out of possibility and necessity of classes of classes of ... sets *ad libitum*. Modal logics have found much disfavour in the works of Quine.[456] But today as there abound a variety of modal logics that find their origins in what we call ontological universals and their connotatives, it looks feasible to work an Einaic sort of modal logic that would assure the possibility of relatively spectrally variegated "truth-probability values" of modals like possibility and necessity based on the Einaic categories.

The ways-of-being interpretation of To Be and universals points to the nature of Einaic Ontology as the meta-science of regional ontologies and the sciences. Anything meta-logical or meta-mathematical does not purport to be totally unconnected to logic or mathematics. Instead, they work at a level that is relatively more *a priori* or theoretically preparatory or axiomatic to logic or mathematics. Even so, the sense in which the term 'meta-metaphysics' is taken today in general may be modified to mean the science 'apriorily' or 'theoretically' preparatory to positive physical ontologies and local ontologies. This, I believe, is one of the most important functions that Einaic Ontology can fulfil. To be accurate, the part of Einaic Ontology that does this is Einaiology, so that General Ontology has relatively greater freedom to feel with regional ontologies and special sciences. But Einaiology and General Ontology are not two separate entities. They play their part together in the most holistic manner for the greater, nay greater and greater, good of human intellectual enterprise.

The "ways of being of processes" interpretation uses what analytic ontology has used as its categories, classes and natural kinds, but shows a way to transcend the particularism of analytic ontology systemically. To that extent, the present work may be considered as an instrument of transition from analytic to scientific ontology in a systemic manner.

---

a proposal to reduce possible worlds to something else." Lewis, *Counterfactuals*, 85. See parts of the same cited and discussed in Brian Skyrms, "Possible Worlds, Physics and Metaphysics", 143. *Reality-in-total may be interpreted as the causally continuous totality of the many necessary worlds*, provided extension-motion is generalized over infinity and eternity in all possible regions of Reality-in-total, and if the individual real possible worlds of the Lewis-type are finitely in extension-motion ensconced and established as real, i.e. based for their ontological and connotative universals on Reality. Except for statements from Lewis above and their likes, the tradition of possible worlds in Lewis and in many other thinkers is in fact at the most akin to Reality-in-general, and many parts of them mere possible worlds.

456   For example, Quine, *Quintessence*, 114.

4. Cosmological-Theological Prospects

What is primarily given to naked veridical perception is the world. To go by Cartesian prescriptions, what is immediately clear is the self. Now, to be able to think causally between the events in the world, we should think also of possible realms of entities. Thus, Einaic Ontology is the study of To Be in terms of Reality and vice versa. Whether there is to Reality anything other than this world and the self belongs to astrophysical, mathematical and philosophical cosmology to unearth – an unfashionable but inevitable conviction that traces the age-old traditions of philosophy. Thus, the problem of the Divine will continue to be important in Einaic Ontology, because the existence or otherwise of the Divine will characterize the nature of objectual causality and its infinity or not in Reality. If the universe proceeds from the Divine, it can do so only infinitely, for any other possibility makes the Divine a purely vacuous non-entity.

In connection with the study of traditional metaphysics, Heidegger describes reifying metaphysics thus:

> Metaphysics thinks of beings as such, that is, in general. Metaphysics thinks of beings as such, as a whole. Metaphysics thinks of the Being of beings both in the ground-giving unity of what is most general, what is indifferently valid everywhere, and also in the unity of the all that accounts for the ground, that is, of the All-Highest. The Being of beings is thus thought of in advance as the grounding ground. Therefore all metaphysics is at bottom, and from the ground up, what grounds, what gives account of the ground, what is called to account by the ground, and finally what calls the ground to account.[457]

I would put the nature of metaphysics slightly differently. The All-Highest Divine is the ground for all traditional metaphysics, since the nominal aspect of To Be and of universals / *abstracta* is taken by it in isolation from their nomic (rule-like) and verbal (processual) aspects. The nomic aspect is the generality, the verbal aspect is the processual. Due to the 2300 years of Western over-emphasis on the nominal aspect in the tradition via Plato, Aristotle, Scholasticism, modern thinkers and linguistic analytic philosophy (and, to mention, the more than 3500 years of Eastern over-emphasis on the nominal), the nomic was taken as the nominal and reified, without accessing the verbal-processual aspect of notions. This caused the famous confusion between To Be and the All-highest Being who is the highest instance of conscious instantiation of To Be. In the process of cosmologically proving the existence of this All-Highest Being, one tended to transpose the To Be of Reality upon the "to be" of the Highest Being, to the detriment of integration of

---

457 Heidegger, *Identity and Difference*, 58.

the verbal with the nomic and the nominal. Moreover, there did not exist much of a truth-probabilistic and processual account of the nomic and the nominal aspects of universals. This added to the ready theoretical reification of nominal which occurs normally in practical life. The concept of Being, conceived always as the verbal concept transposed and equated to the Highest Being, allowed reification of God and the equivocation of Theodicy with Metaphysics. The absence of probabilistically ever more true and ontologically processual account of nomic and nominal universals resulted in concepts of things being made reifications of the nominal and these were borrowed from common usage into philosophy as no other way was found. This long tendency sanctified the other reified parts of metaphysics (cosmology, anthropology, psychology etc.) and thus completed the notion of reified metaphysics. God too was reified, but God could not be this world. Thus God was reified into a pure, absolute and thus vacuous entity.

If there is the non-vacuous Divine, and it cannot be the universe, then it has to be a sort of physique characterized by a non-vacuous and non-this-worldly physical nature. Whatever that physique be, it would have to be infinitely active, since all else is finitely active and the Divine is not absolute vacuum. Such activity could very well be called infinitely conscious activity, infinitely conscious at every given extension-motion region, which alone would make it simultaneously trans-this-world-physical and non-vacuous. The productive God-world relationship presupposed by the nature of such a Divine would be infinite also in all extension-motion. In short, block universes of finite masses and relative extension-motion nature must be taking dependent origin – not by transformation of the Divine physique (for it would still be the Divine), and so *ex nihilo*. This sort of a realistic Philosophy of God does not allow a recurrence of the purely vacuous Divine too, but ends up in the concept of an infinitely concentrate and infinitely realized and actual Divine body, active at infinity at any given extension-motion and *unchangeably* infinitely active / moving / changing. This concept of unchangeability is the only sort of Divine unchangeability thinkable. The infinite movement in the Divine at the locus of any given extension-motion region, summated unto infinite actually possible worlds, is to be understood as the definition of Divine activity or love. In that case, the causality exercised by block universes upon other block universes is finite, and an infinite Divine causality shoud be understood to be active at any given block universe.

What could be the guiding categorial foundation of such Divine activity / love / consciousness? It could not be taking place without the Divine con-spectus of Reality (i.e. the Divine and the universe), at the *species* (way of looking at / mirroring) of infinity and eternity, at any iota of temporal extension of its uniquely non-vacuous activity / physique. For it to be simultaneously uniquely different from vacuum and this-worldly physical matter-energy, it should be capable of the *species* of infinity and eternity. It may easily be found to be best effected within the point of view of the To Be of Reality. Hence, most possibly, Philosophy of God has its probabilistically most systemic expression in Einaic Theology. The active relationship (productive, loving etc.) of the Divine with the universe is Einaic, and so, the infinite Divine determinations upon himself and the universe, along with the finite determinations of block universes upon themselves and a finite number of other block universes would amount to the Providence of God.

Thus, Einaic Theology can affect the very concept of Reality. No amount of direct or indirect philosophical "arguments" has done much in the direction of showing the truth or otherwise of existence of such a Divine. Hence, a cosmological way out of this millennia-old philosophical difficulty is a need. It will characterize Einaic Theology, Cosmology, Anthropology, Ethics etc. If we have a causal way of possibilizing the infinitely creative Divine in connection with the world, we will be in a position to understand the issue of all possible real worlds better, i.e. causally; and the "continuous" concept of Reality will give greater scientific and ontological assurance of validity. I would suggest that we have a General Gravitational Coalescence Cosmology (GGCC) that solves a uniquely clear paradox in the interface of the various cosmological theories of the origin and evolution of the universe. Thus, Einaic Cosmology would base itself on GGCC, which in turn would give a fine foundation to Einaic Theology.

If developed face to face with further philosophical and scientific realities and characteristics brought out by the proposed GGCC, the categories of Einaic Ontology, which are connotatively truth-probabilistic, could be seen to possess a unique universality and tenability. Thus, not only in Einaic Semantics, Epistemology, Ethics etc., but also in Einaic Theology and Cosmology, the priority and universality of sway of Einaic categories may be well perceived if GGCC can show the possibility of there being a creative Divine with a nature different from that of pure vacuum and this-worldly physical existence. Thus, GGCC can be the foundational cosmological *sine qua non* for the development of Einaic Ontology and Einaic Philosophy. Thus, the aim of Einaic Ontology based on the Einaic categories is to develop a system of philosophy that is cosmology-and-physics-compatible and indirectly also humanities-and-religion-compatible.

5. New Einaic Sense and Universality of 'Category'

I would claim that this essay has attempted to alter the philosophical sense of the term 'category' from the sense it has occupied in ontological and logical traditions from the time of Plato and Aristotle.[458] Although the purely classificational sense of categories is missing in the categories developed in this essay, the original meaning of the term 'category', as still useful for discourse, is preserved. The Greek *katēgoria*, "accusation / statement (of something about something else)", is from the verb *katēgorein*, "speak down / speak against", which is from the prefix *kata-*, *kat-*, "down", and *agoreuein*, "speak in the assembly". *Agoreuein* is from *agora*, meaning "market place" or "assembly". It was customary in ancient Greek cities to announce, by royal or civil order, certain matters of information in favour of or against persons and practices, by shouting out in the public market place that 'such-and-such is so-and-so'. This was a sort of classificational statement in public about persons, groups and practices.

The public aspect of 'category' is preserved by 'predication' / 'qualification'. This does not mean that what can be said of something else is a category. Not merely a qualitative term, but even a term that represents an object in one and the same statement can be a category. By its very nature, a predicate is predicable not merely of one but of many – which is why To Be was no predicable in Aristotle. In Aristotle, substance (*ousía*, "that which is / has being") is the category *par excellence*, and all other categories depend on it for their subsistence and meaning, and yet, substance is not predicated of anything in the qualitative sense. "A *substance* – that which is called a substance most strictly, primarily, and most of all – is that which is neither said of a subject nor in a subject, e.g., the individual man or the individual horse."[459] Yet, for Aristotle even 'substance' that merely represents all natural kinds in general is a category a genus, although in its simultaneously (conceptually) generalizing and (ontologically) individuating sense, since it too is able to categorize natural kinds. It is also possible to think of one total substance as such as the one and only one case of the category of substance. But, for this purpose, even 'substance' is not unique enough. Any group of substances is a substance in the immediately total sense. Reality-in-total is most total – it is total of all groups of substances. It is the unique singleton set of that kind. In short, Reality cannot be taken as a special case of substance.

We seek an Einaically acceptable meaning of 'substance'. We will have to admit that it is based on the universals involved in the conceptualisation of the same.

---

458   The Platonic categories (fundamental "forms", *eídē* or *génē*) are: being, motion, rest, identity and difference. Taylor, *Plato: The Man and His Works*, 889. It is easy to see that *Aristotle's ten categories are the result of bifurcation of each of the five Platonic categories into two.*

459   Aristotle, *The Complete Works of Aristotle*, vol. 2, 4 (*Categ.* 2a11). See also Marx, *Introduction to Aristotle's Theory of Being as Being*, 18.

Hence, the only way to get at the genuine meaning of 'substance' is to maximise it by universals, and then make the rest of individuated substances and other predicables possible. This implies that we must always take Reality as the substance *par excellence* and then proceed to ontologize. Reality has thus to be a category – which works as the maximal case – and it ontologically summarizes the genuine meaning of substance beyond its purely scientific and cosmological sense. It includes all possible substances and thus *facilitates predication* beyond a narrow, particularistic sense. But we need to differentiate between the whole and the particular. Reality facilitates predication of the genuine sense of ontological universals and class terms like substance on reality-in-particular.

As just mentioned, the involvement of universals in predication of the genuine sense of 'substance' is very important for discourse. This implies the ideal of all discourse, i.e. Reality-in-general, as the maximal case of connotative universals in consciousness, without which there is no discourse of Reality. This is the genuine meaning of all qualifiers, without which predication and ontological discourse are impossible. Hence, Reality-in-general is also a category *par excellence*, with special reference to the epistemic aspect of ontological and cosmological predication. This category, therefore, is a maximal *facilitator of predication* and of ontological discourse based on Reality. In this sense, this category has the sense of predication as facilitator of predication.

Reality-in-general is the connotative predicate of Reality. So, in discourse on Reality, Reality-in-general works as the conceptually or epistemologically colluding predicate. Ontologically, To Be is its colluding predicate. Reality is not predicated of anything. The other two categories are predicables of Reality separately in the epistemological and ontological senses. Predication in the fullest sense of the activity is with sense if, seemingly contrary to the tradition of anti-metaphysical philosophies, we accept To Be as the ideal case of the objectual facts of laws of nature and predication of ontological universals. Ontological universals have their own paths independent of our epistemic activity. As a matter of the principle of ontological equity, we do not have to predicate To Be to any being other than Reality. Thus, under the purview of Reality, categorially justified systemic thinking is possible only if To Be is ontologically predicated of Reality.

That does not mean that To Be is merely a category. It is the only and highest supra-categorial verbal-processual predicable ever imaginable. Hence, To Be is also a category that *facilitates predication (and instantiation) for reality-in-particular and Reality*. In the sense of facilitators of predication, the Einaic categories of To Be and Reality-in-general remain necessaray. They are of such a nature that they thrive by mutual collusion of and dependence on each other, by basing themselves not merely on *some* possible starting points, but on the deepest and broadest and highest of all starting points, i.e. the To Be of Reality. This fact results in transcendence of *the starting point frenzy*. Gödel's theorem has allowed us to push primitive concepts and find the Einaic commonalities (our three categories) in all sorts of starting points by Gödelian pushing of senses of ontological primitive notions.

This line of thought improves the coherence standpoint of truth, i.e. (1) from merely internal coherence to the conjunction of internal coherence with the dimension of the best possible coherence from the very layer of the system's starting points, (2) culminating in correspondence-level foundation in Reality-in-total accruing at the level of the act of ontological commitment in every system, and (3) led by the pragmatic necessities of pushing axiomatic notions in Reality-in-general, connotative universals, ontological universals and To Be for system-building. Such an epistemology possibilizing ever better truth-probabilities in systems facilitates seeing Einaic Ontology as well integrating ontological coherence with correspondence-level foundation in the conceptual origin of the system in causal-processual actuality of Reality – which is, of course, not merely in reality-in-particular – and with the amount of pragmatism necessary in system-building. That is, particulars are ingredient in Einaic Ontology indirectly over the actuality of Reality, and this fact makes Einaic Ontology free from the shackles of particularism. Although even a regional ontological system thus envisioned may be quite involving (as are many others), the demands of pragmatism are accommodated only to the extent that the demands of ontology in science admit. This further facilitates systemic ontologisation in a more adequate manner than is the case in Aristotelian, Thomistic, analytic, hermeneutic, pragmatic, process philosophical and other senses. I hold that the Einaic categories can thus be the beginning of second generation scientific ontologies. Such a train of scientific ontologies can help ameliorate the particularism rampant in the above forms of ontologies, though my attempt here has mostly criticised the occurrence of particularism in analytic and scientific ontologies.

Due to their new sense of categorial predication, our categories vouch for ontological, natural scientific, epistemological and other priorities at one go in Einaic Ontology, since its categories involve all that exist and all that science can contribute to philosophy. I believe that this is a contribution to the future of scientific ontologies – it being the case that the categories of the various metaphysics that are not science-based ontologies have failed to yield a tenable *synthesis of the truth-probabilities of Reality without access to the science of the cosmos and of all that can exist*. Specifically, this feature of involving all sorts of categorial predication should be a criterion for constructing future categories of scientific ontologies, because metaphysics (in the sense of ontology) without imbibing the science of the cosmos, with its category of Reality, is wary of adequacy of content. Hence, ontology is improper without cosmology or physics. To conclude, Einaic Ontology / metaphysics needs physics. The same may be said also about adequating Einaic Ontology with psychology and other sciences.

The specific set of categories worked out herein is projected as possibly the most fundamental set of such categories on the path to integration of some of the merits of traditional metaphysics, contemporary ontology and anti-metaphysical philosophies with the philosophies of physics and knowledge. The present work is not merely a categorial synthesis of physics with metaphysics. It has also been a search into whether the science of the cosmos is genuinely possible without clarity on and contributions from its ontological foundations. If the present inquiry has been successful to any extent, it is in that it has discovered the manner in which not merely (1) reality-in-particular is imbued with ontological universals, Reality and its To Be, but also (2) discourse about reality-in-particular is imbued with both connotative universals and their acme, i.e., Reality-in-general. If so, physics and the science of the cosmos, which cover the discourse of the physical province of Reality, are in Einaic Ontology imbued with ontological universals, Reality, To Be, connotative universals and Reality-in-general. That is, to conclude, genuine physics is impossible without metaphysics as Einaic Ontology.

The conclusions in the two paragraphs above are not merely ontological but equally well also physical by reason of the mutual implication and possibilization of physics and metaphysics. Similarly, any provincial science and its ontology have a claim on Einaic Ontology, and Einaic Ontology has a claim on those sciences too. In short, all positive sciences and regional ontologies are imbued with their theoretically presupposed Einaic Ontology and its categories, by way of possibilization and acceptance of supporting material from; and Einaic Ontology is beholden to positive sciences and regional ontologies, which naturally instantiate the general categories of Einaic Ontology in the small.

To put the whole thesis of the above work in gist, a continuously causal concept of Reality as such supported by the ultimate dimensional and mutually collusive cosmological, epistemological and ontological categories of thought can show that particular sciences and regional ontologies require the hand of an Einaic variety of ontology for their existence, development and fruition.

Appendix. Beyond Heidegger's Anthropologized Being: Nomic-Nominal, Verbal-Processual, Universal "To Be" in Einaic Ontology

I begin the critique of Heidegger's work with a statement of concern. Anyone who attempts to read this Appendix without first reading my arguments in the text of the whole book and without a critical attitude to Heidegger, is liable to misunderstand my arguments here as misinformed or trivial. I think we should get behind Heidegger's words by chipping his prohibitively poetical and mystifying language off its rhetorically adumbrating shades, in order to get at the senses and implications of his Fundamental Ontology and Being-historical Thinking. True, there is no complete chipping off, nor is there an analysis without already interpreting. Such hermeneutic is basic to all understanding.

This does not mean that we cannot get sufficiently deep into the fundamental implications of his work. I write this Appendix in view of evaluating what I consider as the major ontological imperfection in Heidegger's thought from the point of view of the categorial demands of the history of ontology and scientific ontology, and of the way in which I conceive of the jolts and peaks in such history. Although Heidegger has not given a categorial scheme, he is one of the few twentieth century thinkers of ontological consequence, after Aristotle (in favour of an abstract concept of Being) and Kant (against treating the concept of Being as an attribute), to have dealt extensively with a very special concept of Being and our already interpretive ability to get at To Be. I present here in gist the difference between the *Dasein*-Interpreted concept of Being and the ontologically most widely committed, Einaic Ontological, nomic-nominal and verbal-processual concept of To Be.

Heidegger's famous life-long bemoaning of the oblivion of Being after the pre-Socratic times has done much to restore the importance of To Be in philosophy. While remaining in stunned admiration of his colossal work in favour of restoring ontology to the concept of Being, I have formulated in the work above what I find to be a more adequate and realistic concept of To Be and point out what Heidegger's concept of Being has fallen short of. It is universally admitted that Heidegger's Being is inherently phenomenal, showing itself from, all beings, but it is also verbal in the sense that it represents what is universal of all showings. The strong characterization here is that it is Being only for *Dasein* insofar as *Dasein* receives it as showing and adumbrating itself, opening and closing itself.

This, I believe, is a defective Being, since it does not admit of ontological commitment of any general sort other than the ek-sistence of human as *Dasein*. So, my basic questions to Heidegger are: Could you not accept a Reality, of which alone you could speak of an *a priori* nomic-nominal verbal-processual To Be, in a fully and most widely ontologically committing ontology? Could you not conceive of a To Be, which is not a reifying or absolutising universal?

Aristotle's universals and his Being are purely nominal and abstract absolutes. Contemporary analytic and continental thinkers have been scrupulous against incurring the purely abstract or reifying or onto-theological concept of Being of traditional metaphysics, on the way ostracising both the nominal and the verbal universality of the ontological universal, To Be. Now, it is a sign of unjustified philosophical partiality of thinking in Heidegger, if he accepts only the verbal universality of To Be, but not its nominal universality. Preference for the verbal meaning demonstrates the fear towards the nominal meaning thought to have been overthrown forever. I feel that this has been from trepidation for metaphysics and the consequent distaste for (to some extent, nervousness about) universals. This fear combined with the phenomenalism inherited from Kant and partially from phenomenology has made Heidegger create an overly anthropically verbal concept of Being, which gives itself only to *Dasein*, enowns only *Dasein*, and allows only *Dasein* to be thrown off from the Being-thinking ideal. In Einaic Ontology, this concept of Being is an exclusively connotative one.

First of all, Einaic Ontology does not speak of things any more than to mean processes. Processes are in extension-motion never fully identical to themselves as ontologically circumscribed. Secondly, every process, by its extension-motion expanse proper, is connected with a finite number of other processes and, by its causally extension-motion roots, connected to a finite number of processes in its causal past – be they based in this world, in the different actual worlds or in the infinitely extension-motion level active Divine. This ontological connection of every entity / process to an infinite number of entities / processes is ontologically epitomized by our epistemologically abstract, but ontologically relational and objectually trans-consciousness, entities active in processes and groups of processes, namely, ways of being of beings, called also ontological universals. To Be is the fundamentally causal-processual and relational universality of ontological universals / essences in Reality. Ontological universals are ways of beings, and To Be is *the* universalized Way of being of Reality. The conceptual existence of To Be in consciousness is what Chapter 2 in the present book has called the connotative ideal, Reality-in-general.

Moreover, ontological universals and To Be are never fully (absolutely) circumscribed from within the realities-in-particular that they relate processually, because these universals constantly point beyond themselves by connecting themselves to beings (processes) beyond themselves. Moreover, although To Be is ontologically fully circumscribed in Reality, To Be is in fact the dimension of processual connexity within Reality, and so, it implies reality-in-particular in all its infinitesimal complexity and variegation, which means that To Be is also not connotatively fully circumscribable under Reality, when Reality is considered merely in its totality. To that extent, ontological universals and To Be are naturally probabilistic.

In short, the ways-of-being interpretation of universals and To Be, embraced in Einaic Ontology, moves with universals as dimensions of connexity, not merely as ontologically reified and circumscribed relationalities, nor as existent processual entities. This is what is meant by saying that ways are objectually and causally based in actual processes of reality-in-particular and/or Reality, and are at the same time verbal (processual), thus integrating fluency and staticity in the highest Transcendental, To Be. Einaic Ontology is capable of demonstrating the merit of the Einaic concept of To Be over the absolute universalism of the Aristotelian-metaphysical universal Being. It is capable of juxtaposing itself against the Heideggerian anthropologically aletheial, presencing, enowning (which I would call otherwise as "en-proper-ing") and projecting-open (i.e. throwing *Dasein* off and within) concept of Being that belongs not to Reality but to *Dasein*'s relationship to Being in a question-begging manner – which Being, in turn, is recognized as Being only insofar as it belongs to *Dasein*'s ways of thinking Being and being.

To illustrate and bear out the latter part of this claim in context, I make first of all a short study of the ambivalence between *Dasein* and Reality in Heidegger's anthropologisation of the concept of Being, which should otherwise have been conceived as fully Reality-imbued (i.e. accepted by Reality in its totality, and not in its parts like the Being-thinking human) in ontological commitment to Reality, because I believe that having to make the Transcendental Being accepted only by *Dasein* shows commitment to the Kantian-Husserlian hegemony. I proceed then to show that Einaic Ontology has synthesised nominal universals / essences and the universal concept of Being in metaphysics with the processual-verbal universals, to be's and To Be of ontology, within the To Be of Einaic and truth-probabilistic ontology. If such can be presumed, it brings into question Heidegger's phenomenalistic and phenomenological prejudice that (1) has made everything subject to *Dasein*'s ontologically *conceptual* processes, (2) by doing away with the need to theoretically admit and to theorize about the *Dasein*-independent and trans-consciousness existence of and the need of ontological commitment to reality-in-particular and Reality, (3) which has given rise to the merely anthropically responsible verbal concept of Being in Heidegger.

To begin with, I cite a representative passage from *Being and Time* (earlier Heidegger) and thus bring to focus the above-mentioned philosophical deficiency:

> Descartes has narrowed down the question of the world to that of Things of Nature [*Naturdinglichkeit*] as those entities within-the-world which are proximally accessible. He has confirmed the opinion that to *know* an entity in what is supposedly the most rigorous ontical manner is our only possible access to the primary Being of the entity which such knowledge reveals. But at the same time we must have the insight to see that in principle the 'roundings-out' of the Thing-ontology also operate on the same dogmatic basis as that which Descartes has adopted.
>
> We have already intimated [...] that passing over the world and those entities which we proximally encounter is not accidental, not an oversight which it would be simple to correct, but that it is grounded in a kind of Being which belongs essentially to Dasein itself.[460]

On his way to delving into Being beyond round-out thing-ontology, he says that the conscious encounter with entities is not an oversight, and so he concludes that it is grounded in a kind of Being, which belongs essentially to *Dasein*. Why could Being – in fact the To Be – not be based ontologically on the broader Reality, and only epistemologically and anthropologically on *Dasein*, so that the Being-thinking human is not positable as the guardian of the To Be of Reality? The sort of this "being grounded", the sort of this "Being" (as given in conscious *Dasein*) and the sort of "belonging of Being to *Dasein*" are not paid attention to – as in the concept of Reality-in-general, where To Be does not belong but is only connotatively conceived – and this inattention has resulted from the muddling regarding the concepts of universals and To Be, i.e. from the lack of distinction between ontological and connotative universals. Heidegger forgets that, after the failure of Kant's phenomenalism and his refusal to take Being as a predicate, we have taken to be and To Be as *a priori* to every other *a priori* universal, because theoretically the latter sort of *a priori* are founded on the former by ontological commitment. So, ontological commitment to Being should be to something ontologically most Transcendentally (at the layer of the To Be of Reality) *a priori* to the meagre and anthropic transcendental *a priori* of the *Dasein*-relatedness of beings.

---

460   Heidegger, *Being and Time*, 133 (100). Italics of the German are mine.

For Einaic Ontology, *Dasein*'s sort of Being is at the most Reality-in-general, which is the connotative representation of To Be in consciousness. To Be as such is the ontological Way of being of Reality. Earlier Heidegger calls it "presencing", unaware that it is the presencing of Reality occurring only in *a priori* ontological commitment; and, being overly subjected to the Kantian anti-metaphysics of cognitive exclusiveness of hold over the phenomenal, he speaks of Being as what is (epistemically and consciously) given, presenced and revealed in *Dasein*, by reason of *Dasein*'s being unto Being in thinking, or even in all acts of consciousness.

Without settling this issue of unwarrantedly having to fill the place of the ontological Transcendental with the epistemological Transcendental – which, in my opinion, is that of inattention to the ontologically cosmological totality of Reality and of inattention to the reality of ontological commitment to the nomic-nominal and verbal-processual universal aspects of ontological universals – Heidegger plunges himself into his account of Being insofar as it belongs to presencing in and enowning of *Dasein*. This seems to be under the influence of the Kantian allegation of primacy to consciousness-related phenomena without ontological commitment in the acts of consciousness and of the Kantian allegation of inability to mind to approach noumena. In fact, Heidegger's Being is Reality-in-general, i.e. the *connotative concept* of To Be as is ideally given in the consciousness and life of *Dasein*. Insofar as it is given in *Dasein*, it is phenomenal, and not universal of the phenomena-noumena continuity that we have called Reality. This is the gist of the ontological paradox that he has incurred in the process of anthropologisation of To Be – a grave failure indeed, in my opinion.

Unwarranted conviction in this phenomenalistically influenced anthropologisation (even in later Heidegger's Being which enowns *Dasein* and projects-open *Dasein*) and epistemic verbalization of what our present work understands as To Be (i.e. earlier Heidegger's Being giving-itself to and distancing-itself from *Dasein*, whereby Being is taken as purely verbal so to say of the "being" of *Seinsdenken*) have enabled Heidegger to come down heavily upon the metaphysical, thing-ontological, onto-theological, concept of Being. Criticism of the traditional metaphysical concept of Being should not have been carried out merely by means of the said anthropologisation and epistemologisation of Being. This is because his inability to see the inevitable nomic-nominal aspect of his verbal Being (in fact, the ontologically cosmological verbal-processual aspect of To Be) has resulted in an anthropic Being; and because, without the nomic-nominal To Be, the verbal aspect of To Be should have meant purely the fluent aspect of

Reality, but not merely the *Dasein*-based showing-itself, closing-itself, enowning and projecting-open, of Being.

Commenting on the first part of the afore-given passage, Kovacs says:

> Cartesian ontology considers Being as a being; it reduces Being to a being and leaves unasked the question regarding the 'beingness' (*Seiendheit*) of a being (and of beings in general). This ontology is the source of interpreting God as the most perfect being. The ontological problematic is projected into the notion (and problem) of God. The Cartesian idea of the 'World' results from conceiving Being as something static, self-sufficient, and permanent (eternal). Descartes consolidates the following opinion of traditional ontology: The ontical knowledge (*Erkennen*) of a (particular) being constitutes at the same time the access to the (more primary) Being (the "to-be") of this same being discovered in the ontical knowledge. This means that, in his view, Being is described through the attributes of the particular being(s) (things, mere objects). This ontological presupposition indicates that Descartes is unable to grasp beings like There-being; he is concerned with the qualities of things (of nature).[461]

While attempting to create a concept of Being different from that of Descartes, Heidegger does not transform Descartes' concept of beings and Reality as such into something more fluent and does not – as he should have at the same time –admit the staticity that Beings-in-total partially possess in complementarity to their fluency. This lack makes Heidegger too handicapped to work a concept of the To Be of Beings-in-total. In short, Heidegger does not realize that To Be encapsulates the staticity and fluency of beings-in-particular in their totality as Reality, seen purely ontologically, cosmologically and epistemologically. Instead, he keeps on with interpreting Being as not equivalent to becoming, seeming, thinking and ought,[462] and goes on to regard *Dasein* as the criterion for the meaning of Being as different from the above four: "Being is the fundamental happening, the only ground upon which historical Dasein is granted in the midst of beings that are opened up as a whole."[463] Historical *Dasein*'s grounding is connotative. But the ontological grounding by which Reality is grounded in To Be is forgotten. So, more than the difference of the four concepts with Being, what was needed was to first conceive To Be as encapsulating and subsuming both staticity and fluency and also the remaining four allied realities that he mentions.

Thus, the statement quoted above from Kovacs summarizing the Cartesian admittance of the traditional view, i.e. that "[t]he ontical knowledge (*Erkennen*) of a (particular) being constitutes at the same time the *access to the (more primary)*

---

461 Kovacs, *The Question of God in Heidegger's Phenomenology*, 67.
462 Heidegger, *Introduction to Metaphysics*, 98ff (Chapter 4).
463 Heidegger, *Introduction to Metaphysics*, 215–16.

Being (the "to-be") of this same being discovered in the ontical knowledge" [italics mine], may be attributed to Heidegger too. The anthropic particularism of the ontic being of *Dasein* and of the concept of *Dasein*-based Being in Heidegger allows him to consider *Being as belonging to (as given in) a being – not to the Being of God, but to the Being of Dasein*. This makes him to conclude that *Dasein* alone is in the sense of be-ing, i.e., ek-sists ("stands out") into Being; and Reality simply is, without any "existential" connection with Being! The specialized meaning of "existence" steals into equivalence with Being.

Such pathways of argumentation pave the way for Heidegger to unconsciously allow what we consider as the connotative Reality-in-general as his generalized verbal Being that "enowns" and "projects-open". From then on, it was easy for him to smoothly move off to a *Dasein*-coloured Being, under its *Dasein*-centred-verbal (processual with respect to its giving-itself in *Dasein*, i.e. going on giving itself to *Dasein*) nature, with the support of his Kantian (phenomenalistic) and Husserlian (phenomenological) heritage and in the absence of a taste for philosophical cosmology and a distinguished ontological commitment to the nomic, nominal, verbal and processual "to be" of beings and To Be of Reality. In the afore-mentioned passage from Heidegger, it seems to me that he is citing the fact that (human, conscious) passing over the world and entities is not accidental as the reason why Being belongs to *Dasein*. Everywhere in his *Being and Time*, especially at its beginning, we find this sort of jugglery. The nature of this belonging of Being to *Dasein* is, in my opinion, that of the connotative Reality-in-general, but mistaken as the ontological To Be.

The primordial themes in Heidegger are: (1) Thinking, (2) Being and (3) correlation of Being and thought, as suggested by the terms 'appropriation' and 'mittence'. The Thinking-Being contrast in Heidegger is in fact that between thought and actuality-as-thought – (1) thought in the subjective-objective, non-calculative-technical (the response to the call of Being, i.e. *das besinnliche Denken*) and non-representational sense, and (2) actuality-as-thought within the verbal (for him, in the restricted sense of *Dasein*'s – not of all processes as in Einaic Ontology – processual involvement in Being via non-objectification of beings) sense of the involvement of the objective in the process of the subjective. It is noteworthy that Heidegger does not take Being as the universal verbal-processual aspect of actuality in general, nor does he take it simply metaphysically as the universal with any extra-mental existence. The moment the word 'universal' is uttered, it tastes totalitarian metaphysics for him. He does not realize the first-order, second-order and superior-order ontological, logical and semantic necessity of ontological talk in terms of universals. Instead of making Being "abstract" as in essentialist metaphysics, and instead of reifying Being as Reality, he opts for taking Being as the subjectively objective generality of the processual giving itself of Being to *Dasein*, which I have further generalized cosmologically and ontologically and called Reality-in-general. "The experience of multidimensionality of thinking is a preparation for the discovery of thinking as an event and appropriation of Being (*Ereignis des Seins*), because thinking is the 'thinking of Being.'"[464]

To illustrate this I quote from one of Heidegger's properly metaphysical writings: "For Being, which already predetermines what can be represented, is in the first instance *hypokeímenon*, and in the second instance objectivity which is grounded in a *subiectum*, but in a *subiectum* whose essence is not identical with that of *hypokeímenon*."[465] *This makes it clear that what I have proposed as Reality-in-general, rather than Reality or To Be, comes closer to Heidegger's concept of Being.* The above sort of thinking espoused phenomenology to overcome the metaphysics of equivocation of Being with the Ultimate Being as follows:

> The god who acts here as ground is not thought theologically, but purely ontologically, namely as the highest being in whom all beings and Being itself are caused [....] With this determination of the *facere*, Being's character of production appears in the sense that Being itself is made and effected by a being [....] But within the causal nature of beingness permeating metaphysics everywhere in the most various forms, the exigent nature of Being still becomes determinative in the developed beginning of modern metaphysics. The eminence of the *exigere*, however, does not relinquish the representational character of Being; for this character preserves the tradition of the beginning and primal essence of Being which becomes evident as presencing.[466]

Note that presencing has a processual-subjective sense here. This sense is most reflected in Reality-in-general, not in To Be. The metaphysical concept of *hypokeímenon* is from the verb *hypokeímai*, "to be established", "set before". In the sense of existence, *to ektòs hypokeímenon* is "the external *reality*".

We know that what is meant by Being in Heidegger is not the universal, but the very actuality of things as appropriated at the Being-level, i.e. at the interface of *Dasein*'s thinking and the "giving" of Being. The Being of the actual is nothing but the actuality of the actual in its Enowning of the thusness of things and actualities in the subjective-objective manner. Since Reality-in-general is at the side of the subjective *Dasein*-aspect, I hold that this presupposes the objectual Reality

---

464 Kovacs, *The Question of God in Heidegger's Phenomenology*, 4.
465 Heidegger, *The End of Philosophy*, 46.
466 Heidegger, *The End of Philosophy*, 44–45.

at the side of actuality of the objective Enowning-aspect. *But the dimension of the highest ontological verbal-processual universal To Be, as also the ontological nomic-nominal abstract object or universal involved from the part of Reality, is not admitted in Heidegger.* The Being of Heidegger is, then, the processual actuality-as-such of reality-in-particular accounted for by the Interpretative *Dasein* as our Reality-in-general. Insofar as Heidegger does not at all posit ontological universals, and due to the consequent absence of the ontological universal – To Be – in Heidegger's thought, his Being is very much particularistic from the Einaic Ontological point of view.

Here follows yet another point that Heidegger does not see: Though human epistemological inabilities and *Dasein*'s contextual existence do make our judgments constructed on concepts truth-probabilistic, this fact is not the whole story of the truth-probabilism we are faced with. For that matter, even connotative universals active in concepts, including Reality-in-general, when expressed in judgments, yield only truth-probabilities, though they are based on real ways of processes. It is also primarily a truth-probability that characterises the very judgmental expression of the nature of processes as never fully thing-ontologically themselves within their own factic extension-motion nature. A general statement based on the ontological commitment to the existence of "something" alone is absolutely beyond truth-probabilism regarding the dimensional fact without the knowing mind. But formulations of such facts are always truth-probabilistic. All these mean that each process is connected unto infinite others, which are finite in their connections with their contemporary world and probably infinite in its causal past connections, thus making up a uncircumscribable probable infinity of causal influences. This is the ontological aspect of the truth-probabilism (based on the processually objectual-causal nature of content) concerning statements regarding entities / processes within the context of ontological universals and To Be. Heidegger's not envisioning this ontological (not epistemological) aspect of truth-probability in his thought has motivated some of my disparaging opinions about Heidegger's epistemologization and anthropologisation of To Be.

Later Heidegger is partially aware that To Be belongs to Reality, not merely to *Dasein*. Yet, he is unable to make To Be co-extensive with Reality, which would involve taking To Be as both (1) the ontologically nominal and nomic universal thusness already in the static dimension of Reality *and* (2) the ontologically (not merely in an alleged *Dasein*-centredness) truth-probabilistic verbal and fluent dimensional essence, which is the thusness-process of relationality in Reality. Instead, Heidegger mistakes the conscious, connotative, manner of appropriation of To Be by *Dasein* for the ontological appropriation of To Be. This, I would

345

submit, is crass anthropologisation of To Be. The question of the beingness of Reality as the highest truth-probabilistic universal To Be is complementary to that of the ontologically processual-verbal truth-probabilistic nature of To Be. Only both the characteristics together are capable of ontologically encompassing Reality. So, the question of To Be is to be treated as an objectual-causal-probabilistic and cosmologically ontological question at the level of Reality, not merely at the level of the connotative Reality-in-general as appropriated in *Dasein*. Appropriating To Be at the level of Reality-in-general in consciousness, in actions etc. is not the highest ontological way. The highest ontological way is the Way of To Be in Reality, beyond a phenomenalistic involvement of human consciousness.

Therefore, the genuinely verbal sense of To Be is that of the Greek *phúein*, "grow" and Sanskrit *bhū* "be", both of which derive from an original Indo-European stem. Evidently, it is not merely "emergence" into and "concealment" from *Dasein*, nor is it "emergence" and "concealment" as *Dasein* envisions in Being-thinking, as Heidegger claims it to be.[467] Moreover, the sense of To Be cannot be derived merely etymologically from the ancients. Instead, *To Be is the widest "thusness" of Reality given in the widest possible ontological commitment, which is the admittance of processes and ways of being of processes beyond human*. In earlier Heidegger "emergence" is overly anthropological, and in later Heidegger, it is very much ontological. Along with admitting this improvement in Heidegger, I hold that even the concepts of "emergence / uncovering" (*Alétheia*), which is used in earlier Heidegger, and "Enowning" (*Ereignis* as "letting *Dasein* be within Being") and "projecting open of being / projecting being open" (*Ereignis* as throwing *Dasein* off Being / *Entwurf des Seins*)[468] that are used throughout later Heidegger, especially in his *Contributions to Philosophy (From Enowning)* to encapsulate the verbal aspect of To Be, are insufficient to include the universal, nomic (acausal), objectual-causal aspects.

The reason for this is that the verbal meaning of Being in Heidegger is that of the processual nature of Being -giving-itself and -enowning-*Dasein* in *Dasein*'s being and consciousness, but without the rightful ontological commitments to the *Dasein*-independence of Reality, reality-in-particular, ontological universals and To Be, and with the ontological commitment to connotative universals and Reality-in-general. But ontology with the extent of ontological commitments Heidegger unconsciously admits is no ontology. *The Einaic Ontological To Be is the whole staticity and fluency of Reality, emerging co-extensively at the level of*

---

467 Heidegger, *Introduction to Metaphysics*, 15.
468 Emad and Maly, "Translators' Foreword", xxix.

Reality, and theoretically given in human only as apriorily given in Einaic ontological commitment. It cannot anyway be reduced into what shows itself in *Dasein* at the level of Reality-in-general, which is still the concept of Being even in later Heidegger's *Contributions to Philosophy*. It is the same phenomenalistically and phenomenologically anthropological reduction of Being of earlier Heidegger that has made the later Heideggerian concept of the "Emerging-Enowning" Being ontologically insufficient. Hence, the verbal meaning he assigns to Being is anthropic, not ontological. Later Heidegger has never sufficiently freed himself from the clutches of the earlier's anthropic Being. The Einaic Ontological To Be is not a mere metaphysical universal that is set forever. It is the ever-emerging *phúein*, thusness, of Reality, given to all processes and to human by means of the ever-broadening nature of ontological universals of reality-in-particular, and which is connotatively idealizable in consciousness. Hence, the aspect of idealization in consciousness should also, though vaguely, reflect the concept of To Be. This possibility may be built up in an ontological system only via the concepts of connotative universals and Reality-in-general.

The root cause of inefficiency of the concept of Being in Heidegger may thus be pointed out as his anthropologised concept of Being and lack of distinction between ontological and connotative universals. At least the structure of some of the unthinkable amount of unconscious shifts and sleights of meaning in Heidegger would thus have to be brought out and shown to be based on his anthropologisation of To Be – enough matter for book-length studies. He says, "'*There is*' truth only in so far as *Dasein is[,] and so long as Dasein is*. Entities are uncovered only *when* Dasein *is*; and only as long as Dasein *is*, are they disclosed."[469] Here by the onslaught of concern for truth the ontologically committed thusness of Reality is forgotten. If truth is *Alétheia* ("Uncovering"), it is of Reality via its To Be. This is the totalized ontological commitment to To Be that Heidegger has forgotten due to the over-insistence on *Dasein*'s place in truth.

Uncovering (*Alétheia*) results in Reality-in-general. Hence, "Uncovering" as an anthropic and connotatively verbal-processual notion is not To Be. No one controverts that truth as humans have in phenomena has the element of truth anthropically, i.e. in terms of humans' constructive positing. But it is the Kantian phenomenal seeing-as, divorced from seeing and seeing-that, i.e., without adequating seeing-as by seeing and seeing-that. In the present book, under 1.1.2. in general and 1.1.2.2. in particular, I have demonstrated the partiality in taking seeing-as as equivalent to the wholeness of seeing, seeing-that and

---

469   Heidegger, *Being and Time*, 269 (226).

seeing-as. Strictly, the disclosure Heidegger speaks of (which he wants to take as truth for *Dasein*) is part of the ontological showing itself of Reality in its To Be – i.e., only part of the showing itself of ontological affairs in their ontological (cosmologically objectual-causal) rootedness in the To Be of Reality. That is, purely ontological showing is absolutely cosmically ontological, not given in *Dasein* alone, due to the ontological independence of the To Be of Reality from *Dasein*.

His example for his concept of truth makes clear that for him truth and Being are purely anthropic. I quote: "Newton's laws, the principle of contradiction, any truth whatever – these are true only as long as Dasein *is*."[470] Note that even the principle of contradiction, a methodological category, is included in the list, without reference to ontological commitment. Here truth means 'truth as given in *Dasein*, the Being-thinking human, without the *Dasein*-independent objectual ontological commitment involved in the showing-itself of Reality'. If one were to argue that even what is given in ontological commitment is as it is given in *Dasein*, it is nothing but sophism. Einaic Ontology brings out the truth-probabilistically foundational dimension in discourse, by unearthing the dimensions of ontological foundation in the three categories, which are taken to be relatively more *Dasein*-independent than any other notion.

The absence of these dimensions in Heidegger is a proof of his ambivalence between the *Dasein*-independent and *Dasein*-dependent concepts of To Be, where *Dasein* is the Being-thinking human who is the mediator and that-to-which of the giving of Being. In the passage quoted above, he says also of the trans-*Dasein* quality of what he calls truth. Note the difference between his *Dasein*-specific truth (similar to the Kantian seeing-as) as divorced from ontological commitments (which are via seeing and seeing-that) and our asymptotic approach to truth-probabilities based on ontological commitment to the objectual-causal Reality, To Be, ontological universals, Reality-in-general, connotative universals and token entities:

> Before there was any Dasein, there was no truth; nor will there be any after Dasein is no more. For in such a case truth as disclosedness, uncovering, and uncoveredness, *cannot* be. Before Newton's laws were discovered, they were not 'true'; it does not follow that they were false, or even that they would become false if ontically no discoveredness were any longer possible. Just as little does this 'restriction' imply that the Being-true of 'truths' has in any way been diminished.
>
> To say that before Newton his laws were neither true nor false, cannot signify that before him there were no such entities as have been uncovered and pointed out by those

---

470   Heidegger, *Being and Time*, 269 (226).

laws. Through Newton the laws became true; and with them, entities became accessible in themselves to Dasein. Once entities have been uncovered, they show themselves precisely as entities which beforehand already were. Such uncovering is the kind of Being which belongs to 'truth'.[471]

This kind of Being is the Being of uncovering, based completely on the agent of uncovering, not on Reality. If all Being in an ontology were that of Uncovering, its Being does disown the ontological commitment that is theoretically *a priori* (not merely temporally previous) to the existence of the agent, the process of uncovering, and the showing-itself, Enowning and throwing off of Being with respect to *Dasein*. For any basic ontological commitment, Heidegger lacks the essential ontological concern for Reality, to which To Be properly belongs. This is a major ontological defect in Heidegger.

In the quote given above too, he was unable to transcend the anthropological understanding of To Be, since the very meaning of 'truth' is defined in a purely ontic and anthropocentric manner: as we uncover, so is it uncovered. The Platonic paradox of trans-conscious existence of universals is brushed aside, without attempting to quell the real problem. Instead, he merely equates 'the Being of *Dasein*' (muddled as purely the connotative Being in the probabilistic seeing-as of *Dasein*, beyond which the concept of Being should have been derived by ontological commitment to Reality) with truth: *"Because the kind of Being that is essential to truth is of the character of Dasein, all truth is relative to Dasein's Being."*[472] How bold of him to say so, without ever respecting the theoretical necessity of apriorily positing the being-in-itself of beings and Reality, without realizing that even *Dasein*-centred Being is theoretically a priori to perception. Thus, in Heidegger 'the Being of *Dasein*' is not in fact the same as To Be. Earlier Heidegger has no qualms in identifying Being with the Being of *Dasein*, in the sense of the Being as disclosed in *Dasein*, insofar as it is disclosed by *Dasein* – due to which later Heidegger could not escape from the clutches of later Heidegger. Is this not anthropologisation of Being directly from the inheritance of Kantian phenomenal seeing-as?

The later Heideggerian concepts of Being, *Ereignis* as "Enowning-Eventing-Emerging of Being" and *Ereignis* as "projecting open of Being" resist being discussed at the level of Reality if the context is of so anthropologised Being and if the aspect of idealization of ways of being as nomic-nominal universals does not give rise to idealization of the Way of the To Be of Reality. Enowning should have

---

471 Heidegger, *Being and Time*, 269 (226–27).
472 Heidegger, *Being and Time*, 270 (227).

a meaning only within *Reality, which does the Enowning and the Projecting-open of the ontologically truth-probabilistic Transcendental To Be* that works as nomic, nominal, verbal and processual of Reality. Reality is idealized in the cognizing mind, and that is through a truth-probabilistic (but connotative) universal. What is thus probabilistically carried over for idealization in consciousness is not To Be. To Be is in Reality, and absolutely so. Unluckily, such idealization, which allows To Be to be in Reality, and allows Reality to Enown and Project-open humans and beings, is absolutely absent in later Heidegger. Thus, the following passage will have to be re-interpreted:

> Enthinking is not thinking-out and haphazard invention but rather [is] that thinking that through questioning places itself before be-ing and demands of be-ing that it attune the questioning, all the way through. But in enthinking of be-ing, beings in the whole must be put up for decision every time. In each case this succeeds only in *one* purview and turns out to be all the more needy, the more originarily the hinting of be-ing strikes this thinking. The territory that comes to be through and as the way of enthinking of be-ing is the *between* [*Zwischen*] that *en-owns* Da-sein to god; and in this enownment man and god first become 'recognizable' to each other, belonging to the guardianship and needfulness of be-ing.[473]

The belonging together of Reality under the purview of its To Be is not what is made clear in the statement: "[…] in enthinking of be-ing, beings in the whole must be put up for decision every time." *It simply says that there are beings, and wholly and totally so. This, and this alone, is what easily, in the next possible step of the widest ontological commitment, would have prepared Heidegger to admit a more adequate concept of To Be. But, unluckily, he does not make Being ontologically ownmost to Reality.* This does point to the General Ontological merit of later Heidegger, but it has no Einaiological adequacy to Reality. If so, somehow his concept of Being has to be made adequate to Reality, not merely to the Enowning and Projecting-open of Dasein, beings and god (which is not merely God) by Being in Being-thinking Dasein.

That is, after earlier Heidegger's boldly arguing to the effect that the truth of Being is the truth of uncovering of Being by *Dasein*, later Heidegger does also hold that Reality is beyond *Dasein* and that it is that of all beings, but does not hold that its To Be is also beyond *Dasein*. *This, I believe, is the exact difference and connection between earlier and later Heidegger's concepts of Being.*

To put things in brief: Heidegger is unable to claim ontologically committed processual reality to the To Be of Reality, since he has not created a concept of

---

473   Heidegger, *Contributions to Philosophy*, 60.

Dasein-independent Reality, to which To Be could belong. Instead, he has phenomenalistically made To Be to rest on *Dasein*. This debilitates his ontology to grasp the trans-*Dasein* meaning of To Be as ontologically nomic, nominal, verbal and processual. The one desideratum, very much lacking in the statement above from Heidegger, is the ontologically unavoidable admittance of the fact of the nomic and nominal ideality of the To Be of Reality – which is ontologically independent of *Dasein* – in *Dasein*'s thinking of the *Dasein*-independent Reality – where *To Be and Reality, as given in Dasein, are merely Reality-in-general*, by reason of the distinction between connotative and ontological universals in Einaic Ontology.

To say that all that I have said about Heidegger's concept of Being is nonsense, and that an interpretation of Heidegger requires interpreting him merely from his own context, is to say that he is immune to being scrutinized by alternative visions of To Be, Reality and Thought. Insulating Heidegger within his own framework yields a decidedly highly inequitable and biased evaluation resulting in maintenance of all the muddling present in his conceptual foundations and axioms (or lack of conceptual foundations and axioms). I have been questioning Heidegger's foundations (and lack of foundations), and not directly the results of his foundations. Fell a tree by its roots, and you have many more trees with deeper roots – which too will be felled in the course of history, guaranteeing the birth of more adequate systems.

In the absence, in the following passage (and everywhere else) from *Contributions to Philosophy*, of the possibility of nomic-nominal-verbal-processual idealization by *Dasein* of the concept of To Be, the effect of muddling will be evident in the way Heidegger connects be-ing, as *Dasein*'s (ontic) uncovering, with the truth of be-ing:

> It is necessary here perhaps to say, even somewhat extensively, what is *not* meant with the words *truth of be-ing*. The expression does not mean 'truth' 'about' be-ing, as if it were the conclusion of correct propositions about the concept of be-ing or were an irrefutable 'doctrine' of be-ing. Even if such would be appropriate for be-ing (which is impossible), one would have to presuppose, not only *that* there is a 'truth' about be-ing, but above all of what kind that truth really is, the truth in which be-ing comes to stand. But from where else should what is ownmost to *this* truth and thus to truth as such be determined, except from be-ing itself? And that not only in the sense of a 'derivation' from be-ing, but in the sense of effecting this 'ownmost' by be-ing – such an effecting in terms of which we cannot access be-ing through any 'correct' notions but rather one that belongs solely to the sheltered moments of being-history. But the expression also does not mean 'true' be-ing, as in the unclear meaning of 'true' beings in the sense of true or actual. For here once again a concept of 'actuality' is presupposed and laid at the foundation of be-ing as

a measure, whereas be-ing not only grants to beings what they are but also and primarily unfolds for itself that truth that is appropriate for what is ownmost to be-ing."[474]

If the truth of Being is not any of the former – which are all metaphysical – then what does "the truth in which be-ing comes to stand" mean? This difficulty is clear in his statement above: "[...] from where else should what is ownmost to *this* truth and thus to truth as such be determined, except from be-ing itself?" The truth in which Being comes to stand is not merely *this* truth, but also truth as such. Concepts of truth as such were to arise only in absolutistic metaphysics. To the extent that truth as such is for Heidegger the truth of Being-as-such, it is somehow the highest truth thinkable, but only as showing forth (and Enowning, Projecting-open etc.) and thought-within by *Dasein*. Now, if 'truth' is 'truth as is for *Dasein*', then the 'truth of Being' should also be 'truth of Being as is for *Dasein*', which should be 'truth of truth of Being as is for *Dasein*', and so on *ad libitum*. The infinite regresses that one fights against in reifying metaphysics re-appear in such ways in phenomenalistic ontologies too, which point to their own sophist tendency. This argument of mine would not be sophism, but the result of a free extension of his hermeneutic and phenomenological ontology, which sprang up as slightly phenomenalistically affected responses to Kantian phenomenalism. In short, phenomenalism and phenomenology, devoid of ontological commitment to all necessary types of it, has indirectly shaped Heidegger's concept of Being.

We shall discuss also the categorial and axiomatic status of Heidegger's concepts, in order to bring out the categorially agile nature of Reality. Reality is what is active, and To Be is the processual-relational reason based on the process of Reality-in-total, for such agility. From this point of view too Heidegger is deficient in a cosmologically ontological notion of Reality. Let me explain: It is a fact that both the earlier and later phases of Heidegger consider Being as superior to *Dasein* in content and in Being's "ontologically" more fundamental status. The passage quoted above from *Contributions* is evidence. Whatever the actual reasons for this in earlier and especially in later Heidegger, the implied reason may be found to be that *Being (as To Be) is most extensive*, both as an ontological and *Dasein*-based being-historical fact in processes and as a conceptual / connotative relational entity that is active in the dynamic aspect of the connection between processes (for us, reality-in-particular and Reality) and *Dasein*. It is also a fact that there are conceptual gradational differences in concepts in Heidegger. For example, "Care" is more fundamental than "ready-to-hand". One concept is more fundamental insofar as it is more general with respect to relevance to entities and

---

474 Heidegger, *Contributions to Philosophy*, 64.

thought. Ultimately, the differences of importance pivot around extensions and intensions of terms, not merely on the latter.

We do not have a term with extension alone or intension alone. But we do not find Heidegger speaking of the extension of terms and concepts, nor of the extension-motion expanse that Reality is in, in its physical staticity and fluency. Extension and intension of terms lead us to the relative importance of terms in thinking. Terms may be more relevant to the whole, i.e., they are categorial. Therefore, the mode of analysis in Heidegger is still based on categorial thinking and is at the same time deeply infused in the nature of Being-thinking.

Hence, we should also venture asking if Heidegger's Being permits any more extensively intensive terms, and we find none but Reality as its exactly co-extensive one and Reality-in-general as its exactly co-intensive one. But if this question were interwoven with that of the possible agency of the acts of *Alétheia*, "Uncovering", and *Ereignis*, "Enowning, Throwing-open etc." by To Be, we would have to look for the agent coextensive with To Be (Reality), and not for anything co-intensive. Moreover, To Be is the verbal Transcendental of Reality. It is not any thing. The modes of "doing" by To Be and by *Dasein* differ. This is because *Dasein* is an agent and To Be is not one. Even so, the so-called "acts" (Enowning, Throwing-open etc.) of To Be shall have an agency, and this agent, we find, is Reality, and not *Dasein*. This argument is not so simple as might be pointed out by some Heideggerians. I believe that verbal-nominal universals are ways of processes nominalized, and so, To Be is the verbal-nominal of Reality.

Reality is the Transcendent agent *par excellence*, because that is where all agents of process at the level of the verbal To Be totalise unto, and beyond which there can be nothing, no process. Insofar as Reality is the highest Transcendent, its To Be that is the Transcendental co-intensive with the connotative (Reality-in-general) belonging to the co-extensive Reality can only be a Transcendental. It is so, also for the reason that anything of the order other than "*some* thing" has to be a transcendental, and To Be is the highest at that. As a name, it is a conceptual entity and its ontological ingredient (that "collects" all relations and generalizes) is not a thing. It is the name for the concept of the highest relational-processual To Be of the static-dynamic process of Reality. It is the deepest transcendental and, at its conceptualisation, it becomes Reality-in-general. Hence, we agree to call To Be as the Transcendental *par excellence* devoid of actual activity. It is the ontological relationality which is the reason for the agility of Reality. This difference between To Be and Reality is badly absent in Heidegger. He does not have the latter category in his thought. This fact, I believe, has handicapped his thought beyond repair. I believe that this is why even his later phase could not free itself from *Dasein*-based Being-thinking.

That is, earlier and later Heideggerian expressions like "Enowning by Being", "Being gives", "Being shows itself" etc. (for which no special reference needs to be given, except mentioning that almost every page of his earlier and later works, including *Contributions to Philosophy*, gives many such phrases), gives sickening glosses and blurbs of the anthropological substantivation of the To Be that Heidegger has insisted to be purely verbal and given in *Dasein*. A passive reading through his later *magnum opus*, namely, *Contributions to Philosophy* puts one at a loss as to what he means to accomplish by a sort of onto-anthropologized Being as an agent, except to promise a poetic joy and a rhetorical conquering of intellects in favour of a phenomenal giving-itself of a non-existent entity i.e. Being, to *Dasein*, without the intervention of Reality via reality-in-particular. Einaic Ontology suggests that its maximally classificational categories serve to convert To Be simultaneously into a nomic and nominal universal and a processual-verbal universal, based in Reality. This, I submit, is a more adequate idea of To Be than the anthropic Being of earlier and later Heidegger.

One important reason for my terming Heidegger's Being as anthropic / anthropological, and not merely as epistemological, is that later Heidegger's Being has attempted to transcend its consciousness-relatedness and included it in the Enowning and Projecting-open of *Dasein* by Being, which works at the level of the ontology of authentic existence of humans. The anthropological may very well be thought to include the epistemological and consciousness-related aspect of Being. Similarly, the proper nature of his concept of Being as anthropic may also be thought to include the epistemic, since he has driven the consciousness of Being-thinking humans into the Being of anything, due to his inability to escape the clutches of the Kantian phenomenality of consciousness. Thus, in view of Einaically summarizing both the earlier and the later phases of Heidegger, it is justified to term his Being as anthropic / anthropological.

The failure of Heidegger's Fundamental Ontology is due to the fact that, Einaically, he has only (1) a highly partial Einaiology that deals with Being's giving itself to *Dasein* and its Enowning-projecting-open of *Dasein* in terms of its conscious reflection as if Being were the same as Reality-in-general, and (2) a highly partial General Ontology that studies *Dasein* as Being's child in terms of its reflection of what we call Reality-in-general. We do not find Heidegger talking in terms of any cosmological concepts, at an age when astrophysical cosmology kept on astonishing humans with its discoveries and theories. We do not also find Heidegger discussing determinism in its epistemological and ontological aspects, which could be solved by a Ontological Principle of Excluded Vacuous Middle. Thus, it is clear that Heidegger does not have a relevant cosmology at all. In short, the whole cosmologically ontological dimension of Reality is absent in Heidegger. So, his Fundamental Ontology is not a scientific ontology, nor a scientifically adequate one. This, I would submit, is the reason why he could not create an ontology with an absolutely Reality-imbued and ontologically most widely committing concept of To Be that allows science to flourish in ever better ways and develops science in ways that it ought to have tread.

Bibliography

Ando, Takatura. *Metaphysics: A Critical Survey of Its Meaning*. The Hague: Martinus Nijhoff, 1974.

Antony, Louise. "Semantic Anorexia: On the Notion of "Content" in Cognitive Science" (105–135). In *Meaning and Method: Essays in Honor of Hilary Putnam*, George Boolos, Ed. Cambridge: Cambridge University Press, 1990.

Aristotle. *The Complete Works of Aristotle*, Vols. 1 and 2, Revised Oxford Translation, Jonathan Barnes, Ed. Princeton: Princeton University Press, Bollingen Series LXXI.2, 1984.

Armstrong, David M. *A Combinatorial Theory of Possibility*. Cited in William G. Lycan, "Armstrong's New Combinatorialist Theory of Modality" (3–17). In *Ontology, Causality and Mind: Essays in Honour of D. M. Armstrong*, John Bacon, Keith Campbell and Lloyd Reinhardt, Eds. Cambridge: Cambridge University Press, 1993.

–. *Nominalism and Realism, Universals and Scientific Realism*, Vol. 1. Cambridge, Cambridge University Press, 1995.

–. *What Is a Law of Nature?* Cited in Sfendoni-Mentzou, "The Reality of Thirdness" (55–95). In *Realism and Anti-realism in the Philosophy of Science, Beijing International Conference, 1992*. Robert S. Cohen, Risto Hilpinen and Qiu Renzong. Dordrecht: Kluwer Academic, 1996.

–. "Universals as Attributes" (65–91). In *Metaphysics: Contemporary Readings*. Michael J. Loux, Ed. London: Routledge, 2001.

–. *Truth and Truthmakers*. Cambridge: Cambridge University Press, 2004.

Bambrough, R. "Universals and Family Resemblances" (266–79). In *The Problem of Universals*. Andrew B. Schoedinger, Ed. New Jersey: Humanities Press, 1992.

Barrow, John D. *The World within the World*, Oxford: Oxford University Press, 1994 (reprint with corrections).

Bell, J. S. *Speakable and Unspeakable in Quantum Mechanics*. Cambridge: Cambridge University Press, 1988.

Blackburn, Simon. *Essays in Quasi-Realism*. Oxford: Oxford University Press, 1993.

Bohm, David and Basil J. Hiley, *The Undivided Universe: An Ontological Interpretation of Quantum Theory*. London: Routledge, 1993.

Bohr, Niels. *Atomic Physics and Human Knowledge*. Cited in Mario Bunge. *Treatise on Basic Philosophy, Volume 7, Part I: Formal and Physical Sciences*. Dordrecht: D. Reidel, 1985.

Boole, George. *An Investigation of the Laws of Thought on Which Are Founded the Mathematical Theories of Logic and Probabilities*. New York: Dover, Corrected Edition, 1958. First published in 1854.

Boyd, Richard. "How to Be a Moral Realist" (145–82). In *Beginning Metaphysics: An Introductory Text with Readings*, Geirsson, Heimir and Michael Losonsky, Eds. Oxford: Blackwell, 1998.

Brentano, Franz. *Psychology from the Empirical Standpoint*. Oskar Kraus and Linda L. McAlister, Eds., and Antos C. Rancurello, D. B. Terrell and Linda L. McAlister, Trans. London: Routledge, 1995.

Brown, C. "Internal Realism? Transcendental Idealism?" In *Realism and Antirealism*, P. French *et al.*, Eds. Cited in Qiu Renzong, "How to Know What Rises up Is the Moon? On the Concept of Realism and the Irrelevance of Quantum Mechanics to the Debate on Realism vs. Antirealism" (55–73). In *Realism and Anti-realism in the Philosophy of Science, Beijing International Conference, 1992*. Robert S. Cohen, Risto Hilpinen and Qiu Renzong, Eds. Dordrecht: Kluwer Academic, 1996.

Bunge, Mario. *Treatise on Basic Philosophy, Volume 7, Part I: Formal and Physical Sciences*. Dordrecht: D. Reidel, 1985.

Cambridge Dictionary of Philosophy, The, second edition. S.v. "Counterfactuals", "Formal Logic", "Haecceity", "Identity of Indiscernibles", "Intension", "Intensional Logic", "Is", "Mereology", "Operator", "Philosophy of Science", "Predicables" and "Semantic Holism".

Carl, Wolfgang. *Frege's Theory of Sense and Reference: Its Origins and Scope*. Cambridge: Cambridge University Press, 1994.

Carroll, John W. *Laws of Nature*. Cambridge: Cambridge University Press, 1995.

Charlton, William. *The Analytic Ambition: An Introduction to Philosophy*. Oxford: Blackwell, 1991.

Chellas, Brian F. *Modal Logic: An Introduction*. Cambridge: Cambridge University Press, 1995.

Chisholm, Roderick. *A Realistic Theory of Categories: An Essay on Ontology*, Cambridge: Cambridge University Press, 1996.

Churchland, Paul M. "The Ontological Status of Observables: In Praise of the Superempirical Virtues" (35–47). In *Images of Science: Essays on Realism and*

Empiricism, with a Reply from Bas C. van Fraassen, Paul M. Churchland and Clifford A. Hooker, Eds. Chicago: The University of Chicago Press, 1985.

Coughlan, G. D. and Dodd, J. E. *The Idea of Particle Physics: An Introduction for Scientists*, Second Edition. Cambridge: Cambridge University Press, 1994.

Cushing, James T. *Quantum Mechanics: Historical Contingency and the Copenhagen Hegemony*. Chicago: The University of Chicago Press, 1994.

D'Espagnat, Bernard. *Reality and the Physicist: Knowledge, Duration and the Quantum World*, J. C. Whitehouse and Bernard D'Espagnat, Trans. Cambridge: Cambridge University Press, 1990.

Dictionary of Latin and Greek Theological Terms: Drawn Principally from Protestant Scholastic Theology. Muller, Richard A., Compiler. Grand Rapids: Baker Books, 1985.

Dictionary of Philosophy, 1983 Edition. Dagobert D. Runes, Ed. S.v. "Categorial (Judgment)".

Dilworth, Craig. *The Metaphysics of Science: An Account of Modern Science in Terms of Principles, Laws and Theories*. Dordrecht: Springer, 2006.

Dingguo, Hong. "On the Neutral Status of QM in the Dispute of Realism vs. Anti-realism" (307–316). In *Realism and Anti-realism in the Philosophy of Science, Beijing International Conference, 1992*, Robert S. Cohen, Risto Hilpinen and Qiu Renzong, Eds. Dordrecht: Kluwer Academic, 1996.

Durrant, Michael. *Sortals and the Subject-Predicate Distinction*. Stephen Horton, Ed. Aldershot: Ashgate, 1991.

Edelman, Gerald M. and Giulio Tononi. *A Universe of Consciousness: How Matter Becomes Imagination*. New York: Basic Books, 2000.

Efimov, N. V. *Higher Geometry*. Moscow: Mir Publishers, 1980.

Einstein, Podolsky and Rosen, "Can Quantum-Mechanical Description of Physical Reality Be Considered Complete?" *Physical Review* 47. Cited in Murdoch, *Niels Bohr's Philosophy of Physics*. Cambridge: Cambridge University Press, 1987.

Ellis, Brian. "What Science Aims to Do" (48–74). In *Images of Science: Essays on Realism and Empiricism with a Reply from Bas C. van Fraassen*. Paul M. Churchland and Clifford A. Hooker, Eds. Chicago: University of Chicago Press, 1985.

Emad, Parvis and Kenneth Maly, "Translators' Foreword" (xv-xlv). In Martin Heidegger, *Contributions to Philosophy (From Enowning)*. Parvis Emad and

Kenneth Maly, Trans. Bloomington, Indiana University Press, 1999. (The copy consulted is a pre-print, obtained from a reviewer.)

Encyclopaedic Dictionary of Mathematics (in five volumes), Sunny Sareen. S. v. "Eigenvalue" and "Gödel's Completeness Theorem". New Delhi: Sarup & Sons, 2000.

Feibleman, James K. *Assumptions of Grand Logics*. The Hague: Martinus Nijhoff, 1979.

Feynman, Richard. *The Character of Physical Law*. Cambridge, Mass.: The MIT Press, 1998.

Feynman, Richard, Robert Leighton and Matthew Sands. *The Feynman Lectures on Physics, Vol. 3. Quantum Mechanics*. New Delhi: Narosa, 1989.

Franklin, Allan. "There Are No Antirealists in the Laboratory" (131–148). In *Realism and Anti-realism in the Philosophy of Science, Beijing International Conference, 1992*. Robert S. Cohen, Risto Hilpinen and Qiu Renzong, Eds. Dordrecht: Kluwer Academic, 1996.

Folse, Henry J. "The Bohr-Einstein Debate and the Philosophers' Debate over Realism versus Anti-realism" (289–298). In *Realism and Anti-realism in the Philosophy of Science, Beijing International Conference, 1992*. Robert S. Cohen, Risto Hilpinen and Qiu Renzong, Eds. Dordrecht: Kluwer Academic, 1996.

Frege, Gottlob. *Posthumous Writings*. Cited in Wolfgang Carl, *Frege's Theory of Sense and Reference*. Cambridge: Cambridge University Press, 1994.

Gell-Mann, Murray. *The Quark and the Jaguar: Adventures in the Simple and the Complex*. New York: W. H. Freeman, 1994.

Ginzburg, V. L., "Supplement" (317–60). In V. A. Ugarov. *Special Theory of Relativity*, Yuri Atanov, Trans. from the Russian. Moscow: Mir Publishers, 1979.

Gödel, Kurt. *Unpublished Philosophical Essays*. Francisco A. Rodríguez-Consuegra, Ed. Basel: Birkhäuser, 1995.

–. "What Is Cantor's Continuum Problem?" Cited in Penelope Maddy, *Naturalism in Mathematics*. Oxford: Clarendon Press, 2000.

Goodman, Nelson. *The Structure of Appearance*. D. Reidel, 1977.

Goodman, Nelson and Quine, Willard van Orman. "Steps toward a Constructive Nominalism". Cited in Penelope Maddy, *Naturalism in Mathematics*. Oxford: Clarendon Press, 2000.

Gribanov, D. P. Albert Einstein's Philosophical View and the Theory of Relativity. H. Campbell Creighton, Trans. from the Russian. Moscow: Progress Publishers, 1987.

Gribbin, John. Q Is for Quantum: Particle Physics from A to Z. Hyderabad: Universities Press, 1998.

Guthrie, W. K. C. A History of Greek Philosophy, Vol. 2. Cambridge: Cambridge University Press, 1965.

Haak, Susan. Evidence and Inquiry: Towards Reconstruction in Epistemology. Oxford: Blackwell, 1996.

Heidegger, Martin. Being and Time. John Macquarrie and Edward Robinson, Trans. Oxford: Basil Blackwell, 1967.

–. *Identity and Difference.* Joan Stambaugh, Trans. New York: Harper & Row, 1969.

–. *The End of Philosophy.* Joan Stambaugh, Trans. New York: Harper & Row, 1973.

–. *Hegel's Concept of Experience.* San Francisco: Harper & Row, 1989.

–. *Contributions to Philosophy (From Enowning).* Parvis Emad and Kenneth Maly, Trans. Bloomington: Indiana University Press, 1999. (The copy consulted is a pre-print obtained from a reviewer.)

–. *Introduction to Metaphysics.* Gregory Fried and Richard Polt, Trans. New Haven: Yale University Press, 2000.

Heisenberg, Werner. The Physical Principles of the Quantum Theory. Carl Eckart and Frank C. Hoyt, Trans. New York: Dover, 1949.

–. *Physics and Philosophy: The Revolution in Modern Science.* New York: Harper & Row, 1958.

Hodes, Harold. "Ontological Commitment Thick and Thin" (235–60). *Meaning and Method: Essays in Honor of Hilary Putnam.* George Boolos, Ed. Cambridge: Cambridge University Press, 1990.

Holland, Peter. The Quantum Theory of Motion: An Account of the de Broglie-Bohm Causal Interpretation of Quantum Mechanics. Cambridge: Cambridge University Press, 1993, 1995 (Paperback).

Honderich, Ted, Ed. *The Oxford Companion to Philosophy.* Oxford: Oxford University Press, 1995.

Husserl, Edmund. Formal and Transcendental Logic, Dorion Cairns, Trans. The Hague: Martinus Nijhoff, 1978.

–. *Ideas Pertaining to a Pure Phenomenology and to a Phenomenological Philosophy. First Book: General Introduction to a Pure Phenomenology.* F. Kersten, Trans. Dordrecht: Kluwer Academic, 1982.

Hylton, Peter. "Quine on Reference and Ontology" (115–150). *The Cambridge Companion to Quine.* Roger F. Gibson, Jr., Ed. Cambridge: Cambridge University Press, 2004.

Jiachang, Luo and Hu Xinhe, "Relational Realism on Reform of the View of Physical Reality and Its Logical Manifestation" (359–379). In *Realism and Anti-realism in the Philosophy of Science, Beijing International Conference, 1992.* Robert S. Cohen, Risto Hilpinen and Qiu Renzong, Eds. Dordrecht: Kluwer Academic, 1996.

Johansson, Lars-Göran. "Realism and Wave-Particle Duality" (329–338). In *Realism and Anti-realism in the Philosophy of Science, Beijing International Conference, 1992.* Robert S. Cohen, Risto Hilpinen and Qiu Renzong, Eds. Dordrecht: Kluwer Academic, 1996.

Johnson, Oliver A. *The Problem of Knowledge.* The Hague: Martinus Nijhoff, 1974.

Jordan, Pascual. *Physics of the 20th Century.* Cited in James T. Cushing, *Quantum Mechanics: Historical Contingency and the Copenhagen Hegemony.* Chicago: The University of Chicago Press, 1994.

Jubien, Michael. *Ontology, Modality and the Fallacy of Reference.* Cambridge: Cambridge University Press, 1993.

Kant, Immanuel. *Immanuel Kant's Critique of Pure Reason.* Normal Kemp Smith, Trans. London: Macmillan, 1980.

–. *Theoretical Philosophy, 1755–1770.* D. Walford and R. Meerbote, Trans. Cambridge: Cambridge University Press, 1992.

Kirk, Robert. *Relativism and Reality: A Contemporary Introduction.* London: Routledge, 1999.

Kovacs, George. *The Question of God in Heidegger's Phenomenology.* Evanston, Ill.: Northwestern University Press, 1990.

Lewis, David. *Counterfactuals.* Oxford: Blackwell, 1973.

Lewis, David. *Counterfactuals.* Cited in Brian Skyrms, "Possible Worlds, Physics and Metaphysics" (143–152). In *Analytical Metaphysics: A Collection of Essays, Vol. 5, Necessity and Possibility: The Metaphysics of Modality.* Michael Tooley, Ed. New York: Garland, 1999.

Maddy, Penelope. *Naturalism in Mathematics.* Oxford: Clarendon Press, 2000.

Marx, Werner. *Introduction to Aristotle's Theory of Being as Being*. The Hague: Martinus Nijhoff, 1977.

Matheson, Carl. "Is the Naturalist Really Naturally a Realist?" Cited in Musgrave, Alan. "Realism, Truth and Objectivity" (19–44). In *Realism and Anti-realism in the Philosophy of Science, Beijing International Conference, 1992*. Robert S. Cohen, Risto Hilpinen and Qiu Renzong, Eds. Dordrecht: Kluwer Academic, 1996.

McKeon, Richard. "Experience and Metaphysics" (83–89). In *Experience and Metaphysics*, Proceedings of the XI International Congress of Philosophy, Brussels, August 20–26, 1953, Vol. 4. Amsterdam: North-Holland, 1953.

McMullin, Ernan. "The Problem of Universals". Cited in D. M. Armstrong, *Nominalism and Realism: Universals and Scientific Realism*, Volume 1. Cambridge: Cambridge University Press, 1995.

Moser, Paul K. *Knowledge and Evidence*. Cambridge: Cambridge University Press, 1991.

Muller, Richard A. *Dictionary of Latin and Greek Theological Terms: Drawn Principally from Protestant Scholastic Theology*. Grand Rapids: Baker Books, 1985.

Murdoch, Dugald. *Niels Bohr's Philosophy of Physics*. Cambridge: Cambridge University Press, 1987.

Musgrave, Alan. "Realism, Truth and Objectivity" (19–44). In *Realism and Anti-realism in the Philosophy of Science, Beijing International Conference, 1992*. Robert S. Cohen, Risto Hilpinen and Qiu Renzong, Eds. Dordrecht: Kluwer Academic Publishers, 1996.

Nagel, Thomas and James R. Newman. *Gödel's Proof*, London: Routledge & Kegan Paul, 1976.

Neelamkavil, Raphael. *Causal Ubiquity in Quantum Physics: A Superluminal and Local-Causal Physical Ontology*. Frankfurt: Peter Lang, 2014.

Neurath, Otto. "Unified Science as Encyclopedic Integration" (1–27). In *Foundations of the Unity of Science: Toward an International Encyclopedia of Unified Science*, Vol. 1, Otto Neurath, Rudolf Carnap and Charles Morris, Eds. Chicago: The University of Chicago Press, 1971.

The New Encyclopaedia Britannica: Micropaedia 15th Edition. S.v. "Lorentz, Hendrik Antoon", "Poincaré, Henri" and "Relativity".

The New Shorter Oxford English Dictionary on Historical Principles. S.v. "Eigen-", "Modal", "Strong" and "Weak".

Newton, Isaac. *The Principia: Mathematical Principles of Natural Philosophy*. Bernard Cohen and Anne Whitman, Ed. Berkeley: University of California Press, 1999.

Nietzsche, Friedrich. "The Thing-in-Itself and Appearance, and the Metaphysical Need". In *Immanuel Kant: Critical Assessments, volume 1, Kant Criticism from His Own to the Present Time*. Anthony M. Ludovici, Trans. and Ruth Chadwick, Ed. London: Routledge, 1992.

Niiniluoto, Ilkka. "Queries about Internal Realism"(45–54). In *Realism and Anti-realism in the Philosophy of Science, Beijing International Conference, 1992*. Robert S. Cohen, Risto Hilpinen and Qiu Renzong, Eds. Dordrecht: Kluwer Academic, 1996.

Oxford Latin Dictionary. Oxford: Clarendon Press, 1968.

Peirce, Charles Sanders. *Collected Papers of Charles Sanders Peirce*, Vols. 4, 6, 8, C. Hartshorne and P. Weiss, Eds. Cambridge, Mass.: The Belknap Press of the Harvard University Press, 1933 (Third Printing, 1974).

Penrose, Roger. *Shadows of the Mind: A Search for the Missing Science of Consciousness*. Oxford: Oxford University Press, 1994.

Pickover, Clifford A. *Time: A Traveller's Guide*, Oxford: Oxford University Press, 1998.

Popper, Karl R. *The Myth of the Framework: In Defence of Science and Rationality*, M. A. Notturno, Ed. London: Routledge, 1994.

Psillos, Stathis. *Scientific Realism: How Science Tracks Truth*, Philosophical Issues in Science (Series), W. H. Newton-Smith, Ed. London: Routledge, 1999.

Putnam, Hilary. *Realism, Truth and History*. Cited in Qiu Renzong, "How to know What Rises up Is the Moon? On the Concept of Realism and the Irrelevance of Quantum Mechanics to the Debate on Realism vs. Antirealism" (55–73). *Realism and Anti-realism in the Philosophy of Science, Beijing International Conference, 1992*. Robert S. Cohen, Risto Hilpinen and Qiu Renzong, Eds. Dordrecht: Kluwer Academic, 1996.

Quine, Willard Van Orman. *Word and Object*. Cambridge, MA.: The MIT Press, 1960.

–. *From a Logical Point of View*. Cambridge, Mass.: Harvard University Press, 1961.

–. *Methods of Logic*. London: Routledge & Kegan Paul, 1966.

—. "Facts of the Matter" (155–69). In *Essays on the Philosophy of W. V. Quine.* Robert W. Shahan and Christ Swoyer, Eds. Hassocks, Sussex: The Harvester Press, 1979.

—. *Theories and Things.* Cambridge, MA.: The Belknap Press of the Harvard University Press, 1981.

—. *Quiddities: An Intermittently Philosophical Dictionary.* London: Penguin, 1990.

—. *Ways of Paradox.* Cited in Harold Hodes, "Ontological Commitment Thick and Thin" (235–60). In *Meaning and Method: Essays in Honour of Hilary Putnam.* George Boolos, Ed. Cambridge: Cambridge University Press, 1990.

—. "Epistemology Naturalized" (15–31). *Naturalizing Epistemology.* Hilary Kornblith, Ed. Cambridge, Mass.: The MIT Press, 1994.

—. "Carnap and Logical Truth". Cited in Penelope Maddy, *Naturalism in Mathematics.* Oxford: Clarendon Press, 2000.

—. *Quintessence: Basic Readings from the Philosophy of W. V. Quine*, Roger F. Gibson, Jr., Ed. Cambridge, Mass.: The Belnap Press, 2004.

Renzong, Qiu. "How to Know What Rises up Is the Moon? On the Concept of Realism and the Irrelevance of Quantum Mechanics to the Debate on Realism vs. Antirealism"(55–73). In *Realism and Anti-realism in the Philosophy of Science, Beijing International Conference, 1992.* Robert S. Cohen, Risto Hilpinen and Qiu Renzong, Eds. Dordrecht: Kluwer Academic, 1996.

Rescher, Nicholas. *Conceptual Idealism.* Washington, D.C.: University Press of America, 1982.

Rodríguez-Consuegra, Francisco A. "Realism, Metamathematics and the Unpublished Essays". In Kurt Gödel, *Unpublished Philosophical Essays*, Francisco A. Rodríguez-Consuegra, Ed. Basel: Birkhäuser, 1995.

Rorty, Richard. *Philosophy and the Mirror of Nature.* Princeton: Princeton University Press, 1980.

Rothman, Milton A. *Discovering the Natural Laws: The Experimental Basis of Physics*, New York: Dover, 1989.

Routledge Encyclopaedia of Philosophy, 1998 edition. S.v. "Counterfactual Conditionals" by Frank Döring, "Natural Kinds" by Chris Daly, "Relativity Theory, Philosophical Significance of" by Michael Redhead and "Russell, Bertrand Arthur William" by Nicholas Griffin.

Russell, Bertrand. *The Problems of Philosophy.* Bombay: Oxford University Press, 1980.

–. *Introduction to Mathematical Philosophy*. New York: Simon and Schuster, n.d.

Sachs, Mendel. "On the Elementarity of Measurement in General Relativity: Toward a General Theory" (56–80). In *Boston Studies in the Philosophy of Science*, Volume 3, in Memory of Norwood Russell Hanson. Dordrecht: D. Reidel, 1967.

Schuhmann, Karl. "Brentano's Impact on Twentieth-century Philosophy". In *The Cambridge Companion to Brentano*. Dale Jacquette, Ed. Cambridge: Cambridge University Press, 2004.

Sfendoni-Mentzou, Demetra. "The Reality of Thirdness: A Potential-Pragmatic Account of Laws of Nature" (75–95). In *Realism and Anti-realism in the Philosophy of Science, Beijing International Conference, 1992*. Robert S. Cohen, Risto Hilpinen and Qiu Renzong, Eds. Dordrecht: Kluwer Academic, 1996.

Shapere, Dudley. *Reason and the Search for Knowledge, Boston Studies in the Philosophy of Science*, Vol. 78, Dordrecht: D. Reidel, 1984.

Singh, S. P. and Bagde, M. K. *Elements of Special Relativity*. New Delhi: S. Chand & Co., 1988.

Stenlund, Sören. *Language and Philosophical Problems*. London: Routledge, 1990.

Strawson, Peter. F. *Individuals: An Essay in Descriptive Metaphysics*. London: Methuen, 1984.

–. *Introduction to Logical Theory*, London: Methuen, 1985.

–. "Particular and General" (212–231). *The Problem of Universals*, Andrew B. Schoedinger, Ed. New Jersey: Humanities Press, 1992.

–. "The Theory of Property and the Theory of Reality in Quantum Mechanics". Cited in Qiu Renzong, "How to Know What Rises up Is the Moon? On the Concept of Realism and the Irrelevance of Quantum Mechanics to the Debate on Realism vs. Antirealism" (55–73). In *Realism and Anti-realism in the Philosophy of Science, Beijing International Conference*, 1992, Robert S. Cohen, Risto Hilpinen and Qiu Renzong, Eds. Dordrecht: Kluwer Academic, 1996.

Takeuti, Gaisi. "Work of Paul Bernays and Kurt Gödel" (77–85). In *Logic, Methodology and Philosophy of Science VI*, L. Jonathan Cohen, Jerzy Łoś, Helmut Pfeiffer and Klaus-Peter Podewski, Eds. Amsterdam: North-Holland Publishing Company, 1982.

Taylor, A. E. *Plato: The Man and His Work*. London: Methuen, 1986.

Ugarov, V. A. *Special Theory of Relativity*. Yuri Atanov, Trans. from the Russian. Moscow: Mir Publishers, 1979.

Uspensky, V. A. Gödel's Incompleteness Theorem. Moscow: Mir Publishers, 1987.

Van Fraassen, Bas C. The Scientific Image. Oxford: Oxford University Press, 1980.

Van Fraassen, Bas C. "Essences and Laws of Nature" (189–200). In *Reduction, Time and Reality: Studies in the Philosophy of the Natural Sciences.* Richard Healey, Ed. Cambridge: Cambridge University Press, 1981.

Velarde-Mayol, Victor. On Brentano. Belmont: Wadsworth, 2000.

Wallner and *Peschl,* "Cognitive Science – An Experiment in Constructive Realism; Constructive Realism – An Experiment in Cognitive Science" (103–116). In *Realism and Anti-realism in the Philosophy of Science, Beijing International Conference, 1992,* Robert S. Cohen, Risto Hilpinen and Qiu Renzong, Eds. Dordrecht: Kluwer Academic Publishers, 1996.

Wheaton, Bruce R. The Tiger and the Shark: Empirical Roots of Wave-particle Dualism. Cambridge: Cambridge University Press, 1992.

Whitaker, Andrew. Einstein, Bohr and the Quantum Dilemma. Cambridge: Cambridge University Press, 1996.

Whitehead, Alfred North. Process and Reality: An Essay in Cosmology, Gifford Lectures Delivered in the University of Edinburgh During the Session 1927–28, Corrected Edition. David Ray Griffin and Donald W. Sherburne, Eds. New York: The Free Press, 1978. Originally published by Macmillan, 1927.

Wittgenstein, Ludwig. Tractatus Logico-Philosophicus, German Text with an English Translation *en regard* by C. K. Ogden and Introduction by Bertrand Russell. London: Routledge, 2000.

–. *Notebooks 1914–1916.* G. H. von Wright and G. E. M. Anscombe, Eds. Chicago: The University of Chicago Press, 1984.

Zhengkun, Yin. "Truth and Fiction in Scientific Theory" (266–67). In *Realism and Anti-realism in the Philosophy of Science, Beijing International Conference, 1992,* Robert S. Cohen, Risto Hilpinen and Qiu Renzong, Eds. Dordrecht: Kluwer Academic Publishers, 1996.

Zuoxiu, He Zuoxiu, "On the Einstein, Podolsky and Rosen Paradox and the Relevant Philosophical Problems" (299–305). In *Realism and Anti-realism in the Philosophy of Science, Beijing International Conference, 1992,* Robert S. Cohen, Risto Hilpinen and Qiu Renzong, Eds. Dordrecht: Kluwer Academic Publishers, 1996.

Index

"is" 26

A

a priori 9, 11, 15, 22, 38, 39, 40, 41, 50, 51, 72, 74, 75, 76, 83, 89, 90, 91, 94, 111, 128, 154, 156, 158, 162, 230, 238, 242, 294, 319, 321, 329, 338, 340, 341, 349
absolute 9, 32, 33, 36, 52, 57, 59, 63, 65, 66, 69, 72, 76, 101, 103, 104, 113, 118, 119, 123, 126, 131, 132, 137, 140, 141, 144, 152, 176, 182, 191, 225, 226, 227, 238, 242, 262, 267, 268, 294, 296, 297, 306, 322, 339
absolutisation 14, 44
absolutism 14, 45, 79, 115, 143, 145, 147, 175, 180, 194, 199, 211, 267, 306, 319, 321
absolutistic 142, 143, 206, 352
abstract 22, 26, 35, 36, 38, 39, 40, 74, 77, 152, 154, 157, 158, 160, 161, 181, 182, 184, 189, 199, 203, 212, 221, 223, 225, 230, 234, 244, 248, 249, 250, 251, 252, 254, 256, 257, 258, 259, 260, 263, 264, 265, 266, 267, 268, 270, 271, 272, 274, 276, 277, 278, 279, 283, 285, 286, 287, 290, 293, 295, 298, 303, 304, 305, 306, 308, 309, 310, 318, 320, 321, 322, 324, 338, 343, 345
– abstractly 34, 298, 321
abstractly 36
abstract object 26, 40, 157, 251, 306
accidental 53, 263, 340, 343
acquaintance 80
actual 12, 14, 23, 25, 26, 29, 30, 31, 32, 34, 35, 40, 42, 44, 52, 53, 56, 60, 63, 64, 65, 67, 68, 69, 70, 71, 72, 74, 75, 77, 78, 80, 81, 86, 95, 97, 100, 102, 105, 107, 113, 120, 123, 125, 127, 128, 138, 142, 144, 147, 148, 151, 152, 153, 154, 155, 156, 157, 158, 159, 160, 162, 164, 172, 174, 175, 176, 179, 180, 181, 194, 197, 201, 204, 206, 212, 213, 214, 216, 217, 219, 226, 234, 237, 243, 246, 252, 258, 259, 262, 263, 264, 265, 267, 270, 272, 274, 276, 277, 290, 292, 295, 298, 300, 303, 304, 306, 308, 309, 310, 318, 324, 338, 339, 344, 351, 352, 353
actual entity 31, 32, 41, 53, 219
actuals 27, 31, 32, 64, 81, 127, 156, 157, 160, 213, 217, 219, 299
Alétheia 35, 346, 347, 353
aletheial 14, 47, 321, 323, 339
analysis 12, 15, 25, 39, 46, 106, 126, 145, 148, 158, 179, 199, 210, 213, 229, 246, 249, 251, 252, 264, 314, 337, 353
analytic 6, 26, 29, 161, 163, 184, 193, 194, 196, 197, 198, 202, 203, 205, 218, 238, 241, 246, 248, 250, 253, 266, 268, 277, 278, 286, 291, 294, 295, 313, 314, 322, 335
anthropic 15, 135, 246, 322, 323, 340, 341, 343, 347, 348, 354
Anthropologised 337
antirealism 60, 203
apperception 64
apriorily 40, 74, 75, 97, 127, 130, 230, 318, 329, 347, 349
Aquinas 26, 36, 202, 250, 273, 322, 323
Aristotle
– Aristotelian 6, 11, 26, 36, 49, 164, 185, 242, 250, 270, 305, 322, 323, 324, 325, 333, 337, 338, 357, 363

369

Armstrong 6, 11, 71, 181, 196, 197, 198, 199, 200, 201, 202, 203, 204, 205, 208, 209, 210, 211, 212, 213, 214, 215, 216, 217, 218, 232, 320, 322, 357, 363
Aspect 120, 124, 232
asymptotic 35, 105, 127, 221, 348
attribute 26, 43, 52, 53, 54, 55, 81, 145, 149, 162, 177, 245, 256, 263, 273
axiom 45, 115, 221, 223, 229, 249, 295

B

being 6, 9, 11, 12, 13, 16, 21, 22, 23, 24, 27, 29, 31, 32, 33, 36, 39, 41, 44, 47, 50, 51, 52, 53, 54, 55, 56, 58, 61, 63, 64, 67, 68, 71, 74, 75, 76, 81, 83, 84, 85, 88, 89, 92, 93, 95, 96, 97, 113, 117, 125, 128, 129, 134, 141, 148, 152, 153, 155, 157, 160, 161, 162, 164, 166, 167, 177, 178, 179, 181, 183, 185, 186, 189, 192, 193, 196, 197, 199, 200, 202, 204, 207, 208, 210, 211, 212, 214, 215, 216, 218, 222, 223, 227, 229, 235, 243, 244, 245, 246, 247, 249, 250, 257, 259, 261, 262, 266, 267, 269, 270, 271, 273, 274, 276, 282, 283, 291, 292, 299, 300, 301, 303, 304, 307, 308, 309, 310, 312, 315, 317, 319, 320, 321, 322, 323, 324, 325, 327, 328, 329, 332, 333, 334, 335, 338, 339, 340, 341, 342, 344, 346, 349, 351, 352
be-ing 310, 343, 350, 351, 352
Being 13, 14, 22, 26, 29, 35, 36, 45, 47, 99, 155, 233, 241, 242, 248, 250, 305, 310, 317, 320, 321, 322, 323, 324, 325, 326, 330, 333, 337, 338, 339, 340, 341, 342, 343, 344, 346, 347, 348, 349, 350, 351, 352, 353, 354, 361, 363
Being-historical 323
belief 66, 73, 84, 89, 90, 165, 188, 233, 238, 268, 292, 293, 319

Bell 102, 109, 124, 357
Blackburn 65, 357
Bohm 109, 114, 120, 123, 124, 142, 144, 357, 361
Bohr 11, 77, 78, 79, 87, 88, 89, 92, 102, 110, 111, 117, 118, 119, 121, 123, 128, 132, 151, 317, 358, 359, 360, 363, 367
Boolean 52, 54
bottlenecked 100, 103
bottlenecking 98, 99, 100, 101, 102, 104
bound variable 196, 274, 297, 306, 309
bound variables 83, 257, 277, 278, 283, 297, 298, 303
Brentano 30, 55, 252, 273, 274, 358, 366, 367
bulging 100, 104
Bunge 87, 88, 89, 93, 94, 110, 120, 122, 125, 358

C

Campbell 199, 357, 361
Carnap 54, 241, 265, 267, 268, 269, 274, 291, 363, 365
Carroll 167, 168, 169, 171, 178, 179, 180, 358
categorial 11, 13, 24, 26, 29, 34, 36, 37, 42, 45, 46, 51, 55, 57, 58, 60, 61, 63, 64, 72, 75, 76, 79, 86, 126, 130, 141, 142, 145, 151, 173, 184, 193, 195, 197, 205, 219, 221, 227, 235, 238, 243, 245, 251, 253, 284, 296, 309, 310, 311, 312, 313, 317, 319, 325, 334, 335, 336, 337, 352, 353
categories 6, 7, 11, 12, 13, 14, 15, 16, 21, 24, 25, 26, 27, 28, 29, 30, 34, 35, 36, 37, 38, 41, 42, 43, 44, 45, 46, 49, 50, 51, 55, 56, 57, 58, 59, 60, 61, 62, 63, 64, 65, 69, 72, 74, 75, 76, 77, 79, 80, 83, 84, 86, 100, 102, 128, 129, 130, 139, 140, 143, 145, 147, 151, 155,

156, 163, 164, 192, 195, 197, 205, 211, 218, 219, 220, 224, 225, 227, 228, 231, 235, 237, 238, 239, 241, 242, 247, 251, 253, 259, 265, 282, 283, 285, 287, 288, 289, 292, 294, 296, 297, 301, 302, 310, 311, 313, 314, 315, 316, 317, 318, 321, 322, 323, 324, 327, 328, 332, 333, 334, 335, 336, 348, 354
category 12, 13, 14, 15, 25, 26, 35, 36, 37, 41, 42, 44, 45, 49, 50, 51, 57, 58, 61, 65, 69, 74, 75, 76, 84, 86, 87, 112, 113, 127, 128, 129, 130, 134, 140, 143, 144, 145, 147, 149, 152, 155, 157, 162, 163, 164, 170, 188, 193, 195, 196, 200, 211, 212, 214, 219, 230, 234, 235, 236, 237, 238, 242, 247, 253, 275, 289, 296, 310, 311, 313, 317, 318, 319, 320, 323, 328, 332, 333, 334, 336, 348, 353
causal 12, 21, 27, 30, 36, 50, 51, 56, 57, 68, 70, 74, 75, 86, 91, 92, 96, 97, 98, 100, 101, 102, 108, 112, 113, 114, 115, 116, 117, 119, 121, 122, 123, 124, 125, 129, 130, 136, 139, 142, 143, 144, 152, 157, 160, 162, 163, 164, 171, 173, 174, 175, 176, 177, 178, 179, 180, 183, 187, 188, 189, 191, 193, 194, 195, 196, 197, 198, 200, 201, 203, 204, 205, 206, 207, 208, 209, 210, 211, 212, 213, 214, 215, 216, 217, 218, 219, 220, 225, 231, 235, 236, 238, 241, 243, 245, 246, 247, 252, 253, 259, 260, 261, 263, 265, 270, 271, 273, 277, 282, 283, 285, 286, 287, 288, 289, 292, 293, 295, 298, 302, 303, 304, 310, 311, 312, 319, 320, 321, 332, 338, 344, 345, 346, 348
causality 21, 35, 50, 51, 55, 57, 58, 74, 76, 89, 90, 92, 98, 112, 113, 114, 120, 121, 126, 127, 129, 139, 140, 141, 142, 143, 144, 145, 166, 171, 174, 175, 177, 178, 180, 182, 183, 184, 189, 191, 198, 201, 202, 204, 205, 208, 211, 215, 216, 218, 219, 231, 232, 235, 237, 265, 323, 330
cause 11, 35, 44, 55, 58, 92, 96, 122, 142, 144, 145, 181, 208, 209, 210, 215, 216, 253, 276, 347
Chisholm 52, 53, 54, 55, 56, 149, 176, 257, 317, 322, 323, 358
Churchland 59, 85, 358, 359
class 26, 35, 38, 50, 52, 53, 54, 56, 71, 73, 112, 184, 190, 222, 230, 249, 255, 256, 257, 262, 263, 278, 286, 321, 327, 328, 334
classification 27, 37, 172, 199, 214, 256, 318, 323, 324
classificational 11, 13, 22, 24, 28, 29, 34, 37, 42, 43, 45, 49, 50, 55, 162, 195, 215, 235, 237, 242, 253, 296, 301, 311, 315, 316, 318, 322, 323, 333, 354
classon 89
Clauser 120, 124
co-extensive 15, 163, 250, 345, 353
cognition 30, 51, 58, 60, 61, 64, 90, 147, 195, 232, 235, 271
cognitive 15, 29, 42, 62, 76, 89, 90, 206, 238, 243, 318, 341
cognitive science 26, 28, 357, 367
Cohen 169, 357, 358, 359, 360, 362, 363, 364, 365, 366, 367
coherence 12, 34, 45, 181, 233, 239, 313, 335
coherent 87, 327
coherentism 165, 233
co-intensive 353
collection 52, 123, 200
collude 37, 45, 46, 316, 323
collusive 11, 14, 15, 21, 25, 34, 36, 37, 41, 42, 44, 51, 239, 242, 296, 297, 316, 321
complementarity 97, 102, 151

concept 6, 11, 12, 13, 14, 21, 23, 26, 28, 29, 30, 31, 34, 35, 38, 39, 42, 44, 47, 49, 50, 52, 55, 56, 61, 62, 63, 65, 67, 68, 69, 71, 72, 74, 75, 77, 80, 82, 85, 86, 87, 89, 90, 97, 100, 104, 105, 111, 113, 114, 115, 116, 117, 126, 128, 131, 132, 138, 141, 142, 143, 147, 148, 149, 150, 158, 160, 164, 166, 171, 172, 174, 175, 177, 178, 179, 180, 181, 183, 185, 186, 188, 189, 191, 192, 193, 199, 202, 206, 208, 212, 213, 215, 217, 220, 227, 229, 237, 238, 243, 246, 248, 249, 250, 251, 253, 254, 256, 258, 259, 269, 272, 278, 282, 285, 286, 287, 292, 296, 298, 305, 308, 310, 312, 317, 318, 320, 321, 322, 323, 325, 326, 327, 332, 337, 338, 339, 341, 342, 343, 344, 347, 348, 349, 350, 351, 352, 353, 355

conceptual 12, 21, 22, 23, 26, 27, 29, 31, 32, 33, 34, 35, 36, 37, 38, 46, 56, 58, 59, 60, 62, 64, 65, 68, 71, 72, 80, 97, 134, 145, 149, 153, 154, 157, 158, 159, 162, 174, 179, 186, 192, 193, 201, 202, 206, 211, 218, 226, 233, 234, 237, 238, 247, 251, 257, 259, 262, 263, 264, 265, 267, 277, 279, 280, 282, 288, 290, 299, 300, 318, 321, 325, 338, 339, 351, 352, 353

conceptual scheme 251, 278, 279

concrescence 31, 32

concretism 61, 79, 121, 179

concretistic 51

conflate 62, 128

conflating 64

conflation 62, 64

connective 42, 183, 244, 267, 270, 274, 290, 299

connotation 217, 246, 261, 272, 290, 300

connotative 11, 13, 15, 21, 23, 29, 31, 36, 56, 69, 73, 74, 145, 153, 154, 155, 156, 157, 158, 159, 160, 161, 163, 164, 165, 169, 171, 172, 175, 176, 177, 178, 182, 187, 189, 191, 192, 193, 195, 196, 202, 203, 205, 206, 207, 208, 210, 212, 213, 214, 215, 217, 218, 219, 220, 225, 226, 228, 230, 231, 232, 234, 235, 236, 237, 238, 242, 244, 245, 246, 247, 249, 250, 251, 252, 259, 260, 262, 265, 270, 273, 274, 277, 278, 279, 282, 285, 286, 287, 288, 289, 292, 295, 300, 301, 307, 309, 310, 316, 321, 322, 327, 334, 335, 336, 338, 340, 341, 342, 343, 345, 346, 347, 348, 349, 350, 351, 352, 353

connote 35, 207, 322

conscious 11, 13, 22, 23, 25, 29, 31, 32, 33, 36, 40, 44, 160, 167, 169, 183, 194, 201, 204, 206, 208, 215, 218, 220, 235, 244, 245, 246, 247, 250, 310, 325, 340, 343, 345, 349, 354

consciousness 11, 15, 27, 31, 33, 36, 38, 40, 44, 62, 73, 127, 128, 155, 157, 162, 164, 167, 189, 191, 192, 203, 206, 213, 219, 236, 237, 239, 243, 244, 245, 246, 247, 248, 251, 262, 270, 272, 289, 292, 300, 307, 309, 322, 323, 325, 328, 334, 338, 340, 341, 346, 347, 350, 354

continuity 6, 21, 26, 51, 67, 69, 70, 74, 75, 76, 79, 86, 103, 104, 108, 112, 123, 124, 126, 127, 131, 136, 139, 140, 141, 142, 157, 171, 180, 185, 189, 259, 341

continuous 7, 12, 32, 34, 45, 46, 49, 50, 51, 61, 63, 65, 69, 75, 77, 78, 79, 86, 90, 102, 103, 104, 111, 124, 126, 130, 141, 149, 175, 183, 184, 191, 205, 206, 211, 242, 269, 288, 294, 296, 301, 302, 304, 318, 320, 329, 332

Copenhagen 89, 91, 92, 93, 94, 359, 362

Copernicus 76, 79
corpuscles 87, 102
corpuscular 88, 102
counterfactual 166, 168, 170, 173, 204, 226, 227
counterlegal 173, 176
criterial velocity 135, 136, 143
Cushing 89, 90, 91, 359, 362

D
D'Espagnat 109, 146, 359
Dasein 15, 35, 42, 73, 244, 310, 322, 337, 338, 339, 340, 341, 342, 343, 344, 345, 346, 347, 348, 349, 350, 351, 352, 353, 354
de Broglie 143, 306, 361
deduction 64, 136, 224, 226
deductive 220, 222, 224, 225, 226, 229, 230, 305
deductiveness 224
Denitt 60
denotable 35, 152, 156, 160, 164, 247, 277, 282, 287, 290, 295
Descartes 53, 340, 342
determination 94, 112, 117, 119, 173, 174, 197, 202, 204, 205, 227, 283, 297, 298, 327, 344
determinationism 171, 174, 175, 206, 220
determinism 90, 91, 115, 121, 123, 126, 143, 145, 166, 171, 175, 180, 206, 220, 354
discourse 12, 14, 15, 23, 24, 26, 46, 47, 65, 69, 73, 74, 81, 84, 112, 156, 162, 164, 166, 172, 183, 185, 186, 195, 204, 207, 208, 218, 226, 232, 235, 238, 241, 244, 252, 254, 257, 264, 266, 276, 277, 282, 283, 289, 293, 299, 301, 306, 307, 309, 314, 316, 317, 333, 334, 336, 348
Divine 12, 25, 32, 33, 37, 41, 45, 74, 149, 153, 247, 322, 325, 330, 332, 338

domain 11, 27, 37, 44, 54, 194, 197, 242, 277, 280, 306
double-slit experiment 87, 104, 107
duality 87, 90, 102, 105, 106, 107, 108, 126, 130, 140, 143, 256, 269
Durrant 194, 195, 359

E
eigenfunction 92, 116
eigenstate 92, 130
eigenvalue 90, 92
Einaic 5, 7, 10, 11, 13, 14, 15, 16, 21, 23, 28, 37, 41, 42, 44, 46, 47, 150, 151, 156, 157, 158, 160, 161, 178, 196, 200, 215, 219, 220, 225, 227, 232, 239, 241, 242, 245, 246, 248, 249, 250, 252, 253, 254, 259, 263, 265, 266, 271, 274, 283, 284, 285, 287, 288, 289, 290, 292, 293, 294, 295, 296, 297, 299, 300, 301, 302, 303, 307, 308, 309, 310, 313, 314, 315, 316, 317, 318, 320, 321, 322, 324, 325, 326, 327, 328, 329, 330, 332, 334, 335, 336, 337, 338, 339, 341, 345, 346, 347, 348, 351, 354
Einaic Ontology 14, 21, 37, 156, 157, 228, 232, 265, 271, 292, 294, 300, 301, 314, 315, 316, 317, 320, 321, 324, 326, 327, 329, 336, 339
Einaic Semantics 266, 283, 314, 332
Einaiology 14, 21, 37, 300, 324, 326, 327, 329, 354
Einstein 11, 59, 77, 78, 79, 89, 91, 92, 97, 109, 110, 114, 115, 116, 117, 119, 120, 121, 122, 123, 124, 131, 132, 133, 134, 136, 137, 138, 140, 317, 359, 360, 361, 367
electromagnetic 67, 95, 103, 132, 134, 135, 138, 139, 141
Ellis 66, 359
empiricism 56, 59, 72, 84, 85, 89, 90, 111, 264, 269, 284, 285, 319

373

enown 350
enowning 14, 339, 341, 342, 346, 349, 352, 354
Enowning 35, 36, 344, 346, 349, 353, 354, 359, 361
en-proper-ing 339
entities 11, 13, 22, 23, 25, 27, 29, 31, 32, 34, 35, 37, 40, 42, 43, 44, 52, 53, 55, 56, 57, 58, 59, 60, 63, 64, 65, 66, 67, 75, 77, 78, 80, 81, 83, 85, 86, 88, 90, 108, 109, 111, 112, 123, 125, 140, 145, 149, 150, 151, 152, 153, 154, 155, 159, 160, 161, 164, 167, 172, 175, 178, 180, 183, 188, 191, 192, 194, 195, 197, 201, 208, 212, 213, 215, 217, 218, 219, 236, 237, 239, 243, 244, 246, 251, 252, 254, 257, 258, 259, 260, 262, 263, 264, 265, 266, 267, 268, 270, 274, 276, 279, 283, 286, 289, 297, 298, 299, 300, 302, 304, 305, 306, 307, 308, 310, 321, 322, 324, 325, 328, 329, 338, 339, 340, 343, 345, 348, 352
entity 22, 26, 30, 31, 32, 35, 37, 40, 43, 44, 51, 52, 53, 54, 73, 75, 81, 83, 84, 85, 86, 99, 123, 125, 126, 127, 130, 140, 145, 148, 150, 152, 153, 155, 171, 179, 181, 188, 189, 191, 193, 197, 201, 213, 215, 216, 231, 242, 247, 248, 249, 251, 253, 259, 260, 261, 262, 263, 265, 269, 274, 277, 279, 280, 286, 297, 308, 310, 317, 338, 340, 352, 353, 354
Entwurf des Seins 15, 346
epistemic 15, 36, 44, 66, 69, 72, 76, 78, 79, 86, 132, 163, 169, 172, 173, 175, 178, 179, 180, 183, 185, 188, 189, 193, 194, 204, 205, 218, 232, 233, 235, 237, 246, 248, 249, 250, 255, 285, 287, 296, 321, 334, 341
epistemological 11, 12, 15, 16, 21, 25, 26, 28, 34, 37, 44, 45, 46, 47, 56, 60, 74, 76, 78, 79, 80, 86, 89, 90, 91, 109, 112, 121, 145, 147, 148, 162, 163, 175, 182, 184, 193, 195, 196, 201, 202, 203, 205, 208, 209, 211, 214, 215, 217, 219, 220, 226, 234, 235, 236, 237, 238, 242, 245, 246, 247, 248, 250, 259, 274, 284, 294, 302, 307, 313, 319, 320, 321, 323, 334, 335, 341, 345, 354
EPR 87, 110, 112, 114, 115, 116, 117, 118, 121, 122, 124, 136, 139, 140, 143
Ereignis 35, 36, 344, 346, 349, 353
ersatz 59, 66
essence 35, 39, 42, 75, 137, 167, 168, 169, 170, 171, 172, 174, 177, 178, 181, 183, 193, 216, 261, 262, 270, 290, 301, 305, 320, 321, 323, 344, 345
essential 11, 26, 52, 55, 56, 82, 90, 99, 110, 140, 145, 159, 166, 167, 173, 174, 183, 189, 192, 205, 207, 208, 241, 247, 248, 259, 264, 272, 273, 284, 310, 321, 322, 327, 349
Euclidean 105, 127, 221
event 26, 30, 31, 53, 55, 56, 63, 64, 108, 127, 139, 174, 177, 191, 208, 210, 285, 286, 316, 344
evidential probability 232, 233
exist 26, 49, 55, 56, 60, 68, 70, 72, 74, 75, 82, 83, 106, 143, 144, 150, 152, 166, 170, 177, 181, 210, 216, 222, 231, 244, 260, 261, 266, 270, 273, 274, 275, 293, 303, 311
existence 9, 22, 38, 40, 53, 60, 62, 63, 65, 70, 72, 81, 82, 83, 87, 89, 103, 109, 114, 124, 125, 133, 134, 140, 144, 147, 151, 152, 154, 175, 177, 178, 180, 183, 188, 212, 213, 214, 215, 216, 217, 218, 222, 223, 239, 244, 250, 251, 252, 256, 258, 260, 261, 262, 265, 267, 268, 270, 272, 273, 276, 277, 303, 318, 320, 330, 338, 340, 343, 345, 349

extension 31, 32, 52, 53, 54, 87, 99, 105, 106, 144, 145, 147, 151, 153, 172, 174, 175, 176, 177, 182, 202, 216, 217, 223, 227, 271, 275, 288, 289, 290, 292, 295, 297, 298, 299, 300, 301, 323, 327, 352, 353
extra-mental 30, 38, 40, 59, 67, 343
extra-phenomenal 67, 69, 125, 147

F

fact 9, 12, 14, 15, 30, 33, 38, 42, 43, 44, 47, 49, 50, 55, 56, 66, 69, 73, 76, 77, 78, 79, 82, 85, 90, 91, 92, 96, 97, 98, 100, 101, 102, 103, 104, 108, 109, 110, 112, 117, 118, 119, 121, 122, 127, 130, 134, 135, 136, 139, 148, 159, 165, 166, 167, 171, 173, 174, 176, 177, 181, 183, 184, 187, 188, 189, 190, 196, 198, 204, 205, 206, 207, 208, 209, 210, 212, 218, 219, 224, 230, 231, 235, 236, 238, 239, 244, 245, 258, 261, 262, 263, 264, 267, 268, 269, 271, 272, 273, 274, 276, 277, 283, 287, 289, 292, 293, 295, 296, 297, 301, 302, 305, 308, 311, 313, 317, 318, 321, 327, 335, 339, 341, 343, 349, 351, 352, 353, 354
Feynman 100, 126, 360
Field 188, 189, 292
field theory 66, 123, 141
first-order logic 82, 229, 275, 284, 285
fluency 6, 53, 339, 342, 346, 353
formalized 220, 223
four-dimensional 99, 100, 103, 105
Frege 52, 82, 149, 255, 269, 321, 358, 360
Fundamental Ontology 337, 354

G

Galileian transformations 131, 137, 138
Gell-Mann 114, 115, 119, 360
General Ontology 14, 21, 35, 37, 57, 85, 86, 300, 324, 326, 329, 354
generalities 26, 32, 34, 40, 42, 55, 164, 166, 167, 170, 183, 188, 189, 191, 192, 193, 195, 206, 208, 231, 236, 245, 278, 288, 298, 301, 322
generality 26, 34, 35, 38, 63, 71, 159, 162, 163, 164, 170, 173, 185, 188, 191, 192, 193, 194, 203, 206, 208, 231, 236, 242, 245, 282, 300, 326, 330, 344
generalization 11, 13, 45, 56, 84, 97, 132, 137, 169, 212, 213, 226, 249, 254, 295, 298, 311, 328
Gödel 13, 52, 164, 192, 220, 221, 222, 223, 224, 225, 226, 227, 228, 229, 231, 235, 268, 269, 282, 301, 317, 319, 320, 323, 324, 335, 360, 363, 365, 366, 367
Gödelian 21, 220, 269, 282, 294, 301, 313, 335
gravitation 95, 141
gravitational 95, 100, 103, 104, 131, 173

H

Heidegger 6, 11, 14, 21, 26, 35, 36, 47, 246, 248, 250, 310, 312, 322, 323, 330, 337, 338, 339, 340, 341, 342, 343, 344, 345, 346, 347, 348, 349, 350, 351, 352, 353, 354, 359, 361, 362
Heideggerian 15, 73, 339, 347, 349, 354
Heisenberg 87, 92, 97, 125, 361
hidden variables 86, 120, 124, 144
Hintikka 68
Hodes 258, 307, 361, 365
holism 123, 254, 274, 275, 278, 289, 290, 291, 292, 293, 295, 299, 302, 311, 314
holistic 122, 290, 293, 321, 329
Holland 115, 116, 121, 123, 361, 363, 366
Hume 43, 293
Humean 58, 198, 204, 215

375

Husserl 36, 38, 39, 40, 157, 161, 243, 244, 250, 252, 270, 271, 272, 273, 322, 323, 361
Husserlian 30, 62, 252, 310, 343
hyphenated 59, 66

I

ideal 11, 12, 14, 21, 26, 27, 34, 35, 37, 40, 43, 56, 61, 66, 67, 72, 73, 81, 83, 91, 104, 109, 140, 162, 164, 172, 177, 187, 191, 206, 219, 228, 231, 233, 237, 238, 269, 271, 273, 302, 320, 325, 328, 334, 338
Ideal limit 66
ideal *limits* 66
ideal scientific practice 66
idealism 30, 51, 58, 59, 64, 65, 66, 79, 90, 249, 250
idealistic 12, 69, 165, 311
idealization 65, 150, 152, 162, 178, 237, 298, 347, 349, 351
idealized 24, 71, 151, 162, 191, 211, 235, 259, 316, 317, 350
identity 42, 43, 52, 63, 179, 182, 186, 199, 203, 204, 205, 209, 234, 242, 245, 251, 257, 259, 261, 264, 265, 272, 298, 322, 330, 333
IFO 70, 72
indeterminacy 92, 126, 216
indiscernibles 42, 43, 52, 199
inductive 11, 13, 36, 37, 45, 74, 87, 150, 169, 171, 173, 192, 193, 224, 226, 230, 248, 249, 279, 295
inductiveness 224
infinite 21, 33, 49, 51, 69, 86, 90, 92, 97, 98, 99, 104, 105, 108, 110, 111, 113, 116, 117, 122, 125, 127, 133, 141, 144, 145, 146, 151, 152, 154, 156, 158, 174, 175, 176, 177, 183, 191, 198, 201, 203, 206, 219, 221, 223, 231, 232, 260, 265, 270, 285, 295, 308, 319, 328, 338, 345, 352

infinitely 30, 32, 33, 50, 51, 97, 121, 144, 148, 149, 162, 175, 187, 192, 197, 210, 216, 219, 247, 253, 295, 305, 325, 332, 338
infinitesimal 21, 49, 51, 78, 86, 90, 91, 92, 94, 95, 108, 110, 112, 116, 117, 122, 125, 127, 144, 145, 149, 150, 174, 175, 176, 177, 178, 183, 184, 191, 198, 201, 203, 206, 209, 219, 231, 232, 260, 265, 269, 285, 309, 319, 339
infinitesimally 30, 50, 51, 94, 96, 97, 105, 111, 121, 130, 144, 148, 162, 175, 192, 216, 219, 270
ingress 32, 179, 246
ingression 33, 36, 40, 84, 130, 206, 213, 246, 247, 289
instantiate 173, 189, 198, 200, 201, 202, 206, 248, 311, 336
instantiated 31, 128, 176, 202, 212, 217, 235, 245, 247, 248, 249, 288, 308, 311, 322
instantiation 30, 47, 60, 173, 181, 182, 184, 185, 193, 198, 201, 202, 203, 211, 212, 213, 214, 215, 217, 219, 226, 248, 249, 250, 252, 261, 270, 306, 309, 310, 311, 318, 334
instrumentalism 12, 50, 77, 79, 87, 88, 90, 101, 102, 103, 110, 111, 144, 180, 285
instrumentalistic 12, 21, 50, 66, 80, 91, 110, 126, 143, 151, 180, 183
intension 217, 289, 292, 296, 297, 298, 300, 301, 327, 353
intentional 55, 80, 172, 243, 245
intentionality 252, 292
internal realism 60, 66
Internalism 233
invariance 112, 131, 133, 135, 138, 140

J

Johansson 50, 88, 105, 106, 107, 108, 111, 112, 362

376

Jordan 89, 362
judgment 64, 81, 82, 243, 270, 271, 273, 295, 300, 320

K
Kant 11, 21, 29, 36, 49, 50, 57, 58, 59, 61, 62, 63, 64, 67, 68, 69, 70, 72, 74, 76, 78, 81, 90, 127, 158, 162, 252, 253, 293, 294, 317, 318, 323, 337, 338, 340, 362, 364
Kantian 12, 30, 37, 45, 50, 51, 59, 64, 65, 67, 76, 111, 140, 144, 211, 294, 341, 343, 347, 348, 349, 352
Kennedy-Thorndike experiments 132
Kovacs 342, 344, 362
Kripke 168, 254, 275

L
Lakatos 141, 150
language-laden 277, 282, 285
law of nature 166, 169, 171, 173, 176, 178, 180, 198, 205
laws of nature 47, 138, 164, 166, 167, 168, 170, 171, 172, 173, 174, 176, 178, 179, 181, 184, 197, 198, 201, 202, 205, 213, 214, 215, 218, 219, 224, 235, 236, 237, 286, 288, 334
Lewis 71, 173, 204, 205, 215, 328, 362
linguistic 13, 22, 59, 65, 68, 154, 158, 171, 172, 177, 183, 184, 196, 217, 236, 249, 251, 252, 254, 258, 259, 268, 277, 278, 292, 293, 299, 300, 311, 313, 328
local 34, 45, 50, 113, 116, 121, 123, 124, 136, 139, 143, 144, 145, 152, 230, 251, 257, 312, 314, 315, 329
locality 50, 87, 115, 121, 122, 124, 129, 146
Lorenz 131, 138
Lorenz transformations 131, 134, 138
luminal 121, 136, 141

M
macro- 79, 86, 113, 125, 130, 144, 150, 151
Maddy 265, 266, 267, 268, 269, 360, 362, 365
mathematical 52, 53, 77, 79, 80, 87, 88, 91, 101, 102, 104, 105, 107, 109, 110, 111, 125, 126, 129, 140, 144, 150, 151, 152, 153, 154, 155, 156, 157, 158, 159, 160, 161, 188, 189, 191, 192, 220, 221, 223, 224, 225, 227, 229, 230, 244, 265, 267, 268, 283, 293, 294, 324, 329
Matheson 66, 363
matter waves 306
Maxwell 132, 134, 138, 141
Meinong 273, 274
mereological 71, 322
meso- 79, 86, 87, 91, 113, 125, 130, 137, 144, 150
metaphysical 10, 14, 22, 25, 37, 41, 45, 49, 50, 58, 59, 62, 71, 72, 141, 154, 171, 172, 179, 180, 182, 215, 216, 238, 254, 263, 264, 292, 299, 334, 339, 341, 344, 347, 352
metaphysics 6, 9, 10, 12, 21, 41, 44, 45, 46, 47, 50, 65, 72, 145, 181, 191, 192, 196, 220, 247, 248, 252, 253, 261, 302, 310, 315, 316, 319, 320, 326, 329, 330, 335, 336, 338, 339, 341, 343, 352
Michelson-Morley 132
micro- 51, 79, 86, 87, 91, 93, 101, 113, 125, 130, 144, 150, 151
mind-dependent 61, 70, 75
mind-independent 15, 29, 60, 68, 70, 71, 73, 106, 148, 149
modal 166, 168, 169, 171, 179, 266, 267, 328
model 73, 165, 207, 229
Moser 232, 233, 234, 235, 363
Murdoch 110, 116, 117, 118, 119, 359, 363

Musgrave 59, 60, 62, 64, 65, 66, 67, 363

N

nano- 79, 86, 87, 91, 101, 113, 144, 150
natural kind 163, 164, 168, 181, 182, 186, 242, 277, 280, 285
natural kinds 65, 98, 112, 163, 164, 167, 168, 175, 180, 181, 182, 183, 184, 186, 278, 279, 280, 286, 295
Newton 53, 87, 131, 140, 141, 142, 143, 150, 169, 317, 348, 364
Niiniluoto 68, 70, 71, 72, 73, 364
node 98, 99, 100, 101, 102, 103, 104
nodes 98, 99, 100, 101, 102, 103, 104, 105, 107
nomic 11, 14, 21, 22, 47, 48, 56, 156, 167, 168, 169, 170, 171, 173, 178, 179, 180, 181, 182, 183, 184, 187, 194, 204, 205, 208, 235, 236, 245, 246, 247, 259, 272, 277, 278, 283, 285, 288, 305, 306, 308, 320, 321, 322, 330, 341, 343, 345, 346, 349, 351, 354
nomically 36, 167, 178, 236, 259, 292
nominal 11, 14, 21, 26, 35, 47, 48, 55, 156, 167, 168, 176, 242, 274, 317, 324, 330, 338, 339, 341, 343, 345, 349, 351, 353, 354
non-classificational 49
nonlocal 113, 114, 121, 124
nonlocality 50, 113, 120, 124, 129, 131, 136, 139
nonoccurrent probability 232
nonpropositional experience 233
nonpropositional probability-makers 233
non-vacuous 97, 99, 102, 114
non-veridical 169, 208, 209, 210
noumena 12, 21, 49, 51, 57, 58, 59, 60, 61, 62, 63, 65, 67, 69, 73, 74, 75, 76, 86, 111, 125, 140, 144, 180, 252, 253, 317, 318, 323, 341

noumenal 49, 59, 61, 63, 66, 68, 69, 74, 124, 125, 127, 142

O

objective 30, 31, 38, 51, 73, 88, 140, 141, 151, 166, 171, 175, 185, 187, 189, 204, 214, 217, 225, 249, 251, 252, 255, 257, 260, 262, 263, 264, 276, 285, 292, 293, 310, 343, 344
objectivity 59, 140, 217, 255, 259, 265, 277, 284, 287, 344
objectual-causal 74, 142, 153, 162, 189, 198, 203, 206, 214, 219, 244, 245, 247, 252, 259, 270, 277, 285, 286, 288, 303, 310, 312, 345
observation categoricals 81, 284
observation sentences 81, 255, 256, 283, 284, 285, 288, 296
Oddie 70
ontic 148, 258, 259, 284, 343, 349, 351
ontological 5, 6, 11, 12, 13, 16, 21, 22, 23, 24, 25, 26, 27, 28, 29, 31, 32, 33, 34, 35, 36, 37, 38, 39, 40, 42, 43, 44, 45, 46, 47, 50, 51, 52, 53, 54, 56, 57, 60, 61, 62, 63, 65, 66, 68, 69, 72, 73, 74, 75, 76, 77, 78, 79, 80, 81, 82, 83, 84, 86, 90, 91, 92, 97, 98, 103, 108, 109, 111, 114, 117, 118, 121, 125, 128, 129, 130, 141, 144, 145, 147, 148, 149, 150, 151, 153, 154, 155, 156, 157, 158, 159, 160, 161, 162, 163, 164, 166, 171, 172, 173, 174, 175, 177, 178, 180, 181, 182, 183, 184, 186, 187, 189, 191, 192, 193, 195, 196, 201, 202, 203, 204, 205, 206, 207, 208, 209, 210, 212, 213, 214, 215, 216, 217, 218, 219, 220, 221, 222, 223, 225, 226, 227, 228, 230, 231, 232, 234, 235, 236, 237, 238, 239, 241, 242, 244, 245, 246, 247, 248, 249, 250, 251, 252, 253, 254, 255, 258, 259, 260, 261, 262,

264, 265, 266, 267, 270, 271, 272, 274, 275, 276, 277, 278, 279, 280, 282, 283, 284, 285, 286, 287, 288, 289, 290, 292, 293, 294, 295, 296, 297, 299, 300, 302, 304, 305, 307, 309, 310, 311, 313, 315, 318, 319, 320, 321, 322, 323, 325, 328, 332, 334, 335, 336, 337, 338, 339, 340, 341, 342, 343, 345, 346, 347, 348, 349, 350, 351, 352, 353, 354
ontological commitment 11, 12, 13, 22, 23, 26, 29, 36, 37, 40, 45, 50, 56, 61, 65, 73, 74, 81, 82, 83, 84, 86, 90, 91, 97, 98, 103, 111, 117, 118, 130, 144, 148, 149, 150, 151, 153, 154, 155, 156, 157, 158, 160, 161, 162, 172, 174, 181, 184, 186, 189, 191, 201, 207, 208, 211, 217, 219, 227, 228, 232, 234, 238, 241, 242, 250, 252, 253, 254, 255, 256, 258, 259, 260, 272, 274, 275, 276, 277, 279, 280, 282, 283, 284, 286, 287, 288, 289, 290, 292, 293, 295, 297, 304, 305, 307, 309, 319, 320, 335, 338, 339, 340, 341, 343, 346, 348, 349, 350, 352
Ontological Principle of Excluded Vacuous Middle 12, 142, 157, 220, 260, 265, 355
ontological probabilism 109, 147, 231
ontological universals 6, 11, 13, 14, 23, 27, 29, 41, 47, 73, 74, 153, 155, 157, 159, 160, 164, 182, 189, 192, 203, 206, 207, 210, 212, 213, 217, 219, 225, 230, 231, 234, 236, 237, 239, 246, 247, 250, 251, 259, 262, 263, 265, 266, 270, 277, 279, 285, 287, 288, 295, 300, 301, 302, 307, 308, 311, 321, 335, 336, 338, 345
ontologically committed 83, 116, 156, 231, 283, 350
onto-theology 325, 326

organismic 52, 53
ostensive 254, 279

P

paradox 48, 106, 110, 116, 122, 143, 332, 341, 349
Parsons 252, 266
particle 12, 50, 66, 88, 89, 90, 91, 94, 95, 96, 97, 98, 99, 100, 101, 102, 103, 104, 105, 106, 108, 111, 113, 114, 116, 117, 119, 120, 122, 125, 126, 130, 136, 138, 146, 306, 367
particles 43, 78, 86, 87, 88, 89, 92, 95, 96, 97, 98, 99, 100, 103, 104, 105, 106, 107, 108, 112, 113, 114, 119, 121, 122, 123, 124, 125, 127, 129, 132, 136, 141, 142, 143, 147, 150, 151, 257, 276, 304, 306
particular 11, 13, 22, 23, 24, 26, 27, 29, 30, 35, 36, 42, 44, 45, 51, 56, 67, 70, 71, 75, 77, 80, 81, 83, 85, 92, 114, 115, 125, 128, 130, 141, 148, 149, 153, 158, 159, 160, 163, 174, 175, 178, 179, 181, 182, 184, 185, 186, 187, 188, 189, 190, 191, 192, 193, 195, 196, 197, 199, 200, 201, 202, 203, 204, 205, 206, 207, 208, 211, 214, 216, 217, 218, 220, 225, 226, 227, 231, 234, 236, 238, 241, 243, 245, 246, 248, 249, 250, 253, 262, 263, 264, 265, 267, 269, 270, 272, 274, 277, 278, 282, 285, 286, 287, 289, 291, 293, 295, 296, 297, 299, 300, 302, 303, 305, 306, 307, 308, 309, 310, 311, 313, 314, 315, 318, 319, 321, 322, 323, 325, 327, 334, 335, 336, 339, 340, 342, 345, 346, 347, 352, 354
particularism 6, 11, 13, 21, 24, 49, 50, 57, 65, 150, 155, 156, 161, 165, 185, 189, 193, 195, 196, 198, 199, 201, 202, 207, 208, 235, 237, 241, 250,

253, 259, 261, 262, 265, 267, 274, 277, 278, 283, 290, 292, 293, 295, 296, 301, 305, 312, 320, 321, 335, 343
Peirce 50, 167, 212, 364
perception 30, 51, 58, 64, 68, 72, 74, 90, 141, 152, 153, 169, 178, 179, 180, 192, 201, 203, 208, 209, 210, 214, 218, 232, 295, 349
perceptual 68, 89, 90, 233
perspectival 79, 115, 142, 143, 144, 145, 147, 174, 175, 177, 178, 180, 194, 199, 211, 219
phase speed 104
phenomena 12, 21, 30, 38, 49, 51, 57, 58, 59, 60, 61, 62, 63, 65, 67, 68, 69, 71, 72, 74, 75, 76, 78, 86, 90, 92, 104, 110, 111, 113, 124, 125, 127, 128, 130, 132, 134, 135, 137, 140, 141, 144, 151, 180, 252, 276, 285, 304, 317, 318, 323, 341, 347
phenomenal 15, 25, 39, 49, 58, 59, 61, 62, 63, 65, 66, 67, 69, 74, 76, 78, 124, 125, 126, 127, 128, 142, 318, 337, 341, 347, 349, 354
phenomenalism 56, 58, 59, 65, 69, 74, 79, 86, 103, 109, 110, 111, 143, 175, 338, 340, 352
phenomenalistic 21, 50, 65, 66, 73, 80, 143, 145, 279, 339, 343, 346, 352
photon 87, 89, 104, 114, 119, 124
placing 184, 185, 186, 187, 188, 189, 190, 191, 207, 208, 262, 277, 286
Poincaré 132, 363
Popper 57, 194, 364
possible actual worlds 153
possible worlds 12, 49, 70, 74, 113, 123, 134, 139, 140, 149, 152, 166, 170, 328, 332
postmodern 45, 71, 232, 250
potential 33, 40, 121, 153, 154, 157, 158, 160, 175, 177, 181, 182, 212, 213, 215, 216, 218, 238, 304

predicable 26, 27, 50, 162, 164, 195, 264, 272, 319, 323, 334
predicables 26, 27, 50, 163, 195, 323, 334
predication 26, 27, 50, 258, 262, 272, 273, 333, 334, 335
prehend 32
prehension 79, 86, 239
presencing 339, 341, 344
primitive notions 24, 25, 45, 151, 220, 221, 222, 223, 225, 226, 227, 228, 229, 230, 231, 235, 269, 294, 311, 317, 335
principle of continuity 127, 136, 139, 157
probabilism 16, 21, 87, 109, 145, 147, 148, 151, 173, 188, 192, 220, 231, 298, 313, 319, 345
probabilistic 11, 12, 13, 14, 25, 28, 29, 46, 73, 78, 86, 92, 97, 98, 101, 102, 109, 110, 111, 117, 126, 127, 128, 129, 130, 143, 144, 148, 151, 164, 171, 173, 174, 175, 176, 189, 195, 196, 211, 219, 220, 225, 226, 230, 232, 234, 235, 239, 252, 269, 297, 316, 317, 319, 320, 323, 327, 332, 339, 345, 349, 350
probability-maker 232, 233
process 12, 21, 22, 24, 25, 30, 32, 35, 36, 40, 44, 45, 61, 65, 80, 93, 94, 101, 106, 121, 126, 142, 148, 150, 151, 152, 155, 156, 158, 162, 163, 181, 183, 189, 201, 205, 206, 208, 209, 210, 211, 213, 214, 216, 218, 219, 226, 230, 234, 237, 238, 263, 264, 268, 271, 273, 274, 279, 288, 300, 306, 307, 308, 309, 310, 311, 312, 314, 317, 321, 323, 325, 327, 335, 338, 341, 343, 345, 349, 352, 353
processes 11, 13, 14, 22, 23, 24, 26, 28, 29, 32, 56, 57, 60, 66, 68, 75, 76, 77, 85, 86, 89, 90, 92, 97, 106, 109, 117,

123, 125, 126, 131, 142, 145, 148, 150, 152, 155, 156, 157, 158, 160, 162, 165, 167, 171, 175, 177, 178, 180, 181, 182, 183, 188, 189, 192, 193, 201, 202, 203, 204, 206, 208, 209, 210, 211, 212, 213, 216, 217, 218, 228, 230, 231, 232, 234, 236, 237, 238, 243, 245, 246, 248, 250, 251, 252, 259, 261, 264, 265, 268, 270, 272, 274, 279, 283, 285, 289, 292, 298, 299, 300, 301, 303, 304, 306, 308, 309, 310, 311, 321, 325, 327, 338, 339, 345, 346, 347, 352
processual 6, 11, 14, 21, 26, 31, 35, 43, 48, 55, 56, 120, 142, 152, 155, 156, 159, 161, 175, 193, 199, 204, 206, 218, 237, 245, 246, 248, 250, 270, 274, 275, 278, 279, 283, 308, 309, 310, 311, 320, 321, 323, 328, 330, 334, 338, 339, 343, 345, 346, 347, 350, 352, 353, 354
projecting-open 14, 339, 342, 350, 352, 354
propronouns 297
Psillos 71, 364
pulsation 98, 103
Putnam 60, 168, 241, 251, 266, 357, 361, 364, 365

Q

qualia 11, 13, 52, 56, 83, 84, 85, 127
qualities 11, 26, 53, 55, 70, 97, 127, 140, 173, 184, 189, 190, 192, 199, 206, 210, 247, 250, 282, 298, 304, 305, 325, 342
quality 26, 29, 37, 44, 45, 50, 84, 123, 140, 166, 167, 184, 207, 235, 263, 272, 292, 296, 305, 313, 321, 323, 348
quanta 56, 103, 109, 111, 127, 133, 134, 135, 136, 139, 141, 145, 306
quantification 51, 82, 83, 252, 254, 258, 260, 267, 270, 275, 277, 282, 297, 303, 304, 306, 308
quantified 82, 83, 84, 275, 276
quantify 267, 304, 309
quanton 87, 88, 94, 100, 101, 104, 125
Quantum Mechanics 12, 21, 50, 51, 75, 78, 86, 89, 91, 97, 102, 103, 107, 109, 111, 114, 116, 121, 123, 125, 126, 127, 130, 139, 142, 143, 144, 145, 146, 147, 151, 194, 357, 358, 359, 360, 361, 362, 364, 365, 366
quasi-realism 65
quasi-realist 65
Quine 11, 13, 22, 26, 43, 51, 56, 81, 82, 83, 91, 111, 158, 241, 242, 244, 248, 250, 251, 252, 253, 254, 255, 256, 257, 258, 259, 260, 261, 263, 264, 265, 266, 267, 268, 271, 274, 275, 276, 277, 278, 279, 280, 282, 283, 284, 285, 286, 287, 288, 289, 290, 291, 292, 293, 294, 295, 297, 299, 303, 305, 307, 311, 314, 320, 322, 329, 360, 362, 364, 365

R

realism 29, 54, 60, 65, 66, 67, 68, 69, 72, 76, 77, 78, 79, 85, 87, 89, 90, 102, 103, 109, 110, 111, 112, 115, 116, 120, 124, 128, 141, 143, 148, 158, 174, 188, 197, 198, 199, 203, 214, 217, 219, 265, 269, 284, 287, 322, 328, 357, 358, 359, 360, 362, 363, 364, 365, 366, 367
realistic 12, 26, 29, 45, 49, 59, 60, 65, 84, 89, 90, 91, 92, 98, 107, 109, 112, 117, 124, 130, 143, 144, 145, 147, 214, 270, 313, 320, 323, 328, 337
Reality-in-general 11, 13, 14, 15, 26, 29, 34, 35, 36, 37, 42, 44, 45, 47, 75, 127, 148, 156, 160, 161, 163, 164, 170, 172, 175, 187, 188, 191, 192, 193, 194, 195, 197, 198, 201, 202, 203, 205, 206, 208, 211, 212, 213,

214, 217, 218, 219, 220, 225, 226, 230, 231, 235, 236, 237, 238, 242, 245, 247, 248, 249, 250, 252, 253, 260, 265, 273, 282, 283, 286, 287, 289, 293, 295, 296, 300, 301, 303, 309, 310, 311, 313, 318, 320, 321, 323, 325, 326, 327, 334, 335, 336, 338, 340, 341, 343, 344, 345, 346, 347, 348, 351, 353, 354
reality-in-particular 14, 45, 83, 148, 149, 175, 178, 181, 182, 184, 189, 192, 196, 217, 218, 220, 245, 246, 251, 264, 278, 289, 295, 296, 300, 302, 308, 327, 336, 339
Reality-in-total 6, 11, 12, 13, 14, 15, 25, 26, 29, 30, 31, 32, 33, 34, 35, 36, 37, 41, 42, 44, 45, 46, 47, 49, 50, 51, 56, 57, 61, 69, 74, 75, 76, 77, 79, 80, 83, 84, 85, 86, 87, 98, 111, 112, 113, 126, 127, 128, 130, 134, 140, 141, 143, 144, 145, 147, 148, 149, 150, 151, 152, 153, 156, 157, 158, 159, 160, 161, 162, 163, 164, 167, 170, 172, 173, 174, 175, 177, 178, 180, 182, 183, 184, 186, 187, 189, 191, 192, 193, 194, 196, 197, 198, 201, 202, 203, 204, 205, 206, 208, 211, 212, 213, 214, 216, 217, 218, 219, 220, 225, 230, 231, 235, 236, 237, 238, 239, 241, 242, 247, 248, 249, 250, 253, 260, 261, 264, 265, 269, 273, 274, 277, 278, 279, 282, 283, 285, 286, 287, 289, 295, 296, 298, 299, 300, 301, 303, 305, 306, 307, 308, 309, 310, 311, 312, 313, 318, 319, 320, 321, 322, 323, 325, 326, 327, 328, 329, 332, 333, 334, 335, 336, 338, 339, 340, 341, 342, 343, 344, 345, 346, 347, 348, 349, 350, 352, 353, 354, 355
reductionism 269, 291
reference 13, 23, 79, 80, 81, 82, 84, 85, 91, 119, 131, 132, 134, 135, 136, 137, 138, 153, 157, 158, 159, 160, 161, 163, 172, 178, 179, 181, 182, 185, 186, 195, 201, 216, 227, 251, 254, 257, 258, 259, 260, 261, 263, 264, 270, 272, 275, 282, 288, 295, 297, 298, 303, 311, 313, 322, 327, 334, 354
relative 35, 43, 66, 72, 85, 86, 94, 131, 133, 134, 135, 137, 138, 140, 176, 233, 234, 244, 258, 278, 292, 295, 302, 305, 316, 325, 341, 349, 353
relativism 12, 66, 72, 90, 267
relativistic 51, 79, 110, 131, 134, 136, 139, 140, 143, 144, 239
Renzong 60, 61, 169, 194, 195, 357, 358, 359, 360, 362, 363, 364, 365, 366, 367
Rescher 70, 72, 365
Rorty 80, 172, 365
Russell 5, 52, 54, 80, 149, 222, 228, 241, 249, 256, 263, 266, 269, 274, 276, 297, 324, 365, 366

S

Schrödinger 88, 89, 92, 109, 122
Scientific Criterion of Ontological Commitment 277
Scientific Ontology 317
scientific realism 70, 87
second generation scientific ontology 25, 29, 314, 319, 323
seeing as 63
seeing-that 22, 62, 63, 64, 70, 73, 153, 347, 348
Seiendheit 47, 342
semantics 60, 61, 73, 158, 201, 266, 275, 282, 287, 288, 289, 291, 292, 301, 328
sensation 60, 61, 304, 305
sensibility 49, 57, 58, 59, 60, 61, 62, 64, 74, 76, 318
set theory 52, 55, 75, 154, 228, 252, 283
Sfendoni-Mentzou 197, 198, 202, 203, 212, 213, 214, 215, 216, 217, 357, 366
Shapere 85, 150, 366

simultaneity 138, 144
simultaneous 88, 97, 110, 118, 158, 232, 303, 311
sine qua non 29, 42, 44, 47, 64, 213, 233, 282, 311, 313, 323
sinusoidal 100, 101, 104, 105
solipsism 67, 90, 233
sortal 195, 207
sortals 194, 195
space 11, 25, 35, 46, 55, 56, 58, 64, 70, 78, 80, 88, 93, 100, 105, 108, 109, 116, 122, 125, 129, 132, 133, 136, 139, 141, 142, 143, 145, 154, 155, 175, 176, 197, 212, 214, 216, 217, 221, 261, 266, 273, 298, 303, 305, 306, 310, 323, 328, 329
space-time 56, 105, 145, 261, 298, 323
spatiality 23, 55, 57, 139, 155, 300, 309
spatio-temporal 24, 32, 50, 59, 65, 68, 96, 99, 100, 101, 102, 104, 105, 108, 121, 122, 136, 142, 145, 172, 176, 177, 210, 212, 214, 258, 260, 261, 270, 273, 276, 302, 304, 308, 353
Special Theory of Relativity 12, 99, 123, 131, 360, 366
species 11, 23, 163, 164, 190, 242, 248, 257, 276
static-dynamic 353
staticity 6, 232, 339, 342, 346, 353
statistical interpretation 103
Stenlund 55, 153, 154, 155, 158, 159, 161, 366
Stout 6, 198, 199
Stoutian 198, 200, 202
straight-line motion 99, 105
Strawson 11, 13, 184, 185, 186, 187, 189, 190, 191, 193, 194, 196, 199, 200, 202, 207, 208, 248, 320, 323, 366
subluminal 119, 124, 132, 133, 136, 156
subsistence 83, 260, 261, 262, 265, 270, 274, 277, 333

substance 11, 12, 35, 47, 49, 50, 55, 57, 61, 64, 75, 81, 129, 140, 152, 162, 163, 164, 184, 193, 262, 311, 317, 318, 319, 322, 323, 324, 325, 333, 334
Sudarshan 124
superluminal 50, 75, 103, 105, 114, 116, 121, 122, 123, 124, 129, 132, 135, 136, 138, 139, 140, 144, 149, 156
supervenience 204
synthesis 21, 25, 37, 42, 48, 49, 56, 61, 69, 74, 87, 125, 126, 128, 143, 144, 145, 232, 233, 269, 300, 313, 314, 326, 335, 336
synthetic 29, 34, 50, 87, 103, 162, 239, 248, 254, 268, 291, 294, 295, 314
system 7, 9, 14, 21, 25, 36, 37, 42, 45, 46, 61, 71, 72, 84, 92, 108, 115, 116, 117, 120, 121, 122, 123, 131, 137, 146, 148, 150, 153, 163, 164, 192, 197, 201, 210, 211, 220, 221, 222, 223, 225, 227, 228, 229, 230, 231, 237, 241, 242, 257, 258, 276, 277, 279, 282, 291, 292, 294, 301, 302, 304, 310, 313, 315, 319, 321, 335, 347
system building 45, 46, 192, 220, 313, 335
system-bounded 279
systemic 14, 25, 46, 122, 151, 207, 219, 225, 230, 239, 253, 254, 292, 296, 303, 311, 315, 334, 335

T

Tarski 73, 275
temporality 23, 55, 57, 139, 153, 155, 300, 309, 345
theoretical 12, 24, 28, 37, 41, 45, 50, 51, 60, 65, 67, 68, 73, 76, 77, 79, 81, 85, 86, 87, 90, 94, 98, 107, 115, 127, 128, 141, 153, 159, 170, 177, 182, 189, 194, 195, 201, 230, 233, 238, 244, 245, 251, 256, 258, 260, 264, 265, 267, 269, 279, 286, 287, 288, 292,

293, 294, 295, 296, 299, 301, 302, 311, 316, 321, 349
theory-laden 77, 80, 84, 86, 186, 286
thick particulars 199
thin 199, 241, 307
thing-in-itself 25, 57, 58, 72, 74, 125
things-in-themselves 58, 62, 64, 67, 70, 74, 75, 90, 126, 318
three-dimensional 98, 99, 100, 102, 103, 104, 131, 305
time 11, 35, 55, 56, 58, 59, 63, 64, 67, 68, 70, 76, 78, 80, 88, 89, 90, 95, 96, 98, 100, 101, 105, 108, 109, 118, 119, 120, 126, 129, 130, 133, 134, 135, 137, 139, 141, 142, 143, 145, 146, 154, 155, 156, 175, 179, 197, 206, 209, 210, 212, 216, 236, 237, 243, 246, 256, 257, 261, 266, 272, 277, 287, 288, 294, 298, 303, 306, 308, 309, 312, 316, 323, 325, 329, 333, 339, 340, 342, 350, 353
To Be 6, 11, 13, 14, 21, 26, 29, 31, 33, 34, 35, 36, 37, 41, 42, 44, 45, 47, 50, 151, 156, 160, 161, 162, 164, 182, 187, 191, 192, 193, 196, 203, 208, 211, 214, 218, 219, 220, 225, 231, 235, 236, 237, 238, 241, 242, 243, 245, 246, 247, 248, 249, 250, 251, 253, 254, 260, 261, 265, 269, 273, 274, 275, 277, 279, 282, 283, 285, 286, 287, 288, 289, 293, 295, 296, 297, 298, 299, 300, 301, 302, 303, 306, 307, 309, 310, 311, 312, 313, 314, 318, 320, 321, 322, 323, 324, 325, 326, 327, 328, 329, 330, 334, 335, 336, 337, 338, 339, 340, 341, 342, 343, 344, 345, 346, 347, 348, 349, 350, 351, 352, 353, 354, 355
token 11, 26, 37, 42, 44, 57, 73, 75, 84, 85, 148, 164, 182, 189, 205, 215, 216, 231, 242, 247, 250, 251, 254, 260, 265, 269, 272, 273, 277, 278, 279, 280, 285, 287, 289, 292, 295, 298, 305, 311, 319, 348
tokens 11, 13, 23, 26, 27, 50, 51, 81, 160, 164, 168, 181, 182, 184, 186, 187, 190, 199, 210, 225, 226, 247, 250, 259, 261, 265, 268, 270, 274, 275, 277, 278, 279, 285, 286, 287, 288, 290, 292, 293, 295, 299, 300, 305, 307, 311
totalise 46, 128, 149, 150, 211, 237, 289, 353
totalization 11, 13, 36, 37, 45, 74, 87, 150, 151, 152, 193, 225, 279, 295, 311
totals 11, 27, 160, 191, 226
Transcendent 11, 27, 29, 32, 33, 35, 36, 37, 44, 45, 49, 50, 51, 75, 86, 127, 128, 130, 145, 149, 152, 156, 157, 158, 162, 170, 173, 174, 183, 193, 235, 237, 247, 275, 283, 289, 293, 294, 300, 314, 315, 316, 319, 321, 322, 325, 326, 353, 361
transcendental 26, 31, 32, 33, 34, 37, 45, 50, 59, 64, 86, 149, 157, 162, 163, 196, 232, 235, 238, 243, 247, 282, 298, 323, 340, 353
Transcendental 11, 13, 15, 27, 29, 31, 33, 35, 36, 37, 44, 45, 50, 51, 61, 62, 127, 128, 130, 145, 149, 156, 162, 163, 184, 187, 191, 193, 203, 232, 235, 237, 238, 241, 245, 246, 247, 251, 254, 261, 271, 277, 282, 283, 286, 288, 289, 293, 294, 309, 311, 312, 314, 315, 316, 319, 321, 322, 323, 325, 326, 339, 341, 350, 353, 358
transcendentally transcendent 162, 235
Transcendentally Transcendent 162, 183, 326
transcendentals 11, 26, 27, 31, 32, 33, 34, 36, 44, 45, 46, 282
Transcendently Transcendental 37, 237, 247, 294, 326

transcendents 11, 26, 27, 31, 32, 33, 34, 44, 45, 46, 300
trough 98, 99, 100, 102, 103, 104, 105, 107
truth 6, 10, 11, 13, 29, 42, 43, 45, 51, 55, 59, 60, 66, 69, 70, 72, 73, 74, 76, 77, 78, 82, 83, 90, 92, 102, 118, 151, 164, 166, 168, 169, 172, 173, 176, 184, 187, 194, 197, 201, 208, 218, 220, 223, 224, 225, 226, 227, 230, 231, 232, 233, 234, 235, 238, 242, 246, 247, 249, 253, 258, 266, 269, 275, 279, 283, 286, 288, 291, 292, 294, 301, 302, 307, 313, 323, 327, 329, 335, 347, 348, 349, 350, 351, 352
truthmakers 197, 204, 205, 215, 274

U
UFO 70, 72
Ugarov 360, 366
ultra-quantal 86, 100, 103, 109, 113
ultra-Quantum Mechanics 103
unbound variable 283
Uncertainty 92, 219
Unconcealment 35
uncovering 346, 348, 349, 350, 351
universal-bounded 30, 42, 125, 184, 279, 280
universals 6, 11, 12, 13, 15, 21, 22, 23, 25, 26, 27, 28, 29, 30, 31, 32, 35, 36, 37, 38, 40, 41, 42, 44, 47, 56, 69, 72, 74, 75, 81, 82, 83, 84, 85, 87, 125, 127, 128, 141, 148, 152, 153, 154, 155, 156, 157, 159, 160, 163, 164, 168, 169, 171, 172, 173, 175, 176, 177, 178, 181, 182, 183, 184, 187, 188, 191, 192, 193, 194, 195, 196, 197, 198, 199, 200, 201, 202, 203, 204, 205, 206, 207, 208, 210, 211, 212, 213, 214, 215, 216, 217, 218, 219, 220, 225, 226, 228, 230, 231, 232, 234, 235, 236, 237, 238, 242, 245, 246, 247, 248, 249, 250, 252, 253, 259, 260, 261, 262, 264, 265, 266, 267, 269, 270, 271, 272, 274, 277, 278, 279, 282, 283, 285, 286, 287, 288, 289, 295, 296, 298, 299, 300, 302, 303, 304, 305, 306, 307, 308, 309, 311, 313, 316, 317, 320, 321, 322, 326, 327, 328, 329, 330, 333, 334, 335, 336, 338, 339, 340, 341, 343, 345, 346, 347, 348, 349, 351, 353

V
Van Fraassen 84, 166, 167, 170, 173, 174, 177, 317, 367
verbal 11, 14, 21, 26, 29, 34, 35, 47, 48, 156, 242, 245, 274, 275, 283, 300, 309, 310, 320, 321, 323, 324, 326, 328, 330, 334, 338, 339, 341, 343, 345, 346, 347, 351, 353, 354
veridical 68, 169, 208, 209, 223

W
Wartofsky 128, 129
wave 12, 78, 87, 88, 89, 90, 91, 94, 95, 96, 97, 98, 99, 100, 101, 102, 103, 104, 105, 106, 107, 108, 111, 113, 114, 126, 127, 130, 133, 140, 141, 143, 144
wavelength 85, 87, 98, 104, 126, 143
wave-particle 87, 90, 105, 106, 108, 126, 127, 130, 140, 143, 144
wave period 104
wavicle 90, 91, 92, 95, 96, 97, 98, 99, 100, 101, 102, 104, 106, 111, 115, 116, 117, 122, 130, 175
wavicles 78, 91, 92, 93, 94, 95, 96, 97, 98, 99, 100, 102, 103, 104, 109, 113, 114, 116, 119, 122, 130, 144, 149, 151, 156, 306
ways of being 22, 29, 52, 56, 152, 161, 167, 210, 247, 250, 274, 283, 300, 304, 310, 321, 322, 327
weak force 95

Whitehead 5, 6, 31, 36, 40, 41, 52, 53, 55, 79, 84, 85, 109, 149, 222, 228, 246, 266, 306, 312, 316, 322, 323, 324, 367

Wittgenstein 68, 184, 214, 250, 252, 253, 367

Y

Young 87

Z

zero rest mass 105, 133, 135, 136, 143